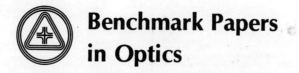

Benchmark Papers
in Optics

Series Editor: Stanley S. Ballard
University of Florida

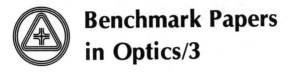

Benchmark Papers
in Optics/3

A BENCHMARK® Books Series

LIGHT IN THE SEA

Edited by

JOHN E. TYLER
Scripps Institution of Oceanography

Dowden, Hutchinson
& Ross, Inc.
Stroudsburg, Pennsylvania

LIBRARY OF CONGRESS CATALOGING IN PUBLICATION DATA

Main entry under title:
Light in the sea.
   (Benchmark papers in optics; 3)
   Includes indexes.
   1. Optical oceanography—Addresses, essays, lectures.
I. Tyler, John E.
GC179.L53   551.4'601   77-468
ISBN 0-87933-265-4

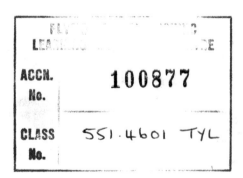
Exclusive Distributor: **Halsted Press**
A Division of John Wiley & Sons, Inc.
ISBN: 0-470-99143-7

SERIES EDITOR'S FOREWORD

Optics, pure and applied, comprises a very broad field. Its roots are ancient; its modern applications are novel and exciting. Hence, a difficult and far-reaching task is faced by the series editor of "Benchmark Papers in Optics." The editor must consider the classical fields, so rich in history and so necessary for the understanding of all optical phenomena, and must bring to the scientific audience current information on the burgeoning areas of applied optics. And what should be done in the several inter- or multidisciplinary fields that include optics?

For all subjects, the classical or benchmark papers must be included, be they a century or only a decade old, and in English or another language. The papers must be placed in proper perspective by the commentaries written by knowledgeable and skillful volume editors. Thus, the overall impact of the book will be more satisfying than the sum of the original papers that are presented in facsimile. These volumes should be of value to scholars, teachers, applied physicists, engineers, and practical opticists. All this constitutes a courageous goal; we can only hope that it is being at least partially achieved.

The subject of hydrologic optics seems ideal for a Benchmark volume. The literature on this broad subject is widely dispersed among many nations, as are the scientists working in this field. There are no complete textbooks. Few authoritative papers cover large portions of the subject, and there are relatively few persons with long-term, intimate experience in the field. One of these is the volume editor, John Tyler, who has studied hydrooptics problems at the Scripps Institution of Oceanography for some twenty-five years. Through his participation in many international congresses and symposia he has also become acquainted with workers from other countries and has discussed with them problems of mutual interest. Thus his selection of the important papers in this field must be regarded as authoritative in all ways.

The published history of hydrologic optics covers a span of 150 years. It reveals the evolution of this interdisciplinary science from early, qualitative observations to the present sophisticated theory and experiment. Mr. Tyler's Preface, historical Introduction, and his commentaries on the seven groups of papers will assist the reader in recognizing and appreciating the great forward strides that have been made in understanding the phenomenology of light in the sea.

STANLEY S. BALLARD

PREFACE

Manifestations of the importance of light in the sea appear in many fields of science. In biological oceanography, light is a fundamental prerequisite of photosynthesis and hence of all life in the sea. Light is a factor in the predatory and escape tactics of fish and in their schooling habits. For man in the sea, light is an indispensable requirement for seeing and for the application of photography or television. Light is also a means for quantitatively describing water and for numerically differentiating clear water from water containing additives or impurities. It thus provides a useful, though not always a reliable, method for detecting and tracing the movement of ocean water masses.

Specialists in many fields have applied their skills to the study of light in the sea, and their results have appeared in diverse journals, often in expedition reports or privately printed documents that are difficult to obtain. This volume attempts to bring together much of the significant published material and offer it in a form that can be most readily applied to problems encountered in these fields of science.

JOHN E. TYLER

CONTENTS

Contents

Contents

CONTENTS BY AUTHOR

INTRODUCTION

The systematic study of light in the sea, otherwise known as optical oceanography or hydrologic optics, apparently received its initial stimulus in the nineteenth century with the independent investigations of O. E. Kotsebu, a Russian naval officer, and P. A. Secchi, an Italian astronomer. Both of these men made visual observations of water transparency by lowering a flat white disc into the ocean and noting the depth of its disappearance. Kotsebu's investigations are reported in the book, *Voyage to the Southern Ocean and Bering Strait to find a Northeastern passage; Undertaken on the ship RYURIK in 1815–1818,* published in St. Petersburg in 1821. Secchi's elaborate experiments were published in a papal navy report by Cmdr. Alessandro Cialdi entitled "Sul Moto Ondoso del Mare e su le correnti di esso specialmente su quelle littorali" (Second Edition, Rome 1866). The subtitle of the section by P. A. Secchi is, "Relazione delle esperienze fatte a bordo della pontificia pirocorvetta l'Immacolata Concezione per determinars la trasparenza del mare."

This simple visual experiment cannot by itself yield data on any single ocean optical property nor accurately distinguish various ocean water types from one another (J. E. Tyler, 1968). The Secchi disc has nevertheless been widely used by oceanographers and Secchi disc observations are frequently found in the literature of biological oceanography.

In 1889 a man named Regnard made the first use of a photoelectric (selenium) cell for measuring natural light in the ocean. This unbelievable experiment, undertaken only thirteen years after Werner Siemens made the first actual selenium cell [1876], was conducted with a galvanometer, on which the readings were taken, located in the laboratory of the Prince of Monaco while the cell was lowered into the harbor at the end of a long cable. Although this experiment produced no useful data, it nevertheless alerted oceanographers to the superiority of photo-electric devices for exploring light parameters in the sea.

The first major application of a photoelectric cell (a Kunz cell) in underwater measurements was made by V. E. Shelford and F. W. Gail (1922). These authors were interested in the relationship between the penetrating light and the distribution of marine plants. They made measurements of underwater "light" as a function of sun altitude and water depth.

The work of Shelford and Gail (1922) stimulated considerable interest in the relationship between radiant energy and marine biology. A short time after their paper appeared, the British scientists, W. R. G. Atkins, of Plymouth, and H. H. Poole, of Dublin, began a series of investigations that continued for thirty years and greatly enriched the knowledge of the optical properties of ocean water and the distribution of light in the sea.

In 1934 Professor Hans Pettersson, at Göteborg's Högskolas Oceanogr. Inst., Sweden, published the first of a series of seven definitive papers on light in the sea. In the brief period between 1934 and 1938 he devised instrumentation and made measurements of all the basic optical properties required for a complete optical documentation of an ocean water type.

On May 13, 1936, the International Council for the Exploration of the Sea held a special conference for the discussion of light in the sea. The participants included W. R. G. Atkins and F. S. Russell (Plymouth, England), G. L. Clarke (Woods Hole, Massachusetts, U.S.A.), C. L. Utterback (Seattle, Washington, U.S.A.), H. H. Poole (Dublin, Great Britain), and H. Pettersson (Göteborg, Sweden), all of whom were actively engaged in research on light in the sea, and A. Ångström (Stockholm, Sweden), who at the time was President of the International Radiation Commission.

At this meeting a deliberate effort was made to standardize the instrumentation for the measurement of light in the sea. An instrument for measuring beam transmittance and another for determining the vertical penetration of natural light were described in detail. Proposals were made for the standardization of optical color filters and opal glass for use with these instruments. Every detail of the instruments, the galvanometers, the cables, the watertight construction, the units of measurement, etc., was discussed and duly recorded in the meeting report (Atkins et al., 1938).

A further proposal was made for the examination and standardization of photocells to be used for the measurement of light in the sea. A perplexing feature of the final Committee report is the recommendation regarding the unit of measurement for biological studies. On this matter, the report reads:

> It is evidently desirable, however, and it may be necessary for some purposes, to be able to obtain results in definite units, and it would be desirable to express them in terms of the radiant power, say in milliwatts per square centimetre, or gram-calories per minute per square centimetre in the given spectral bands.
>
> Until the standardization of cells with filters in energy units, either by the solar method suggested by Ångström, or by the use of a source of known colour temperature, has been more fully worked out, it seems advisable to employ some definite (though arbitrary) unit, in which to measure the illumination as recorded by the cell with and without the various colour filters. When the cell is used with plain opal but without a colour filter we may call this unit the "Lux," although we must remember that it will differ somewhat from the true Lux, as estimated on the visual scale. On the other hand, the use of the term Lux for a light-intensity unit measured by the cell with a colour filter is entirely misleading, so we may content ourselves by expressing the readings of the selenium cell with opal and the colour filters RG1, VG9, or BG12 in "red" units, "green" units, or "blue" units respectively. (Atkins et al., 1938, p. 50)

Thus the photometric unit, "Lux" was to be used for a measurement that admittedly would not yield true lux, nor would it necessarily be closely correlated with the absorption of radiant energy by marine phytoplankton.

It is a curious circumstance that technological change has since made obsolete all of the hardware recommendations set forth in the Committee report. But, the use of lux by biologists has persisted and has been a detriment to a clear understanding of the relationship between the radiant flux penetrating into the ocean and the response to radiant energy by the oceanic phytoplankton.

Shortly after the publication of the Committee report, Professor Yves Le Grand, Sous-Directeur du laboratoire de Physique appliquee au Museum d'Histoire Naturelle, made a careful and exhaustive study of the literature on light in the sea and in October 1939 published a major paper entitled, "La Pénétration de la Lumière dans la mer." His Résumé et Conclusions, together with his annotated bibliography, Paper 1, provide an excellent appraisal of the state of the science of light in the sea through the end of 1938.

Research interest in the physical aspects of light in the sea was greatly stimulated by the Committee report and by Le Grand's paper. Le Grand had concluded that "La loi véritable de décroissance de l'éclairement avec la profondeur, compte tenu de la diffusion, diffère de la loi exponentielle classique. Le "coefficient de'extinction vertical" qui suppose cette loi n'est donc qu'approche." (The true law for the decrease of light with depth, taking scattering into account, differs from the classical exponential law. The vertical coefficient of extinction, which presumes this law, is thus only approximated.)

Pettersson had also recognized the need for detailed directional measurements of the radiant energy in the underwater light field. He stated:

> In studies of subsurface illumination one generally measures the light falling on a plane horizontal surface, either that of the lightsensitive organ itself, say of a photo-electric cell exposed behind a pressure-proof glass window, or else on an interposed opal, as recommended by Atkins and Poole. In most cases the exposed surface faced upwards, thus receiving light from the upper hemisphere only, of which light the more oblique rays had their effect much reduced, according to the cosine law. Occasionally, however, the light-receiving surface was turned sideways or downwards. C. L. Utterback has of late studied the relation between transmitted light, i.e., of downward direction, and upward scattered light, of three different spectral bands, by sending down his photometer case first in its normal position and then inverted. Another type of photometer case, enclosing two different cells, one facing downwards and the other upwards, and provided with a three-stranded cable, allowing measurements from both photometers to be made in rapid succession, has for some time been used onboard the Swedish research ship, the "Skagerak."
>
> For certain purposes it is important to know the total flux of light impinging from all directions on a small object suspended in the water. This quantity which, for the sake of brevity, will in the following be denoted by "total illumination," I_T, has to be considered when dealing with the heating or the chemical effects of submarine light, as, for instance, in the photosynthesis of unicellular plants, the phytoplankton. It must obviously be larger than the intensity, I, measured in the usual manner, and the ratio between the two quantities, $R_r = I_T/I$, may be expected to vary both with angular distribution of the light falling on the water surface, with the depth, and with the transparency of the water, both above and, to a certain extent, also immediately below the point considered.
>
> The following paper gives a brief review of the results of work planned to find the angular distribution of subsurface light and, incidentally, the value of R_T at different depths. A special technique was developed at Bornö Station for the purpose mentioned. A detailed report by two of the Commission's assistants, Fil. lic. N. G. Johnson [also known as N. G. Jerlov] and Fil. kand. G. Liljequist, will be published in a coming issue of Svenka Hydrografisk-Biologiska Kommissionens Skrifter. (Pettersson, 1938)

The report by Johnson (Jerlov) and Liljequist (1938), which is anticipated in Pettersson's paper, gives additional data on the distribution of radiance in the natural lightfield underwater. The authors used ocean depth, color, sun zenith angle, and surface lighting conditions as parameters, and from their data they calculated, by integration, some of the quantities that would later be tied into a general theory of radiative transfer in the sea.

The change in shape of the underwater radiance distribution that was observed as a function of depth and wavelength stimulated the curiosity of both biologists and physicists. Atkins and Poole (1933) had considered that "the average obliquity (of the radiance distribution) at the deeper levels was chiefly governed by a balance between the effects of absorption, which by filtering out the more oblique rays, reduces the average obliquity, and the scattering, which increases it," and Poole was clearly thinking in terms of an "equilibrium" distribution of radiance in the deeper water. He said:

> It is evident that further work on the subject of the angular distribution of submarine light under various conditions of surface and of surface illumination is very desirable, but pending the results of such work we should appear to be justified in assuming that the chief effect of surface disturbance is to increase the average subsurface obliquity, and hence the vertical extinction coefficient in the shallower layers, the obliquity gradually returning to its equilibrium value in deeper water, as determined by the balance of the effects of absorption and scattering by such suspended matter as may be present. (Poole, 1938)

In the United States, L. V. Whitney, who had been studying natural radiant energy in Wisconsin lakes, had obtained radiance distribution data at various depths in both ocean and lake water. In 1941 he published two theoretical papers in which he predicted the existence of an equilibrium radiance distribution at great depth. He said:

> A study of the angular distribution of light intensity at different depths shows an increase of scattered light in deeper water, and the tendency for the angular distribution pattern to become symmetrical about a vertical axis with the maximum intensity from the zenith. Two opposing factors tend to produce a limiting angular distribution pattern for diffuse light at great depths: the more rapid absorption of slanting rays which tends to make the light more directional, and the scattering by particles which tends to make the light less directional. (Whitney, 1941a)

In 1945, Poole, who was apparently unaware of Whitney's paper, published an independent theoretical analysis that similarly predicted the condition of equilibrium radiance distribution at great depth. In his conclusions Poole stated:

> 1) Experimental results show little if any regular variation with depth in the vertical extinction coefficient or in the ratio of the

5

horizontal to the vertical illumination, if the measurements are made for approximately monochromatic light in homogeneous water. The extinction coefficient seems to be only slightly affected by the state of the water surface or by the angular distribution of the daylight.

2)   This had led to the suggestion by more than one worker that the angular distribution, and hence the average obliquity, tend to be determined by a balance between the opposing effects of absorption and scattering, the first decreasing the mean obliquity and the second increasing it.

3)   An attempt is here made to present the theory of such an equilibrium distribution on the assumption that the scattering of a beam of light by the particles in suspension takes place equally in all directions. The polar surface of field brightness so found is an ellipsoid of revolution about a vertical axis, the point of observation being the lower focus.

4)   It is shown that, on this hypothesis (isotropic scattering), the equilibrium value of the vertical extinction coefficient is always less than the extinction coefficient (including all scattering) for a parallel beam. Where only a comparatively small part of the extinction is due to scattering, the difference between the two coefficients is very small. This would mean that equilibrium would only be approached very slowly. (Poole, 1945)

By the end of 1938, the true complexity of the interaction of radiant energy with the ocean had been recognized, the required basic measurements had been separated, and prototype instrumentation had been devised for making these measurements.

In 1947–1948, the Swedish Deep-Sea Expedition (*MS Albatross*), under the leadership of Dr. Hans Pettersson, circumnavigated the earth. Optical measurements were, for the first time, an important part of the scientific program of a major oceanographic expedition. The optical program, conducted by N. G. Jerlov, included measurements of the upward and downward components of the daylight penetrating the ocean; measurements of the penetrating "antirachitic ultraviolet daylight"; *in situ* measurements of water transparency by means of a beam of light; and laboratory measurements of the scattering at 45° from a beam of light. In all cases, measurements were obtained for a variety of spectral regions and for various depths in the ocean. These data, taken at selected stations in the mid-latitudes around the world, demonstrated the range of values to be expected in the optical properties of the world's oceans and the variability of these properties with wavelength and depth.

In 1950–1952, the Danish Deep-Sea Expedition Round the World, under the leadership of Dr. Anton Fr. Bruun, undertook a major research effort to measure the organic productivity of important ocean areas and to relate photosynthesis to measurements of available radiant energy underwater. The program, conducted by E. Steemann Nielsen and E. Aabye Jensen, was innovative and highly successful.

In September 1956 the French vessel *Calypso,* under the direction of Professor Y. Le Grand, undertook a series of measurements to determine the spectral transmittance of coastal waters in the Mediterranean. Mlle. J. LeNoble (1958), who participated in the research, reported on measurements of radiance in the ultraviolet region of the spectrum. M. Alexander Ivanoff (1958), who also participated, reported radiance measurements in the visible portion of the spectrum and on polarization underwater.

During the International Geophysical Year, 1957–1958, the vessels *V.F.S. Gauss* and *Anton Dohrn,* under the direction of Dr. Joachim Joseph, Deutsches Hydrographisches Institut, Hamburg, undertook extensive simultaneous measurements of temperature and transmittance in the North Atlantic with the objective of preparing vertical contour profiles of these two parameters for a study of the interaction of different water masses.

REFERENCES

Ångström, A. 1936. On the unit for radiation in oceanic research. *Perm. Int. Cons. Explor. Mer Rapp. P.-V. Réun.* **101** (II):Precis #5.[a]

Atkins, W. R. G. 1932. Solar radiation and its transmission through air and water. *Perm. Int. Cons. Explor. Mer J. du Conseil* **7** (2):171–211.

Atkins, W. R. G., and H. H. Poole. 1933. The photoelectric measurement of the penetration of light of various wavelengths into the sea, and the physiological bearing of the results. *Philos. Trans. R. Soc. London* **B 222:**129–164.

———. 1940. A cubical photometer for studying the angular distribution of submarine daylight. *R. Dublin, Soc. Sci. Proc.* **24** (4):29–42.

Atkins, W. R. G., G. L. Clarke, H. Pettersson, H. H. Poole, C. L. Utterback, and A. Ångström. 1938. Measurement of submarine daylight. *Perm. Int. Cons. Explor. Mer J. du Conseil* **13** (1):37–57.

Clarke, G. L. 1936. Light penetration in the western North Atlantic and its application to biological problems. *Perm. Int. Cons. Explor. Mer Rapp. P.-V. Réun.* **101** (II):Precis #3.[a]

Ivanoff, A. 1958. Contribution a l'étude des propriétés optiques de l'eau de mer en bretagne et an corse, et a la théorie de la polarisation sous-marine. *Résultats Scientifiques des Campagnes de la "Calypso,"* section III, 309–336, Masson et Cie, Editeurs, 120 Boulevard Saint-Germain, Paris (VIe).

Jerlov, N. G. 1951. Optical studies of ocean water. In H. Pettersson, ed., *Rep. Swedish Deep-sea Exp., 1947–1948* **3** (1):1–69.

[a]Précis are paginated independently.

Johnson, N. G., and G. Liljequist. 1938. On the angular distribution of submarine daylight and on the total submarine illumination. *Sven. Hydrogr. Biol. Komm. Skr.* Ny Serie: Hydrografi XIV.

Joseph, J. 1961. Trübungs- und temperatur-verteilung auf den stationen und schnitten von V.F.S., "Gauss" sowie bathythermogramme von F.F.S. "Anton Dohrn" und V.F.S. "Gauss" im *Internationalen Geophysikalischen Jahr 1957–1958.Dtsch. Hydrogr. Z.* **B4°** (5), suppl.

LeNoble, J. 1958. Pénétration du rayonnement ultraviolet dans la mer. *Résultats Scientifiques des Campagnes de la "Calypso,"* section III, 298–308, Masson et Cie, Editeurs, 120 Boulevard Saint-Germain, Paris (VIe).

Pettersson, H. 1934. A transparency-meter for sea-water. *Göteborgs K. Vetenskapsakad. Vitterh. Samh. Handl.* (5), Ser. B, **3**, (8), (*Högsk. Oceanogr. Inst. Göteborg Medd. 7*):1–17.

———. 1934. Scattering and extinction of light in sea-water. *Göteborgs K. Vetenskapsakad. Vitterh. Samh. Handl.* (5), Ser. B, **4**, (4), (*Högsk. Oceanogr. Inst. Göteborg Medd. 9*):1–16.

———. 1935. Submarine daylight and the transparency of sea water. *Int. Counc. Explor .Sea J. du Conseil* **10** (1):48–65.

———. 1936. Das licht im meer. *Bioklimat. Beibl. Meteorol. Z. Wein,* (11).

———. 1936. The transparency of sea water. *Perm. Int. Cons. Explor. Mer Rapp. P.-V. Réun.* **101** (II):Precis #6.[a]

———. 1936. Transparency of sea-water. *Nature* **137** (3454):68.

———. 1938. Measurements of the angular distribution of submarine light. *Perm. Int. Cons. Explor. Mer Rapp. P.-V. Réun.* **108** (2):7–12.

Pettersson H., and H. H. Poole. 1937. Measurements of submarine daylight. *Göteborgs K. Vetenskapsakad. Vitterh. Samh. Handl.* (5), Ser. B, **5**, (5), (*Högsk. Oceanogr. Inst. Göteborg Medd. 13*):1–26.

Poole, H. H. 1936. The photo-electric measurement of submarine illumination in offshore waters. *Perm. Int. Cons. Explor. Mer Rapp. P.-V. Réun.* **101** (II):Precis #2.[a]

———. 1938. The effect of surface conditions on the intensity and angular distribution of submarine daylight. *Perm. Int. Cons. Explor. Mer Rapp. P.-V. Réun.* **108** (2):3–6.

———. 1945. The angular distribution of submarine daylight. *R. Dublin Soc. Sci. Proc.* **24** (4):29–42.

Russell, F. S. 1936. Submarine illumination in relation to animal life. *Perm. Int. Cons. Explor. Mer Rapp. P.-V. Réun.* **101** (II):Precis #1.[a]

Shelford, V. E., and F. W. Gail. 1922. A study of light penetration into sea water made with the Kunz photo-electric cell with particular reference to the distribution of plants. *Puget Sound Biol. Stn. Publ.*3 (65):141–176.

Steemann Nielsen, E., and E. A. Jensen. 1957. Primary oceanic production, the autotrophic production of organic matter in the oceans. In A. Fr. Bruun, Sv. Greve, and R. Spärck, eds., *Galathea Report, Volume 1: Scientific Results of the Danish Deep-Sea Expedition Round the World, 1950–1952.*

Tyler, J. E. 1968. The Secchi disc. *Limnol. Oceanogr.* **13** (1):1–6.

Utterback, C. L. 1936. Spectral bands of submarine solar radiation in the North Pacific and adjacent inshore waters. *Perm. Int. Cons. Explor. Mer Rapp. P.-V. Réun.* **101** (II):Precis #4.[a]

Whitney, L. V. 1941a. A general law of diminution of light intensity in natural waters and the percent of diffuse light at different depths. *Opt. Soc. Am. J.* **31:**714–722.

————. 1941b. The angular distribution of characteristic diffuse light in natural waters. *J. Mar. Res.* **4:**122–131.

Influential Books and Monographs

Books

Gibbs, R. J. (ed.). 1974. *Suspended solids in water.* Marine Science, No. 4. Plenum Press, New York, 320 pp. Describes current optical methods for locating and studying particulate matter in the ocean.

Goulard, R., S. M. Scala, and R. N. Thomas (eds.). 1968. *Radiative energy transfer.* Pergamon Press, New York, 618 pp. Proceedings of the symposium on interdisciplinary aspects of radiative energy transfer, Philadelphia, Pa., Feb. 24–26, 1966.

Hill, N. M. (ed.). 1962. *The sea—Ideas and observations on progress in the study of the seas, Volume* 1, section IV, chapters 8, 9, and 10, pp. 397–468. Interscience Publishers, New York, 864 pp. A popular review of the theory and practice of radiative transfer in the ocean and the application to underwater visibility and animal life.

Ivanoff, A. 1975. *Introduction a l'oceanographie, tome II.* Proprietes Physiques et chimiques des eaux de mer. Librairie Vuibert, Boulevard Saint-Germain, 63, Paris, 340 pp.

Jerlov, N. G. 1968. *Optical oceanography.* Elsevier Publishing Co., Amsterdam, 194 pp. Extensive review of the literature on the optics of the sea, with abundant references.

Jerlov, N. G., and E. Steemann Nielsen (eds.). 1974. *Optical aspects of oceanography.* Academic Press, London, 494 pp. Collected papers presented at the Symposium on Optical Aspects of Oceanography held at the Institute of Physical Oceanography, Copenhangen, Denmark, June 19–23, 1972.

Monin, A. C., and K. S. Shifrin. 1974. *Hydrophysical and hydrooptical investigations in the Atlantic and the Pacific oceans.* Publishing House "NAUKA," Moscow, 328 pp. (in Russian). Reviews Russian work in optical oceanography and discusses the interdisciplinary application of this data. Includes radiative transfer theory and recent data.

Preisendorfer, R. W. 1976. *Hydrologic optics.* U.S. Govt. Printing Office Order No. C55.602:H99, Washington, D.C. A mathematical treatment of radiative transfer in ocean or lake water. Six volumes.

Shifrin, K. S. (ed.). 1972. *Optics of the ocean and the atmosphere.* Publishing House "NAUKA," Leningrad, 231 pp. (in Russian with English summaries). Contains original scientific papers by many of the outstanding scientists in this field. Emphasizes theory and measurement of properties with some applications to other fields.

Steemann Nielsen, E. 1975. *Marine photosynthesis.* Elsevier Oceanography Series, 13. Elsevier Publishing Co., Elsevier, New York. Discusses the problem of measuring the light available for photosynthesis, the units to be used, and the rate of photosynthesis as a function of the measured irradiance.

Tyler, J. E., and R. C. Smith. 1970. *Measurements of spectral irradiance underwater.* Gordon and Breach Science Publishers, New York, 103 pp. Describes instrumentation in detail and publishes detailed spectral irradiance data as a function of depth in a variety of water types.

Monographs

Electromagnetics of the Sea. 1970. AGARD Conference Proceedings #77 (AGARD-CP-77-70), North Atlantic Treaty Organization, about 500 pp.[b] Contains forty-seven papers about half of which are directly concerned with optical oceanography and applications to imagery underwater.

Jerlov, N. G. (ed.). 1961. *Union geodesique et geophysique internationale.* Monographie No. 10. A publication of the papers given at the *Symposium on Radiant Energy in the Sea* held in Helsinki, August 4–5, 1960. Available from I.U.G.G. Publications Office, 39ter Rue Gay-Lussac 76, Paris V[e], France (as of June 1971). Contains papers on theory, experiment, and instrumentation with some applications to biological processes and descriptive oceanography.

Kozljaninov, M. V. (ed.). 1965. *Investigations in hydrooptics.* Publishing House "NAUKA," Moscow. Academy of Sciences of the USSR, Transactions of the Institute of Oceanology, Vol. 77, 137 pp. Descriptions of instruments and theoretical work by Russian scientists (in Russian with English abstracts).

Oceanology No. 1. 1971. National Scientific Committee on Oceanic Research of the Polish Academy of Sciences, 108 pp. Contains a review article on radiative transfer in the ocean by Jerzy Dera (in Polish) as well as the paper, "Irradiance in the Euphotic Zone of the Sea" by Jerzy Dera (in English).

Oceanology No. 2. 1973. National Scientific Committee on Oceanic Research of the Polish Academy of Sciences, 243 pp. An analysis of underwater visibility conditions in the sea, based on the examples of *The Gulf of GDANSK* by J. Olszewski (in Polish with English summary).

Optics of the Sea (Interface and In-Water Transmission and Imaging). 1973. AGARD Lecture Series #61 (AGARD-LS-61) North Atlantic Treaty Organization, about 400 pp.[b] About two-thirds of this monograph is devoted to the activities of an American Company interested in underwater imagery. Seven additional papers are devoted to theory and experimental results relating to various aspects of optical oceanography.

[b]AGARD Publications are available in the United States from:
National Aeronautics and Space Administration (NASA)
Langley Field, Virginia 23365
Attn: Report Distribution and Storage Unit

1

Reprinted from p. 432 of *Ann. Inst. Océanogr.* [*Monaco*] **19**:393–436 (1939)

LA PÉNÉTRATION DE LA LUMIERE DANS LA MER

Y. Le Grand

[*Editor's Note:* In the original, material precedes this excerpt. An English translation of this excerpt follows.]

VIII. — RÉSUMÉ ET CONCLUSIONS

Les principaux résultats de cette étude sur la pénétration de la lumière dans la mer me semblent les suivants :

1° L'effet des réflexions et diffusions à la surface de l'eau est négligeable. Par contre, j'ai mis en évidence un effet purement instrumental d'agitation, effet important qui concorde avec les pertes apparentes que l'on attribuait jusqu'ici à une couche absorbante superficielle.

2° La mer est un milieu trouble où absorption véritable et diffusions se mélangent intimement. J'ai défini les coefficients correspondants et calculé théoriquement les valeurs de la diffusion des molécules d'eau et de celle des particules en suspension.

3° L'absorption vraie de l'eau distillée est importante vers l'extrémité rouge du spectre. Celle de l'eau de mer purifiée s'en rapproche beaucoup. Par contre, les eaux naturelles révèlent un effet considérable des matières en suspension dans la mer.

4° J'ai étudié les méthodes mesurant l'absorption de la lumière en insistant sur la technique la plus usuelle (photomètre photoélectrique immergé). La loi véritable de décroissance de l'éclairement avec la profondeur, compte tenu de la diffusion, diffère de la loi exponentielle classique. Le « coefficient d'extinction vertical » qui suppose cette loi n'est donc qu'approché.

5° La diffusion a donné lieu à quelques recherches, d'où il résulterait que le rôle essentiel est celui des matières en suspension.

6° J'ai décrit un photomètre universel de ma construction, permettant la mesure exacte d'effets qui se mélangent dans les techniques ordinaires.

7° La couleur de la mer n'est pas due, comme on l'admet parfois, à une diffusion analogue au bleu du ciel. C'est une couleur vraie, résultant de l'absorption du rouge par l'eau pure et de la présence de substances en suspension ou en solution.

8° Enfin l'étude de la visibilité des corps immergés m'a conduit à la règle pratique permettant de comparer le disque de Secchi aux autres méthodes : le produit du coefficient d'extinction vertical par la distance limite de vision du disque varie entre 1,5 (eau trouble) et 4 (eau très limpide).

Cette étude n'est nullement exhaustive : c'est plutôt un essai de mise au point de problèmes anciens et difficiles à résoudre à cause de la superposition de multiples phénomènes. La mer est un milieu vivant qui ne se laisse ni schématiser ni brutaliser. Il faut l'étudier sur place et patiemment. Nous commençons à peine à la connaître, et l'avenir reste largement ouvert à tous ceux qui veulent pénétrer ses secrets.

11

1

THE PENETRATION OF LIGHT IN THE SEA

Y. Le Grand

*This excerpt was translated by the volume editor from "La penetration de la lumiere dans la mer," Ann. Inst. Oceanogr. [Monaco] **19**:432–436 (Oct. 17, 1939), with the permission of the Institut Oceanographique and the author.*

SUMMARY AND CONCLUSIONS

The principal results of this study on the penetration of light into the sea, seem to me as follows:

1) The effect of reflection and scattering at the surface of the water is negligible. On the other hand, I have demonstrated a purely instrumental effect of perturbation (of the light field), an important effect that agrees with the apparent losses that, until now, have been attributed to a shallow absorbing layer.

2) The sea is a turbid medium where true absorption and scattering are intimately mixed. I have defined the corresponding coefficients and theoretically calculated the values for the scattering of the molecules of water and for the particles in suspension.

3) The true absorption of distilled water is important in the extreme red region of the spectrum. That, for purified sea water, is in close agreement. By contrast, natural waters reveal a considerable effect from the suspended material in the sea.

4) I have studied the methods for measuring the absorption of light, stressing the most common technique (submarine photoelectric photometry). The true law of attenuation of light with depth, taking into account the scattering, differs from the classic exponential law. The "vertical coefficient of extinction," which infers this law, is therfore only approached.

5) Scattering is the occasion for some research from which it should result that the essential role is that of the suspended materials.

6) I have described a universal photometer of my construction that permits the exact measurement of the effects, which are mixed together, when measured by ordinary techniques.

7) The color of the sea is not due, as previously supposed, to scattering similar to the blue of the sky. It is a true color resulting from the absorption of red light by pure water and from the presence of substances in suspension or in solution.

8) Finally, the study of the visibility of submerged objects has guided me to the practical rule permitting a comparison of the Secchi disc with other methods: The product of the vertical extinction coefficient by the limiting distance of vision of the disc, varies between 1.5 (turbid water) and 4 (very clear water).

This study is by no means exhaustive: it is rather an attempt to put old problems into focus and resolve difficulties resulting from the superposition of multiple phenomena. The sea is a living medium that allows itself to be neither systematized nor brutalized. It is necessary to study it in place and patiently. We are hardly beginning to know it, and the way is open to all those who wish to penetrate its secrets.

REFERENCES

From the enormous literature concerning the penetration of light into the sea we cite only the accounts effectively used for our work. A complete bibliography of the question is set to appear soon in the Journal du Conseil. For historical works, the reader should consult references 7, 8, and 11 below.

1. CABANNES (J.).—La diffusion moléculaire de la lumière. Les Presses universitaires de France, Paris, 1929.
   Theoretical and experimental work on the molecular scattering in gas and liquids. We have used the general formula from pages 39 and 213 and those given for pure water, pages 187 and 197.
2. FABRY (CH.).—Introduction générale à la Photométrie. Editions de la Revue d'Optique, Paris, 1927.
   Classic work of admirable clarity. I have followed closely pages 101 and following at the beginning of my chapter IV.
3. HARDY (A. C.).—Handbook of Colorimetry. Cambridge, Mass., 1936.
   Definitions and colorometric tables. I have used tables VI and VII for my definition of sunlight.
4. MAURAIN (CH.).—Etude pratique des rayonnements solaire, atmosphérique et terrestre. Gauthier-Villars, Paris, 1937.
   Many references condensed in a small volume. One should consult Chapt. XIII for daylight.
5. MIDDLETON (W. E. K.).—Visibility in Meteorology. The University of Toronto Press, Toronto, 1935.
   Studies in meteorological optics, more advanced than those of the sea; forms an excellent preface. On the transparency of haze, see pages 11 to 20. On the minimum perceptable contrast, see page 49.
6. KNUDSEN (M.).—On Measurement of the Penetration of Light into the Sea. Publ. n° 76, 1922.
   Interesting spectrographic method studied in detail in my account.
7. ATKINS (W. R. G.).—A quantitative Consideration of some Factors concerned in plant growth in water. T. 1, pp. 97–126, 1926.
   Account of old research on the transparency of the sea.
8. ATKINS (W. R. C.).—Solar Radiation and its transmission through Air and Water, T.7, pp. 171–211, 1932.
   Interesting historical account. Certain informations concerning the sun's radiation are incorrect.
9. ATKINS (W. R. G.), CLARKE (G. L.), PETTERSSON (H.), POOLE (H. H.), UTTERBACK (C. L.) et ÅNGSTRÖM (A.).—Measurement of Submarine Daylight. T. 13, pp. 37–57, 1938.
   Fundamental article wherein are assembled and discussed procedures of submarine photometry. Practical recommendations.
10. PETTERSSON (H.).—Submarine Daylight and the Transparency of Sea water. T. 10, pp. 48–65, 1935.
    Contains the description of remarkable apparatus invented by the author and some results.

11. UTTERBACK (C. L.) et JORGENSEN (W.).—Absorption of Daylight in the North Pacific Ocean. T. 9, pp. 197–209, 1934.
    Research on the spectral variation of the extinction coefficient. Bibliography

12. UTTERBACK (C. L.) et MILLER (E. K.).—Variations in Components of Submarine Daylight for 1935 and 1936. T. 12, pp. 305–310, 1937.
    Seasonal variation of the extinction coefficient.

13. ÅNGSTRÖM (A.).—On the Unit for Radiation in Oceanographic Research. T. 101, II-5, 1936.
    Spectral curves of sun radiation. Necessity for employing energy units and not visual units.

14. CLARKE (G. L.).—Observations on the Penetration of Daylight into mid-Atlantic and coastal Waters. T. 85, III, pp. 47–51, 1933.
    Measurements of transparency in the Atlantic. For a comparison of my theoretical curve with experiment, see Fig. 4 of the author, curves 11 and 12 relative to Sargasso Sea.

15. CLARKE (G. L.). Light Penetration in the Western North Atlantic and its application to Biological Problems. T. 101, II-3, 1936.
    Resume of the author's research on submarine photometry.

16. GRAHAM (M.).—Trial of Methods of measuring Transparency of Sea Water. T. 108, II, pp. 13–17, 1938.
    Studies of transparencies in the North Sea by simultaneous measurements with a deck photometer, a submerged photoelectric photometer, and a Secchi disc.

17. PETTERSSON (H.).—Measurements of the Angular Distribution of Submarine Daylight. T. 108, II, pp. 7–12, 1938.
    Principle of measurement by rotating a diaphragm cell, and some results.

18. POOLE (H. H.).—The Photo-Electric Measurement of Submarine Illumination in off-Shore Waters. T. 101, II-2, 1936.
    Summary outline of methods of submarine photometry by means of practical cells. Various corrections. Precision of measurement.

19. POOLE (H. H.).—The Effect of Surface Conditions on the Intensity and Angular Distribution of Submarine Daylight. T. 108, II, pp. 3–6, 1938.
    Contains an interesting analysis of the research of Powell and Clarke (reference 24 below).

20. CLARKE (G. L.) et JAMES (H. R.).—Laboratory Analysis of the Selective Absorption of Light by Sea Water. T. 29, pp. 43–55, 1939.
    Very important work on the transparency of distilled water and of sea water, filtered or not. Role of suspended particles.

21. DAWSON (L. H.) et HULBURT (E. O.).—The Absorption of Ultra-violet and visible Light by Water. T. 24, pp. 175–177, 1934.
    Photographic measurements of the absorption of ultraviolet light by distilled water, with water paths up to 2.7 meters.

22. DAWSON (L. H.) et HULBURT (E. O.).—The Scattering of Light by Water. T. 27, pp. 199–201, 1937.
    Verification of Rayleigh's law for the scattering of pure water. There is an error of 10 in the author's calculations, but that does not change the validity of the verification, which depends only on the behavior of the curve.

23. HULBURT (E. O.).—The Penetration of Ultra-violet light into pure Water and sea Water. T. 17, pp. 15–22, 1928.
    Photoelectric measurements on small thicknesses of solutions of the salts which exist in sea water. Poor precision.

24. POWELL (W. M.) et CLARKE (G. L.).—The Reflection and Absorption of Daylight at the Surface of the Ocean. T. 26, pp. 111–120, 1936.
    Photoelectric research on the effects of the surface. I have studied in detail this interesting work in Chapter II.

25. STEPHENSON (E. B.).—Absorption of Light by Sea Water. T. 24, pp. 220–221, 1934.
    Brightness measurements through the windows of a submarine chamber.

26. UTTERBACK (C. L.) et JORGENSEN (W.).—Scattering of Daylight in the Sea. T. 26, pp. 257–259, 1936.
    Measures of horizontal lighting with depth at various depths and for three colors.

26a. YOUNG (R. T. Jr.).—On the Calculation of Absorption Coefficients of Daylight in Natural Waters. T. 28, pp. 95–99, 1938.
    Calculation of the theoretical relationship between the vertical extinction coefficient and the total coefficient, c, in the case of homogeneous non-scattering water under a clear sky.

27. ATKINS (W. R. G.) et POOLE (H. H.).—The Photo-Electric Measurement of the Penetration of Light of various Wave-lengths into the Sea and the Physiological Bearing of the Result. *Philos. Trans. Roy. Soc. London,* B, t. 222, pp. 129–164, 1933.
    Joint discussion on the transparency of the sea, based on research by the authors. Effect of reflections (p. 138). Employment of the vertical extinction coefficient. Bibliography.

28. BLUMER (H.).—Die Farbenzerstreuung an kleinen Kugeln. *Zeits. f. Physik,* t. 39, pp. 195–214, 1926.
    Conclusion of a series of articles on the calculation of scattering by spherical particles, after the formulas of Mie. It is there that I have borrowed the numerical results of my Table 4.

29. CABANNES (J.).—Sur l'absorption atmosphérique. *Sciences* (Revue de l'Ass. fr. pour l'avancement des Sc.), t. 65, pp. 308–334, 1937.
    Clear and condensed exposition by a specialist on the subject.

30. DARBY (H. H.), JOHNSON (E. R. F.) et BARNES (G. W.).—Studies on the Absorption and Scattering of Solar Radiation in the Sea; Spectrographic and Photoelectric Measurement. *Pap. Tortugas Lab., Carnegie Inst. of Washington,* t. 31, pp. 193–205, 1937.
    Measurements with a watertight quartz spectrograph in shallow water. Study of the penetration of ultraviolet light and of scattering. Interesting results but above all qualitative.

31. HELLAND-HANSEN (B.).—Physical Oceanography and Meteorology. *Rep. Michael Sars N. Atlantic Deep-Sea Exp. 1910,* t. 1, p. 43. Bergen, 1931.
    Photographic measurements in the Sargasso Sea to 1700 meters of depth.

32. JAMES (H. R.) et BIRGE (E. A.).—A Laboratory Study of the Absorption of Light by Lake Waters. *Trans. Wis. Acad. of Sc.,* t. 31, pp. 1–154, 1938.
    Important work, contains, in particular, some measurements of the transparency for 1 meter of distilled water.

33. KALLE (K.).—Zum Problem des Meereswasserfarbe. *Ann. der Hydrographie und maritimen Meteorologic,* t. 1, pp. 1–13, 1938.
    Interesting exposition of the question of color of the sea.

34. KING (L. V.).—Absorption Problems in Radioactivity. *Philos. Mag.,* 6° série, t. 23, pp. 242–250, 1912.
    Contains, on page 245, a table of numerical values for the non-calculable integral which we have encountered in Chapt. IV.

35. KING (L. V.).—On the Scattering and Absorption of Light in gaseous Media, with application to the Intensity of Sky Radiation. *Philos. Trans. Roy. Soc. London,* A, t. 212, pp. 375–433, 1913.
    Problem of multiple scattering treated with considerable mathematical elegance.

36. LE GRAND (Y.).—Mesure photographique de Pagitation de la mer. *C. R. du 59° Congrès de l'Ass. fr. pour l'avancement des Sc.,* p. 219, 1935.
    Resume of a paper presented to the Congress of Nantes, July 1935. This resume is found mixed with that of another communication by the same author on the ripples of sand of the sea shore.

37. LE GRAND (Y.).—Appareil pour la mesure photographique des propriétés diffusantes de l'eau de mer. *Bull. Lab. marit. de Dinard,* fasc. 18, pp. 53–56, 1938.
    First model of the photometer described in Chapt. VI.

38. MAC LENNAN (J. C.), RUEDY (R.) et BURTON (A. C.).—An investigation of the Absorp-

tion Spectra of Water and Ice, with Reference to the Spectra of the major Planets. *Proc. Roy. Soc. London,* A, t. 120, pp. 296–302, 1928.
Absorption spectra for a 21.5 meter path of pure water, in the visible.

39. POOLE (H. H.) et ATKINS (W. R. G.).—On the Penetration of Light into Sea Water. *Journ. Mar. Biol. Assoc.,* t. 14, p. 177, 1926.
Interesting photoelectric research. Study of surface losses.

40. POOLE (H. H.) et ATKINS (W. R. G.).—Further Photoelectric Measurements of the Penetration of Light into Sea Water. *Ibid.,* t. 15, p. 455, 1928.
Continuation, notably of comments on scattering and on the interpretation of measurements made with a Secchi disc.

41. RAMAN (C. V.).—On the Molecular Scattering of Light in Water and the Colour of the Sea. *Proc. Roy. Soc.,* A, t. 101, pp. 64–80, 1922.
Interesting memoir, whose premises are badly in error. Molecular scattering plays a role more restrained that supposed by the author.

42. SHELFORD (V.) et GAIL (F. W.).—A Study of light Penetration into Sea water made with the Kunz photo-electric Cell. *Publ. Puget Sound Biol. Sta.,* t. 3, p. 141, 1922.
First important work on the use of photocells for the measurement of the transparency of the sea.

43. STRATTON (J. A.) et HOUGHTON (H. G.).—A Theoretical Investigation of the Transmission of Light through Fog. *Phys. Review,* t. 38, pp. 159–165, 1931.
Theoretical calculation of the apparent absorption coefficient by scattering, after the formula of Mie, for spherical particles of relative index 1.33. Comparison with the trial experiment for clouds.

44. SWEITZER (C. W.).—Light Scattering of aqueous Salt Solutions. *Journ. of Phys. Chem.,* t. 31, pp. 1150–1191, 1927.
Experimental research on the molecular scattering of water and of aqueous solutions.

45. TSUKAMOTO (K.).—Transparence de l'eau de mer pour l'ultra-violet lointain. *C. R. Acad. Sc. Paris,* t. 184, pp. 221–223, 1927.
Laboratory study of sea water by spectroscopy. The absorption towards 0.2 micron is attributed by the author to bromides in solution.

Part I

THE GLASS-BOTTOMED BOAT

Editor's Comments
on Paper 2

2   **DUNTLEY**
     Light in the Sea

   As early as 1944 S. Q. Duntley at the Massachusetts Institute of
Technology undertook a theoretical and experimental study of light in
the ocean and in lake water. An objective of this study was to develop a
quantitative approach to image deterioration along paths of sight
underwater.
   In October 1961, the Optical Society of America awarded the Fre-
deric Ives Medal for Distinguished Work in Optics to Dr. S. Q. Duntley.
His Ives Medal address (Paper 2), which follows, reviews his research
on light in the sea and relates this research to the work of others, many
of whom were associated with him.

2

Reprinted from *Opt. Soc. Am. J.* **53**(2):214–233 (1963)

Light in the Sea*

SEIBERT Q. DUNTLEY

Visibility Laboratory, Scripps Institution of Oceanography, La Jolla, California

(Received 27 August 1962)

Light in the sea may be produced by the sun or stars, by chemical or biological processes, or by man-made sources. Serving as the primary source of energy for the oceans and supporting their ecology, light also enables the native inhabitants of the water world, as well as humans and their devices, to see. In this paper, new data drawn from investigations spanning nearly two decades are used to illustrate an integrated account of the optical nature of ocean water, the distribution of flux diverging from localized underwater light sources, the propagation of highly collimated beams of light, the penetration of daylight into the sea, and the utilization of solar energy for many purposes including heating, photosynthesis, vision, and photography.

INTRODUCTION

AN interest in the aerial photography of shallow ocean bottoms prompted the author to begin, nearly 20 years ago, a continuing experimental and theoretical study of light in the sea. Some of the principles discovered or extended and generalized by the author and his colleagues are summarized in this paper. Early discussions with E. O. Hulburt and D. B. Judd as well as publications by many investigators[1] provided a valuable starting point. By 1944 the author was using a grating spectrograph, specially designed by David L.

MacAdam, in a glass-bottomed boat off the east coast of Florida to obtain the spectroradiometric data shown in Fig. 1; the presence of reefs and sandy shoals show clearly in the green region of the spectrum.[2] When the spectrograph was flown in an airplane 4300 ft above the same ocean locations, the radiance spectra shown in Fig. 2 were obtained.[3,4] The data in Figs. 1 and 2, displayed in colorimetric form by Fig. 3, exhibit many intricate and beautiful phenomena which are manifestations of some of the physical principles discussed in this paper.

The importance of light in the sea is apparent when it is recalled that solar radiation supplies most of the energy input to the ocean and supports its ecology

---

* Most of the investigations described in this paper were supported by the Office of Naval Research and the Bureau of Ships of the U. S. Navy. Grants from the National Science Foundation have also aided the work. At certain times in the past the research was supported by the National Defense Research Committee and by the U. S. Navy's Bureau of Aeronautics.

[1] See E. F. DuPré and L. H. Dawson, "Transmission of Light in Water: An Annotated Bibliography," U. S. Naval Research Laboratory Bibliography No. 20, April, 1961 for abstracts of 650 publications by over 400 authors in more than 150 Swiss, German, French, Italian, English, and U. S. journals and other sources from 1818 to 1959.

[2] S. Q. Duntley, *Visibility Studies and Some Applications in the Field of Camouflage*, Summary Tech. Rept. of Division 16, NDRC (Columbia University Press, 1946), Vol. II, Chap. 5, p. 212.

[3] See J. G. Moore, Phil. Trans. Roy. Soc. (London) **A240**, 163 (1946–48) for a method of using such data to determine depth and attenuation coefficients of shallow water.

[4] See G. A. Stamm and R. A. Hengel, J. Opt. Soc. Am. **51**, 1090 (1961) for data on the spectral *irradiance* incident on the underside of an aircraft flying above the ocean.

through photosynthesis. The biological productivity of an acre of ocean has been estimated to be, on a world-wide average, comparable to that of an acre of land. Most of the surface of our "water planet" is covered by seas and its atmosphere contains great quantities of water in the form of vapor and clouds. Light in the sea enables the native inhabitants of the water world to find their food and to evade attack. Nowhere in nature is protective coloration more perfectly or dramatically displayed than in the feeding grounds of the sea. Man and his cameras may view underwater scenes by means of daylight or with the aid of artificial lighting devices. Many biological organisms, including some living at very great depth, produce their own light at or near the wavelength for which water is most transparent, presumably both for vision and for signaling. All of these

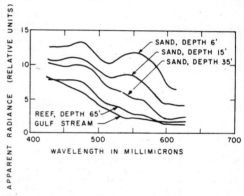

FIG. 2. Spectroradiometric curves of light from the nadir reaching a spectrograph in an airplane 4300 ft above the same ocean locations as in Fig. 1. Spectral resolution: 7.0 mμ; spatial resolution: 3.2×10⁻⁶ sr.

duced by artificial underwater light sources, for any optical input to the water may be represented by an appropriate superposition of highly collimated, monochromatic beams. The following paragraphs describe a variety of experiments which have been made by using a collimated, underwater light source, shown schematically in Fig. 4, at the Visibility Laboratory's Field Station at Diamond Island, Lake Winnipesaukee, New Hampshire.

Attenuation of a Collimated Beam

If a collimated beam of monochromatic light is injected into macroscopically homogeneous water by

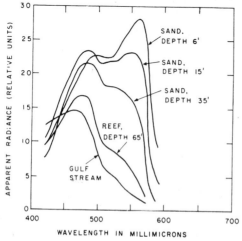

FIG. 1. Spectroradiometric curves of light from the nadir reaching a spectrograph mounted in a glass-bottomed boat off Dania, Florida (March 1944). Spectral resolution: 7.7 mμ; spatial resolution: 2.0×10⁻⁶ sr.

aspects of light in the sea can be treated by describing the optical nature of ocean water, the distribution of flux diverging from localized underwater light sources, the propagation of highly collimated beams of light, and the penetration of daylight into the sea. An integrated account of these topics is the subject of this paper.

OPTICAL NATURE OF OCEAN WATER

Most of the optical properties of ocean water as well as many of the principles which govern the propagation of light in the sea can be studied by injecting a highly collimated beam of monochromatic light into otherwise unlighted water and measuring all aspects of the resulting distribution of flux. This investigative approach even provides a basis for understanding the distribution of daylight in the sea and the submarine lighting pro-

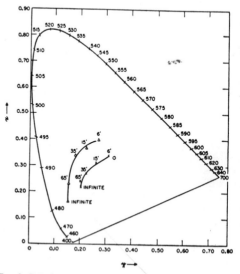

FIG. 3. CIE chromaticity diagram showing loci of the colors of ocean shoals as seen from an altitude of 4300 ft (shorter curve) and from a glass-bottomed boat (longer, upper curve). The points were calculated from the spectral radiance data in Figs. 1 and 2. The circled point represents CIE source C.

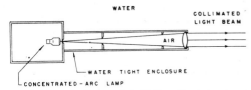

FIG. 4. Schematic diagram of the highly collimated underwater light source represented by a cross-hatched block in Figs. 5, 6, 7, 13, 20, and 21. This source was used in obtaining part or all of the data presented in Figs. 9, 10, 12, 17, 18, 20, and 22. Interchangeable 2, 10, 25, and 100 w zirconium concentrated-arc lamps in a water-tight air-filled enclosure produce nominal total beam spreads of 0.010°, 0.046°, 0.085°, and 0.174°, respectively, when used with a Wratten No. 61 green filter and a specially constructed air-to-water collimator lens having an effective first focal length of 495 mm. This lens, designed for the author by Justin J. Rennilson, is a cemented doublet 55 mm in diameter having radii $r_1 = 269.75$ mm, $r_2 = r_3 = 102.60$ mm, $r_4 = -325.0$ mm and axial thicknesses $t_1 = 3.0 \pm 0.2$ mm, $t_2 = 6.5 \pm 0.2$ mm. The first element is of Hayward LF-2 glass ($N_D = 1.5800 \pm 0.0010$; $\nu = 41.0$) and the second is of Hayward BSC-1 ($N_D = 1.5110 \pm 0.0010$; $\nu = 63.5$). The free aperture is 50.0 mm. The first back focal length of the doublet with its last surface in water is 493.88 mm. The air–glass surface was treated for increased light transmission. The achromatization is such that with the 2-W concentrated-arc lamp the extreme ray divergence is 0.0031°, 0.0039°, and 0.0109° at 480, 520, and 589 mμ, respectively, when the lamp is used in fresh water having a temperature of 20°C. A Wratten No. 61 green filter was used during all of the experiments with this lamp, but it does not appear in Fig. 4 because it was always incorporated in the photometer or the camera. An external circular stop (not shown) can be mounted in the water close to the lens whenever a smaller beam diameter is desired.

means of an underwater projector, as suggested by Fig. 5, it is found that the residual radiant power $P_r{}^0$ reaching a distance r *without having been deviated by any type of scattering process* is

$$P_r{}^0 = P_0 e^{-\alpha r}, \qquad (1)$$

where P_0 represents the total flux content of the beam as it leaves the projector. The zero superscript on $P_r{}^0$ denotes the *zero scattering order*, i.e., nonscattered radiant power. The *spectral volume attenuation coefficient* α, defined by Eq. (1), has the dimension of reciprocal length and can be expressed in *natural log units per meter* (ln/m), natural log units per foot (ln/ft), etc; it is a scalar point function of position which may vary along any underwater path of sight if the water is macroscopically nonhomogeneous.

The attenuation of a beam of light by water results from two independent mechanisms: *scattering* and *absorption*. Scattering refers to any random process by which the *direction* of individual photons is changed without any other alteration. Absorption includes all of the many thermodynamically irreversible processes by which photons are changed in their nature or by which the energy they represent is transformed into thermal kinetic energy, chemical potential energy, and so on. Transformation of photon energy into thermal kinetic

energy of the water is the major absorption mechanism in the ocean. Photosynthetic conversion of light into chemical potential energy is, of course, measurable and vital to the existence of life in the sea. Visible light fluorescence and transpectral effects are ordinarily too minute to be detected in ocean water. The volume attenuation coefficient α is the sum of the *volume absorption coefficient* a and the *total volume scattering coefficient* s: thus $\alpha = a + s$.

Wavelength dependence. The attenuation coefficient of all water (pure, distilled, or natural) varies markedly with wavelength. Typical data are summarized in Table I, wherein the reciprocal of the volume attenuation coefficient, called *attenuation length*, has been tabulated rather than attenuation coefficient for three reasons: (1) a distance is easier to visualize and to remember than a reciprocal distance; (2) visibility calculations and many experiments by swimmers show that any large

TABLE I. Attenuation length of distilled water at various wavelengths.[a–e]

Wavelength mμ	Attenuation length $(1/\alpha)$ meters/ln
400	13.
440	22.
480	28.
520	25.
560	19.
600	5.1
650	3.3
700	1.7

[a] E. O. Hulburt, J. Opt. Soc. Am. 35, 698 (1945).
[b] For ultraviolet attenuation data see L. H. Dawson and E. O. Hulburt, J. Opt. Soc. Am. 24, 175 (1934).
[c] For near infrared attenuation data see J. A. Curcio and C. C. Petty, J. Opt. Soc. Am. 41, 302 (1951).

dark object (such as a dark-suited swimming companion) is just visible at a horizontal distance of about 4 attenuation lengths when there is sufficient underwater daylight; (3) many physicists like to characterize any absorbing–scattering medium (such as water) by the mean free path for a photon in the ordinary kinetic theory sense; this is the attenuation length $1/\alpha$. The term, "20-meter water," signifying water having an attenuation length of 20 m/ln, facilitates verbal discussions.

Water possesses only a single important window, the peak of which lies near 480 mμ unless it is shifted toward the green by dissolved yellow substances. Such yellow solutes, usually prominent in coastal waters, consist of humic acids, melanoidins, and other compounds which result from the decomposition of plant and animal materials. Clear ocean water is so selective in its absorption that only a comparatively narrow band of blue-green light penetrates deeply into the sea[5] (see Fig. 1) but this radiation has been detected at depths greater than 600 m with a multiplier phototube photometer.[6]

[5] J. E. Tyler, Limnology and Oceanography 4, 102 (1959).
[6] S. Q. Duntley, Natl. Acad. Sci.—Natl. Research Council Publ. 473, 79 (1956).

FIG. 5. Illustrating the geometry of Eq. (1). The cross-hatched block represents the collimated underwater light source (projector) shown schematically in Fig. 4.

21

Many have wondered whether there exists any fine structure in the volume attenuation function which was beyond the spectral resolution available to the investigators whose results are summarized by Table I. Is there, for example, a narrow-band window of high transmission? It is the concensus of most physicists that the atomic and molecular structures involved in water provide no reason to expect any significant fine structure in the spectral attenuation function. A careful spectroscopic examination of the region from 3750 to 6850 Å with a resolution of 0.2 Å and sensitivity sufficient to detect a variation of 0.02 ln/m in the attenuation coefficient has been reported by Drummeter and Knestrick.[7] They detected no fine structure, i.e., no narrow-band window.

Water Clarity

The clearest body of ocean water of large extent is reputed to be in the Sargasso Sea, a vast region of the Atlantic Ocean east of Bermuda. Jerlov has reported very clear water between Madeira and Gibraltar,[8] as

TABLE II. Attenuation length of the Atlantic Ocean for wavelength 465 mμ at various depths in the vicinity of Madeira and Gibraltar.[a]

Depth meters	Attenuation length $(1/\alpha)$ meters/ln
0–10	19
10–25	20
25–50	18
50–75	15
75–90	16

[a] N. G. Jerlov, Kgl. Vetenskap. Vitterh. Handl. F.6, Ser. B, BD8.N:o11 (1961).

summarized by Table II. Although clearer water was found at 10 m depth than at 90 m at this location, the reverse is often true elsewhere. Optical oceanographic data are not numerous. Jerlov's measurements during the Swedish Deep Sea Expedition of 1947–48 are classical examples. Table III shows some of these data selected to typify certain indicated locations.[9]

DuPré and Dawson[1] give many references to water-clarity data; users of published data should note carefully whether the attenuation coefficients reported are expressed in ln/m or in log/m and whether the values refer to the attenuation coefficient α for nonscattered light, as in a collimated beam, or to some form of *diffuse attenuation coefficient* K, discussed later in this paper. No single number can adequately specify the clarity of any natural water because two independent mechanisms, absorption and scattering, govern water

[7] L. F. Drummeter and G. L. Knestrick, U. S. Naval Research Laboratory Rept. No. 5642 (1961).
[8] N. G. Jerlov, Kgl. Vetenskap. Vitterh. Handl. F.6, Ser. B, BD8. N:o 11 (1961).
[9] N. G. Jerlov, Reports of the Swedish Deep Sea Expedition of 1947–48 (1951), Vol. III, p. 49, Table 27.

TABLE III. Attenuation length of ocean water for wavelength 440 mμ at various locations.[a]

Location	Attenuation length $(1/\alpha)$ meters/ln
Caribbean	8
Pacific N. Equatorial Current	12
Pacific Countercurrent	12
Pacific Equatorial Divergence	10
Pacific S. Equatorial Current	9
Gulf of Panama	6
Galapagos Islands	4

[a] N. G. Jerlov, Reports of the Swedish Deep Sea Expedition of 1947–48 (1951), Vol. 3, p. 49, Table 27.

clarity. Even for monochromatic light, at least two coefficients, such as α and K, are required, and a more complete specification requires data on the volume scattering function $\sigma(\vartheta)$, defined in the paragraphs which follow.

Daylight, abundant in the mixed layer near the surface, supports the growth of phytoplankton in the biologically productive regions of the oceans. These, in turn, feed a zooplankton population. The transparent planktonic organisms, ranging in size from microns to centimeters, scatter light and thereby produce optical attenuation. Settling of the plankton, particularly after death, tends to produce a high concentration of these scatters just above the thermocline which ordinarily exists at the lower boundary of the mixed layer in the sea.[10] Below the thermocline lies clearer water which may be optically uniform for tens or hundreds of meters before some different water mass is encountered. Interestingly, the optical structure of the ocean resembles, in a sense, that of the atmosphere if depth is considered as analogous to altitude and a proper allowance is made for the decrease of atmospheric density with height.

Scattering

Scattering of light in the sea is predominantly due to transparent biological organisms and particles large compared with the wavelength of light. The magnitude of the scattering is, therefore, virtually independent of wavelength.[11] The variation of attenuation length with

[10] Multiple thermoclines often form in the upper portion of the sea; the maximum optical attenuation is associated with the maximum vertical temperature gradient and frequently falls on a secondary thermocline. Internal waves shift the scattering layer vertically. See E. C. La Fond, E. G. Barham, and W. H. Armstrong, U. S. Navy Electronics Laboratory Rept. 1052 (July 1961), p. 15. Also see J. Joseph, Deut. Hydrograph. Z., Nr. 5 (1961).
[11] Scattering is also contributed by fine particles, by molecules of water, and by various solutes, but these contributions are usually quite minor and often difficult to detect. Even in very clear, blue ocean water scattering by water molecules produces only 7% of the total scattering coefficient and is dominant only at scattering angles near 90°, where it provides more than 2/3 of the scattered intensity (see reference 8); although the magnitude of this small component of scattering varies inversely as the fourth power of wavelength (λ^{-4}), it is so heavily masked by nonselective scattering due to large particles that total scattering in the sea is virtually independent of wavelength. The prominent blue color of clear ocean water, apart from sky reflection, is due almost entirely to selective absorption by water molecules.

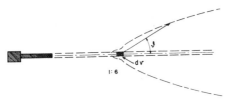

Fig. 6. Polar diagram illustrating Rayleigh scattering by pure water. The ratio of the light scattered into the rear hemisphere to that scattered into the forward hemisphere is 1 to 1. The cross-hatched block represents the collimated underwater light source shown schematically in Fig. 4.

wavelength (see Table I) is due almost wholly to selective absorption.

In the blue region of the spectrum, centering at 480 mμ, approximately 60% of the attenuation coefficient of clear, blue ocean water is due to scattering and 40% is due to absorption; e.g., $s=0.030$ ln/m and $a=0.020$ ln/m.[8] In all other spectral regions absorption is overwhelmingly predominant in very clear water.

Since scattering is virtually independent of wavelength its detailed nature is best revealed by means of experiments conducted at or near the wavelength of minimum absorption. This means experiments with blue light in clear, blue ocean water and experiments with green light in greenish coastal and lake waters.

Scattering by pure water. Consider a scattering experiment performed in pure water, that is, in water molecules containing no dissolved or particulate matter whatsoever. As in Fig. 6, consider an element of volume dv receiving collimated, nonpolarized, monochromatic irradiance H to act as source of scattered light, producing radiant intensity $dJ(\vartheta)$ at scattering angle ϑ. Scattering by the water molecules will be Rayleighian, with $dJ(\vartheta) \sim \lambda^{-4}$ and with the shape of the intensity function $dJ(\vartheta)$ characterized by $(1+0.835 \cos^2\vartheta)$ (see reference 12). Since even the most elaborately prepared distilled water samples show particulate matter when examined in a light beam, scattering by truly pure water has probably never been measured.

Scattering by distilled water. A colleague, John E. Tyler, has performed scattering experiments in many samples of commercial distilled water[13]; Fig. 7 shows a typical result. Obviously, the scattering produced by this sample of distilled water is very different from that

predicted for pure water. The predominant forward scattering is caused by a comparatively few large particles. The dotted curve may be regarded either as a polar plot of the radiant intensity $dJ(\vartheta)$ or of the *volume scattering function* $\sigma(\vartheta)$, defined by the equation $dJ(\vartheta)=\sigma(\vartheta)\,Hdv$, where H is the irradiance produced by the collimated lamp on the volume dv. The dimension of $\sigma(\vartheta)$ is reciprocal length; typical units are reciprocal steradian-meters or reciprocal steradian-feet. The polar curve in Fig. 7 is not complete; it begins at $\vartheta=22\,1/2°$ and stops at $\vartheta=165°$. All conventional scattering meters designed to be used *in situ* possess the limitation that they cannot measure scattering at small angles. Fortunately, the total scattering coefficient s, defined by the relation

$$s=2\pi \int_0^\pi \sigma(\vartheta)\, \sin\vartheta d\vartheta,$$

is insensitive to the magnitude of small-angle forward scattering. Unfortunately, however, the propagation of highly collimated light does depend importantly on small-angle scattering.

Small-angle scattering. The author has devised a special (coaxial) *in situ* scattering meter to supply the missing forward part of the curve. Figure 8 is a schematic diagram of the instrument. It shows the optical system adjusted to measure the volume scattering function at a scattering angle of 1/2 deg. Such a datum was obtained with the coaxial scattering meter at the Diamond Island Field Station and determines the upper end of the upper curve in Fig. 9. This may be the first *in situ* measurement of small-angle scattering by natural water. The very large scattering found at small scattering angles is believed to have been caused primarily by re-

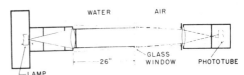

Fig. 8. Coaxial scattering meter for *in situ* measurement of the volume scattering function at small scattering angles. In this schematic drawing the vertical scale has been exaggerated five times over the horizontal scale in order to illustrate the principle of the device more clearly. The collimated underwater light source shown in Fig. 4 is used with the addition of an external opaque central stop which results in the formation of a thin-walled hollow cylinder of light. This traverses 26 in. of water to a high-quality glass window behind which, in air, is a photoelectric telephotometer with a 2° total field of view. The light source and the telephotometer are coaxial, but the latter is equipped with an external stop small enough to exclude the hollow cylinder of light so that only light scattered by the water is collected. The cylindrical scattering volume is indicated by cross-hatching. The upper limit of the scattering angle is determined by the field of the telephotometer and the lower limit is set by the size of its external stop, i.e., by the entrance pupil. A detailed geometrical analysis of the configuration depicted above shows that the scattering is measured at 0.47 deg±0.15°; this datum is used as the volume scattering function for 1/2° scattering angle in Figs. 9 and 10. Photometric calibration of the scattering meter is achieved by removing the external stop on the telephotometer.

Fig. 7. Polar diagram illustrating measured scattering by a typical sample of commercial distilled water. The ratio of the light scattered into the rear hemisphere to that scattered into the forward hemisphere is 1 to 6 for this water sample. Data are by Tyler (see reference 13). The scale of this polar plot is smaller than that used in Fig. 6.

[12] L. H. Dawson and E. O. Hulburt, J. Opt. Soc. Am. **31**, 554 (1941).
[13] J. E. Tyler, Limnology and Oceanography **6**, 451 (1961).

fractive deviations produced by the passage of the collimated light beam through transparent plankton having an index of refraction close to that of water. The curve shape at small scattering angles is chosen to suggest that the magnitude of the volume scattering function may merge tangentially with that of the irradiating beam at vanishingly small angles.

Chemists have, for many years, made laboratory measurements of very small-angle scattering from tiny volumes of scattering materials.[14] Koslyaninov[15] has reported volume scattering measurements at angles down to 1 deg by means of a shipboard laboratory apparatus using water samples brought on board for measurement. Figure 10 shows the data of Koslyaninov for the East China Sea superimposed upon the lake data from Fig. 9 after normalization at a scattering angle of 90°, as denoted by the small circle in the figure. The forward-scattering portions of the curves are similar in shape.

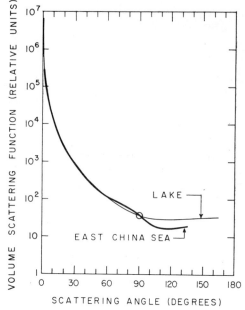

FIG. 10. Comparison of the shape of the *in situ* volume scattering function data for Lake Winnipesaukee, New Hampshire, from Fig. 9 with the shape of the *in vivo* of a curve representing the *in vivo* scattering data obtained by Koslyaninov (see reference 15) using a shipboard laboratory apparatus and a sample of water taken from the East China Sea. The curves have been normalized at a scattering angle of 90° (circled point) for purposes of shape comparison. Koslyaninov used blue light isolated by means of an absorption filter having an effective wavelength of 494 mμ; he reported data at scattering angles of 1, 2.5, 4, 6, 10, 15, 30, 50, 70, 110, and 144 deg. The curves are similar in shape for scattering angles less than 60°.

Comparison with distilled water. Figure 11 shows a comparison of *in situ* scattering measurements by Tyler[13] of commercial distilled water and clear Pacific water. Ocean water scatters more light than does distilled water but the similarity of the shape of the curves is striking and interesting in its implication of the predominant role of large particle scattering.

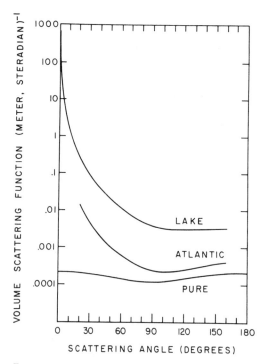

FIG. 9. Volume scattering function curves for pure water (Dawson and Hulburt, see reference 12), the Atlantic between Madeira and Gibraltar (Jerlov, see reference 8), and the Diamond Island Field Station, Lake Winnipesaukee, New Hampshire. The upper curve (lake) represents *in situ* measurements at 5° intervals between scattering angles 20° > θ > 160° by means of a conventional type, pivoted-arm scattering meter and a single datum at θ = 0.5° obtained *in situ* with the coaxial scattering meter shown schematically in Fig. 8; the data are of 20 August 1961; and are for green light isolated by means of a Wratten No. 61 filter.

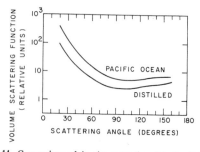

FIG. 11. Comparison of *in situ* scattering data by Tyler (see reference 13) in clear Pacific ocean water near Catalina with comparable data for a typical sample of commercial distilled water. Both curves were obtained with the same pivoted-arm scattering meter and are in the same relative units. The data are for green light isolated by means of a Wratten No. 61 filter.

[14] H. F. Aughey and F. J. Baum, J. Opt. Soc. Am. 44, 833 (1954).
[15] M. V. Koslyaninov, Trudy Inst. Okeanol. Acad. Nauk S.S.S.R. 25, 134 (1957).

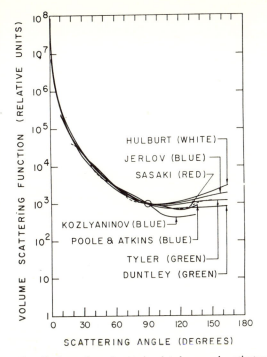

FIG. 12. Comparison of scattering data by seven investigators using dissimilar instruments in seven different parts of the world. All curves are superimposed at a scattering angle of 90°, as indicated by the circled point. Gross similarity in curve shape is apparent in the forward ($0 < \vartheta < 90°$) scattering directions despite major differences in water clarity (2 m/ln $< 1/\alpha < 20$ m/ln), spectral region, geographical location, instrumental design, and experimental technique. Most of the scattering in natural waters is caused by transparent organisms and particles large compared with the wavelength of light. The scattering is believed to result chiefly from refraction and reflection at the surfaces of the scatterers. As a consequence, scattering at small forward angles predominates and polarized light tends to preserve its polarization. To the extent that all scattering curves have identical shapes the scattering by natural waters can be specified in terms of some single number, such as the total volume scattering coefficient s or the volume scattering function at some selected angle.

Comparison between natural waters. A comparison of the scattering properties of natural waters is afforded by Fig. 12, which shows a superposition of measurements by seven different investigators using seven dissimilar instruments in seven different parts of the world. Three of the measurements were made with blue light, two were made with green light, the dashed curve was obtained with red light, and one investigator employed white light. The attenuation lengths of the waters ranged 2 m/ln for the author's lake data to 20 m/ln in the case of Jerlov's data for the Atlantic. It appears that the shape of the forward portion of the volume scattering function is remarkably similar in all of these natural waters, but that significant differences occur in the character of the backscattering they produce.

Although it is a useful first-order concept that natural waters are somewhat similar in the shape of their

volume scattering functions, it is important to note therefore that measurable differences apparently exist and that ocean water masses might therefore be identified by their scattering function curves.

Multiple scattering. The propagation of light in the sea is complicated by multiple scattering. Consider, as in Fig. 13, a plane surface irradiated at normal incidence by the collimated lamp shown in Fig. 4. Every point on the plane receives scattered light from every volume element within the light beam. It receives, moreover, multiply scattered light from every elementary volume of water near the beam. In fact, every volume element within the sea is irradiated by every other volume element both inside and outside the beam. The figure illustrates how irradiation is produced throughout the plane by second-, third-, and fourth-order scattering.

Although theoretical treatments of the effects of multiple scattering on the distribution of light in the sea both from underwater sources and from daylight have been undertaken with partial success by several workers, no fully practical solution has yet been evolved. Some derivations include only secondary scattering and neglect higher-order effects. Others, following the practice of neutron physics, assume the scattering to be virtually isotropic, that is to say, the shape of the volume scattering function is assumed to be spherical or nearly so; this is, of course, highly unrealistic. Four patterns of approach characterize the theories: (1) *Multiple integration* using the volume scattering function, the attenuation coefficient α, and the inverse square law; these treatments suffer from complexity, are never complete, and may neglect sizeable components of flux but some useful approximate solutions have been achieved in special cases. (2) *Diffusion theory.* This applies rigorously only to isotropic or very mildly nonisotropic scattering systems which are not found in the sea; nevertheless, considerable success has been achieved in the prediction of *irradiance* at long ranges; diffusion theory is, however, unable to yield much information concerning the directional characteristics of the underwater light field. (3) *Radiative transfer.* This method is based upon equations of transfer, sometimes in vector form; these integro-differential equations are solved in practice by iterative procedures on the largest electronic computers. (4) *Monte Carlo procedures.* These also require the use of large electronic computers. Al-

FIG. 13. Illustrating the irradiation of an object by multiply scattered light at arbitrary points inside and outside the light beam. The dotted curve associated with each cross-hatched volume element has the shape shown in Fig. 7 and represents a polar plot of the volume scattering function. The need for additional scattering data at small forward angles is obvious.

though, *in principle*, either of the two latter approaches appears to be capable of handling all underwater light propagation problems, neither has thus far achieved appreciable practical success in the treatment of point source or collimated beam geometrics, for the calculations are too massive for even the largest of electronic computers. Success has, however, been achieved for the case of daylight in the sea,[16] wherein the development of theory and the evolution of practical computation procedures followed quickly after experimental explorations of underwater daylight radiance distributions had produced a body of data, described later in this paper, from which valid assumptions could be made and against which predictions could be checked. This experience prompted the author to begin a program of experimental explorations of the distribution of light produced by submerged divergent light sources and by collimated lamps underwater. These explorations are still in progress, but some of the conclusions reached thus far are summarized in the following section.

DIVERGENT LIGHT IN THE SEA

Marine organisms which emit nearly hemispherical flashes of light are found at virtually all depths in the sea. Underwater lighting for vision, television, or photography is often accomplished by means of incandescent lamps or flash tubes which approximate point sources and emit divergent flux. Quantitative prediction of the irradiation produced by such lamps at the object, on its background, and throughout the observer's path of sight can enable optimum lighting arrangements and camera positions to be planned in advance and exposure to be predicted with sufficient accuracy to permit high-contrast photographic techniques to be employed effectively.

Apparent Radiance at the Object

Every underwater object and every elementary volume of water irradiated by a submerged divergent light source is lighted by an apparent radiance distribution which depends upon the radiant intensity distribution of the lamp, the optical properties of the water, and the lamp distance. This radiance distribution can be seen, photographed, and measured by an observer stationed at the position of the object. To such an observer a receding, uniform, spherical lamp appears to be surrounded by a glow of scattered light which becomes proportionately more prominent as lamp distance is increased, until at some range, often 18 to 20 attenuation lengths, the lamp image can no longer be discerned and only the glow is visible. The glow, however, may be seen for a considerably greater distance, depending upon the radiant intensity of the source and the ambient level of light in the sea.

Apparent radiance of the lamp. Densitometric meas-

[16] W. H. Richardson and R. W. Preisendorfer, Scripps Inst. Oceanog., Ref. **60-43** (1960).

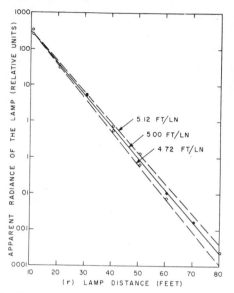

FIG. 14. Apparent radiance of a uniform, spherical underwater lamp at various distances, illustrating the exponential nature of the attenuation of apparent lamp radiance with distance. Photographic photometry was employed using a Wratten No. 61 filter and Eastman Plus X 35-mm film (Emulsion No. 5061-64-16A) developed to unity gamma in D-76. Exposure time at $f/1.5$ varied from 1.75 msec at a lamp distance of 10.5 ft to 180 000 msec when the lamp was 80 ft from the camera. The source of light was a 1000-W incandescent "diving lamp" (No. MG25/1) manufactured by the General Electric Company. The 3-in. spherical lamp envelope was sprayed with a white gloss lacquer in order to produce a uniform translucent white covering which gave the lamp the same radiant intensity in all directions (to within ±7%) except toward the base, which was turned away from the camera. Two or more exposure times differing by 5- or 10-fold were used at each lamp distance. Open circles represent data from a single time of exposure; solid points indicate that identical values of apparent radiance were obtained from negatives made with two different exposure times. A solid straight line, representing an attenuation length $1/\alpha = 5.00$ ft/ln, has been drawn near the points. Dashed lines corresponding to attenuation lengths of 4.72 ft/ln and 5.12 ft/ln, respectively, represent values measured by means of a light-beam transmissometer before and after the all-night experimental session. Cooling of the water during the night correlated with the observed increase of attenuation length, presumably due to plankton shrinkage. Data are of 26 August 1959 at Diamond Island Field Station.

urements of the lamp images in a series of photographs of a receding spherical underwater light source produced the results shown in Fig. 14, wherein the close fit of the data to the solid straight line shows that the apparent radiance of the lamp is attenuated exponentially, as the equation

$$N_r = N_0 e^{-\alpha r}, \tag{2}$$

where N_r is the apparent radiance at distance r, N_0 is the inherent radiance of the lamp surface, and α is the attenuation coefficient for apparent radiance. The dashed lines, constructed from data secured with a light-beam transmissometer designed to conform with the requirements of Eq. (1), provide evidence that numerically identical attenuation coefficients α apply in

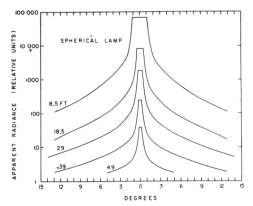

FIG. 15. Angular distribution of apparent radiance produced by a uniform, spherical, underwater lamp at distances of 8.5, 18.5, 29, and 39 feet. The lamp was identical to the one described in connection with Fig. 14. The photometry was by means of an automatic scanning, photoelectric, telephotometer having a circular acceptance cone 0.25° in diameter and with its spectral response limited by a Wratten No. 61 filter. Attenuation length was 5.1 ft/ln. Data are of 3 August 1961 at the Diamond Island Field Station.

Eqs. (1) and (2), indicating thereby that images are formed by photons transmitted without being scattered and that the contribution of scattered light to the exposure of the image portion of the negative was negligible.

Apparent radiance of the glow. Distributions of the apparent radiance of the glow surrounding the distant lamp were obtained by densitometry of the same series of photographs, but more accurate results have been achieved by means of an automatic scanning photoelectric telephotometer which was more free from stray light than was the camera. Distributions of apparent radiance as measured photoelectrically from the target position are shown in Fig. 15. The irradiance on any surface of the target facing the lamp can be computed from these curves and, if the reflectance and gloss characteristics of the target surfaces are known, the inherent radiance of the target in any specified direction can be calculated. If, moreover, the volume scattering function of the water and its attenuation length are known, calculations of inherent background radiance, path radiance, and apparent target contrast can be made from Fig. 15.

Irradiance at the Object

The surface of any underwater object is irradiated by (1) direct (nonscattered) light from the lamp and (2) scattered light. The *nonscattered* or *monopath irradiance* $H_r{}^0$ produced at normal incidence by a lamp radiant intensity J at distance r is given by the relation

$$H_r{}^0 = Je^{-\alpha r}/r^2. \qquad (3)$$

In addition to $H_r{}^0$, the object is irradiated by the scattered or *multipath irradiance* $H_r{}^*$. Thus the *total*

irradiance $H_r = H_r{}^0 + H_r{}^*$. Since H_r can be measured (see Fig. 16) and $H_r{}^0$ can be calculated by means of Eq. (3), $H_r{}^*$ can be found by subtraction; thus, $H_r{}^* = H_r - H_r{}^0$.

Diffusion theory[17,18] based upon the assumption of isotropic scattering suggests that

$$H_r{}^* = JKe^{-Kr}/4\pi r, \qquad (4)$$

where K is an attenuation coefficient for scattered light. If this K is given a value numerically equal to the attenuation function for daylight scalar irradiance k, as discussed later in the portion of this paper devoted to daylight in the sea, Eqs. (3) and (4), when summed, fit the data of Fig. 16 within experimental uncertainty both at short and at long lamp distances; between 10 ft (2 attenuation lengths) and 70 feet (14 attenuation lengths), however, the measured total irradiance is as much as twice the predicted values. A semiempirical modification of Eq. (4) which, added to Eq. (3), fits the data of Fig. 16 within experimental error is

$$H_r{}^* = 2.5(1 + 7e^{-Kr})JKe^{-Kr}/4\pi r. \qquad (5)$$

Effect of beam spread. Underwater sources of divergent light are seldom completely spherical in their radiant intensity distribution. Many underwater lamps

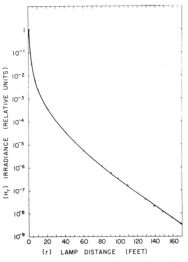

FIG. 16. Total irradiance produced at various distances by a uniform, spherical underwater lamp at the Diamond Island Field Station. The solid curve was passed through the data points by means of a least-squares procedure. The lamp was identical with the one described in connection with Fig. 14. The photometry was by means of an underwater photoelectric irradiance meter facing directly toward the lamp. The spectral response of the irradiometer was limited by means of a Wratten No. 61 green filter. The attenuation length of the water was 5.0 ft/ln. Data are of 26 August 1959.

[17] S. Glasstone and M. C. Edlund, *Elements of Nuclear Reactor Theory* (D. Van Nostrand and Company, Inc., Princeton, New Jersey, 1952), p. 107.
[18] R. W. Preisendorfer (private communication).

emit roughly conical patterns of flux 20° or more in total angular extent. Monopath irradiance is, of course, unaffected by the beam spread, and the effect on multipath irradiance is not large unless the lamp produces a highly collimated beam. Experiments with an underwater light source having a continuously variable beam spread down to 20° resulted in an empirical modification of Eq. (5) to the form

$$H_r{}^* = (2.5 - 1.5 \log_{10} 2\pi/\beta)$$
$$\times [1 + 7(2\pi/\beta)^{1/2} e^{-Kr}] J K e^{-Kr}/4\pi r, \quad (6)$$

where β is the total beam spread. Equation (6) should not be used for beam spreads less than 20°.

Equations (4), (5), and (6) have been tested by the author only at the Diamond Island Field Station, but because of the similarity in the shape of the volume scattering functions of natural waters, as illustrated by Fig. 12, they may have nearly universal applicability as approximations for engineering purposes.

COLLIMATED LIGHT IN THE SEA

Underwater projectors producing beam spreads small compared with 1° exhibit distinctive properties. When seen from the position of the irradiated target, the head-on appearance of a distant, highly collimated lamp is remarkably similar to that of a broad-beam lamp at some lesser range. Thus, the bright disk-shaped image of the lamp is surrounded by a glow of scattered light, having an apparent radiance distribution like that shown in Fig. 17. Although it is difficult to distinguish a distant collimated lamp from a distant divergent source when each is observed from within its beam, radiance distribution measurements reveal subtle differences, the nature of which can be seen by comparing Figs. 15 and 17.

The appearance presented by a moderately distant, slightly averted collimated lamp is, however, very different from that of its divergent counterpart because the intense small-angle scattering, common to all natural waters, produces a readily visible, sharply defined, nearly cylindrical luminous column extending toward the observer from the collimated lamp. Near the lamp and on the axis of this column the monochromatic monopath irradiance normal to the beam at distance r is $H_r{}^0 = H_0 e^{-\alpha r}$, where the irradiance H_0 in the water at the lens of the projector is given by $H_0 = J\psi^2 D^{-2}$ in terms of radiant intensity J, total beam spread ψ, and diameter D of the light beam. Beyond the distance $r' = D/\psi$, at which the lens replaces the source as the aperture stop of the irradiating system, $H_r{}^0$ is given by

$$H_r{}^0 = Je^{-\alpha r}/r^2 = H_0 e^{-\alpha r}(D/\psi r)^2 = H_0 e^{-\alpha r}/(r/r')^2 \quad (7)$$

if diffraction is negligible.

The dashed lines in Fig. 18 illustrate the foregoing relations applied to the case of three collimated lamps having a divergence of 1/6° and exit pupil diameters of 1/300, 2/300, and 8/300 of an attenuation length,

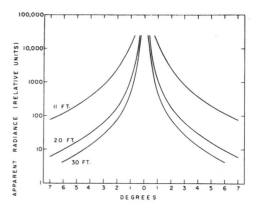

FIG. 17. Apparent radiance produced by scattering from the beam of the highly collimated underwater lamp shown in Fig. 4. The photometry was by means of an automatic scanning, photo-electric telephotometer having a circular acceptance cone 0.25° in diameter and with its spectral response limited by a Wratten No. 61 filter. The beam from the lamp had a divergence of 0.01°; it was directed toward the telephotometer and filled the entrance pupil of that instrument at all times. Lamp distances of 11, 20, and 30 ft were used. Tests of the telephotometer showed that the data in Fig. 17 are free from stray-light effects. Attenuation length of the water was 6.7 ft/ln. The data are of 11 August 1961 at the Diamond Island Field Station.

respectively. For these three lamps the distances r' are 1.15, 2.30, and 9.20 attenuation lengths. The points at $r = r'$, beyond which Eq. (7) applies, lie within the diagram for both of the two smaller lamps and are indicated by triangles. In all cases, diffraction will lower the dashed curves.

The total irradiance H_r on the axis of a collimated beam exceeds the monopath irradiance $H_r{}^0$ by the multipath contribution $H_r{}^*$; i.e., $H_r = H_r{}^0 + H_r{}^*$. This is illustrated by the experimental data points shown in Fig. 18 and the solid curves which have been fitted to them. In the case of the two smaller lamps the multipath contribution was not detected at ranges shorter than r', indicated by the triangle points, but this is not true in the beam from the large-diameter lamp where $H_r{}^*$ and $H_r{}^0$ are approximately equal throughout much of the range of distances covered by the data. The steadily increasing separation of the solid and dashed curves in each of the lower pairs implies that multipath irradiance becomes dominant at large lamp distances.

Data such as those in Fig. 18 can be used to calculate the ratio of monopath to multipath irradiance; i.e., $H_r{}^0/H_r{}^*$. This ratio, independent of the intensity of the lamp or its radiant power output, is a measure of the beam content of the light; it is the ratio of image-forming light transmitted by the water path to the non-image-forming (scattered) light arriving at the irradiated object. Applications dependent on the retention of narrow-beam geometrical characteristics, of coherence, or of single-valued transmission time may require that some usable fraction of the irradiance consist of nonscattered (monopath) light. Figure 19 is a

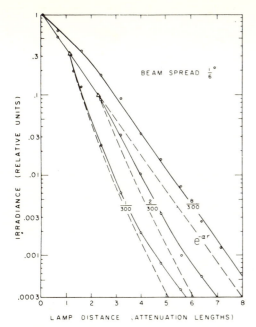

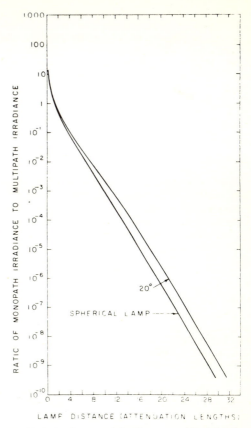

FIG. 18. Irradiance normal to the axis of the beam of light having a divergence of $1/6°$ produced by a collimated underwater lamp (Fig. 4) at distances up to 8 attenuation lengths is shown by the data points and the solid lines for beam diameters of $1/300$, $2/300$, and $8/300$ of an attenuation length. The data are of 14 August 1961 at the Diamond Island Field Station; attenuation length $1/\alpha = 6.3$ ft/ln. Dashed lines represent the monopath irradiance in each case computed from Eq. (7). Geometrical divergence reduces the axial monopath irradiance at all lamp distances beyond the points marked by triangles, which occur at 1.15 and 2.30 attenuation lengths for the two smaller lamps and at 9.20 attenuation lengths (not shown) for the largest lamp. Spreading of the beam by diffraction also reduces the monopath irradiance at all lamp distances, often dramatically. In a plot involving dimensionless lamp distance (such as Fig. 18), the dashed lines cannot be drawn to include the potentially major effect of diffraction because the wavelength of light is independent of the attenuation length, but they should be appropriately lowered when the figure is interpreted in terms of actual dimensions. The vertical separation between the dashed and the solid curves in each pair is a measure of the multipath irradiance. *Caution:* The data in this figure relate only to the axis of an aplanatic underwater projection system having a beam spread $\psi = 1/6°$; they should *not* be scaled by the ratio D/ψ; they do *not*, for example, apply to the case of $\psi = 1/60°$ and lamp diameters $D = 1/3000$, $2/3000$, or $8/3000$ attenuation length.

FIG. 19. Ratio of monopath irradiance to multipath irradiance produced by a uniform spherical lamp (lower curve) and by the same source mounted within a blackened enclosure (box) which limited its emittance to a circular cone $20°$ in total angular diameter (upper curve). In producing these curves, monopath irradiance H_r^0 was calculated by means of Eq. (3) and multipath irradiance H_r^* was obtained by subtracting H_r^0 from the total irradiance data given by Fig. 16 for the unrestricted spherical lamp and from corresponding data for the $20°$ case.

Refractive Deterioration of High Collimation

No discussion of the properties of highly collimated underwater light or image-forming rays would be complete without mention of certain commonly encountered refractive effects which limit the resolution of fine detail and tend to destroy high collimation. Natural waters often contain refractive nonhomogeneities of two kinds: (1) small scale point-to-point variations in refractive index due, for example, to temperature differences; and (2) transparent biological organisms (plankton) which may range in size from microns to centimeters. The effects of these optical nonhomogeneities has been observed by allowing the beam from the 2-in.-diameter $0.01°$ divergent lamp shown in Fig. 4 to fall on an underwater viewing screen after traversing any convenient water path or by photographing the effect with an underwater camera having no lens, in the manner

plot of H_r^0/H_r^* for divergent sources. It shows that for a beam spread of $20°$, $H_r^* = H_r^0$ at 1.4 attenuation lengths and that multipath irradiance predominates at large lamp distances. Experiments now in progress with light beams of small diameter and high collimation may produce corresponding curves for collimated lamps.

Irradiance near a highly collimated beam. All of the foregoing discussion has concerned irradiance produced on the axis of a collimated beam. Measurements of irradiance outside the light beam at various distances from the collimated lamp are shown in Fig. 20.

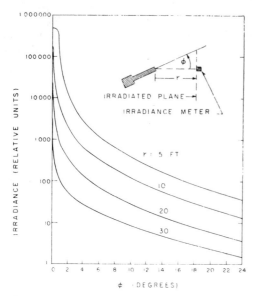

FIG. 20. Irradiance outside a collimated beam of light. Beam divergence: 0.046°; beam diameter: 2-in. Filter: Wratten No. 61. Attenuation length 4.8 ft/ln; Diamond Island.

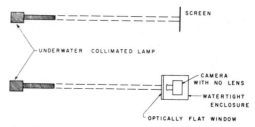

FIG. 21. Techniques for observing (upper figure) and recording (lower figure) the effects of refractive inhomogeneities on the transmission of a highly collimated beam of light through natural water.

suggested by Fig. 21. If such a photograph is made in well-mixed distilled water, only a uniform white field is recorded, but if the distilled water is allowed to stand, a pattern of shadows appears as thermal structures develop. If transparent plankton are added, their refractive shadows are superimposed.

Figure 22 is a photograph of the pattern obtained when such a picture was taken in the clear, natural water at the Diamond Island Field Station in Lake Winnipesaukee, New Hampshire. In this case the light beam passed through 10 ft of lake water. The circular shadows were caused by transparent plankton somewhat less than 1 mm in size whose refractive index differed only slightly from that of water. No effects due to thermal tubulons have been identified in this picture. The light beam was horizontal and 30 in. beneath the surface of the water. A shutter speed of 1/50 second was used because the pattern was in constant restless motion, primarily due to slight wave action, but also due to plankton movements and possibly to thermal drifts.

Loss of resolution. Wavefronts passing through natural waters are distorted by these refractive effects. The edges of objects appear blurred and the apparent contrast of small objects is reduced. Thus, resolving power is impaired and fine details are obliterated. It is said that in some clear, south-sea waters the concentration of transparent plankton is so great that a swimmer cannot distinguish his toes even though his foot is clearly visible at high contrast. Conditions are much less severe at the Diamond Island Field Station, where

magnification is necessary to make the loss of resolution obvious.

An experimental study of this loss of resolution was performed several years ago at the Diamond Island Field Station and a theoretical treatment of the effect was evolved.[19,20] At Diamond Island the loss of resolution was comparable to that caused by the on-axis aberrations of a flat water-to-air window of 1/4-in.-thick commercial plate glass when 10 ft of water separated the object from the camera. The angular magnitude of the blur increases as the square root of the object-to-camera distance, and the apparent contrast of fine details is decreased inversely as the third power of the distance in macroscopically uniform water.[20]

DAYLIGHT IN THE SEA

Most of the light in the sea is from the sun and the sky. In sunny weather each square meter of the water surface may be irradiated by as much as one kilowatt of solar power. Approximately 95% of this power en-

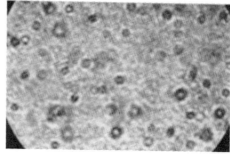

FIG. 22. Photograph of the light distribution from the collimated underwater lamp (Fig. 4) after traversing 10 ft of water in the manner shown schematically in Fig. 21. Camera: Contax without lens. Exposure time: 1/50 sec. Film: Eastman Plus-X. Development: normal, D-76. Beam spread: 0.01°. Beam diameter: 2 in. Attenuation length: 5.6 ft/ln; Diamond Island; 22 August 1961. The diameter of the outer black circular border (caused by the opening in the camera body) measured 1.3 in. on the negative.

[19] S. Q. Duntley, W. H. Culver, F. Richey, and R. W. Preisendorfer, J. Opt. Soc. Am. **42**, 877(A) (1952).
[20] S. Q. Duntley, W. H. Culver, F. Richey, and R. W. Preisendorfer, J. Opt. Soc. Am. (to be published).

ters the water and is absorbed somewhere beneath the surface. Daylight is the principal source of energy for the sea, supplying it with heat and supporting its ecology through photosynthesis. Nearly half of the irradiation is infrared, most of which is absorbed within a meter of the surface. As much as one-fifth of the daylight may be ultraviolet and this can penetrate somewhat more deeply if the concentration of dissolved organic decomposition products ("yellow substance") is low. Fortunately, the peak of the solar spectrum is not far from the wavelength (480 mμ) of greatest transparency in clear ocean water. Blue-green light, representing less than one-tenth of the total incident solar power, penetrates so deeply into the sea that it has been detected photoelectrically below 600 m. Visibility, important to inhabitants of the underwater world, is possible chiefly because of this blue-green light.

Directional Distribution of Daylight Underwater

Sunlight entering at the surface becomes progressively more diffuse with depth until a state of diffusion is reached which (1) is characteristic of the water mass, (2) is independent of the solar altitude and the prevailing sky condition, and (3) is invariant with further increases in depth unless optically different water is encountered. This behavior of daylight in water, a subject of conjecture for more than 30 years, was probably first definitively postulated by Whitney[21],[22] in brilliant speculations based neither upon adequate radiance distribution data nor upon a valid theoretical analysis but chiefly upon insightful interpretations of irradiance measurements. Whitney's hypothesis could not be confirmed until 1957, when an eight-year experimental program, initiated by the author and conducted in its later stages chiefly by several of his colleagues, culminated in the definitive radiance distribution data of Tyler.[23] These data were obtained with superlative equipment representing nearly a decade of apparatus development. The experiments were conducted in a mountain lake containing optically uniform water of very great depth. This lake (Pend Oreille, Idaho) was used only after many futile attempts had been made to find sufficiently uniform, deep water at sea and in other lakes. Even at Pend Oreille optical uniformity occurs only for a few days during the spring of each year. The Pend Oreille data show an unmistakable, systematic trend toward the formation of a characteristic (or *asymptotic*) distribution of underwater daylight radiance. A series of figures developed from Tyler's tabulated Pend Oreille data[23] and described in the section which follows summarize this experimental evidence for the *asymptotic radiance distribution* hypothesis and illustrate the progressive transformation of the light field from the sunny condition near the surface to the characteristic diffuse distribution which prevails at great

depth. A theoretical proof of the existence of characteristic diffuse light (asymptotic radiance distribution) in natural waters has been given by Preisendorfer[24] and confirmatory experimental data in other natural waters have been obtained by Jerlov and Fukuda[25] and Sasaki.[26]

Depth Profiles of Underwater Radiance

The most usable graphical representation of the distribution of daylight radiance in the sea is a family of radiance distribution profiles like those in Fig. 23. Con-

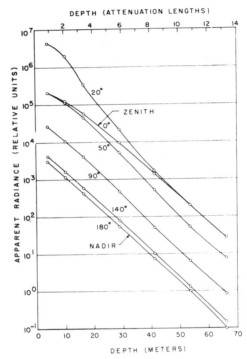

FIG. 23. Depth profiles of underwater apparent radiance for several paths of sight (i.e., zenith angles) in the plane of the sun on a clear, calm, cloudless, sunny day (28 April 1957) at Pend Oreille, Idaho. The circles denote data by Tyler (see reference 23). The solar zenith angle was 33.4°. The submerged photoelectric radiance photometer measured blue light by means of an RCA 931A multiplier phototube equipped with a Wratten No. 45 filter; its field of view was circular and 6.6° in angular diameter. The water was nearly uniform in its optical properties; i.e., the attenuation length (as measured by means of a light beam transmissometer having a tungsten source, an RCA 931A phototube, and a Wratten No. 45 filter) was 2.52 m/ln just beneath the surface and increased very slightly at a steady rate to 2.62 m/ln at a depth of 61 m; that is to say, the change in attenuation length with depth was barely detectable. Additional families of radiance profiles in vertical planes at other azimuths can be constructed from Tyler's tables, which also provide corresponding data for overcast conditions. All such sets of profiles are remarkably similar at great depth. Parallel profiles signify that the radiance distribution has its asymptotic form.

[21] L. V. Whitney, J. Marine Research **4**, 122 (1941).
[22] L. V. Whitney, J. Opt. Soc. Am. **31**, 714 (1941).
[23] J. E. Tyler, Bull. Scripps Inst. Oceanog. **7**, 363 (1960).

[24] R. W. Preisendorfer, J. Marine Research **18**, 1 (1959).
[25] N. G. Jerlov and M. Fukuda, Tellus **12**, 348 (1960).
[26] T. Sasaki, Bull. Japan. Soc. Sci. Fisheries **28**, 489 (1962).

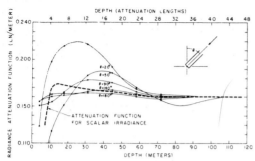

FIG. 24. The solid curves are radiance attenuation functions (i.e., slopes) of the depth profiles of apparent radiance in Fig. 23. The circled points are from Tyler's attenuation function tables (see reference 23). The dashed curve is the attenuation function for scalar irradiance; i.e., the slope of the depth profile of scalar irradiance, a radiometric quantity proportional to the response of a spherical diffuse collector such as that at the top of the instrument pictured in Fig. 25. The transformation of the light field to its asymptotic form is illustrated by the convergence of the radiance attenuation functions to a common, steady value at sufficient depth.

ceptually, each curve represents the results of lowering vertically into the sea a radiance photometer having a fixed zenith angle and azimuth. The unique utility of such profiles arises from the fact that the contrast transmittance of any path of sight in the day-lighted sea is given by the ratio of the apparent background radiances at the *terminals of the path* multiplied by the beam transmittance of the path [see Eq. (8)]. This important general theorem is rigorously true despite any degree of stratification or nonhomogeneity possessed by the water and despite any amount of nonuniformity in the lighting throughout the path of sight. Radiance distribution profiles like those in Fig. 23 enable the apparent background-radiance ratio to be read for any pair of terminal points regardless of the shape of the profile.

In Fig. 23 each curve is nearly, but not quite, straight and nearly, but not quite, parallel with its fellows. When, at sufficient depth, all of the profiles are parallel, the asymptotic radiance distribution prevails.

Radiance Attenuation Functions

The inverse slope of the semilogarithmic underwater radiance distribution profiles in Fig. 23 is called the *radiance attenuation function*. It is symbolized by $K(z,\theta,\phi)$, where z refers to depth, θ specifies the zenith angle of the radiance photometer, and ϕ denotes its azimuth. Figure 24, developed from similar ones by Preisendorfer,[27,28] is a plot of the radiance attenuation functions (slopes) of the radiance profiles shown in Fig. 23. The curves in Fig. 24 have been extrapolated beyond the greatest depth explored by Tyler's measurements in order to illustrate the asymptotic radiance distribution

[27] R. W. Preisendorfer, Scripps Inst. Oceanog. Ref. **58-59**, (1958).
[28] R. W. Preisendorfer, Scripps Inst. Oceanog. Ref. **58-60**, (1958).

concept more completely. Differential equations for the radiance attenuation functions have been evolved by Preisendorfer.[27]

Attenuation Function for Scalar Irradiance

The slope of a vertical profile of scalar irradiance $h(z)$, a radiometric quantity measurable by means of a spherical diffuse collector, is called the *attenuation function for scalar irradiance* at depth z and is denoted by $k(z)$. This function is shown by the dashed curve in Fig. 24. The limiting value $k(\infty)$ of $k(z)$ is a convenient experimental parameter for describing the optical properties of the sea because (1) $k(z)$ approaches its asymptotic value at less depth than do the radiance attenuation functions, and (2) it is easier to measure. Figure 25 shows a water-clarity meter proposed by the author and constructed

FIG. 25. Water-clarity meter for measuring depth profiles of scalar irradiance $h(z)$ and attenuation coefficient $\alpha(z)$ at sea. The hollow, translucent, white sphere at the top of the instrument is the collector for the measurement of scalar irradiance. Attenuation is measured by means of a highly collimated beam of light, produced by a projector in the lower compartment, which travels upward to a photoelectric telephotometer in the upper chamber. Baffles are used to minimize the effect of daylight in near surface measurements. The use of multiplier phototubes enables this equipment to produce profiles of scalar irradiance at depths greater than 10 attenuation lengths. A pressure transducer is incorporated in the instrument to indicate its depth. Due to the spherical nature of the irradiance sensor, the orientation of the instrument is not important; it can, if desired, be oriented horizontally (see reference 29).

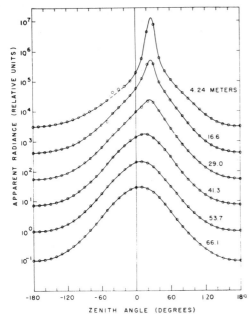

FIG. 26. Underwater radiance distributions in the plane of the sun on a clear, sunny day at depths of 4.24, 16.6, 29.0, 41.3, 53.7, and 66.1 m, respectively. The circles denote data by Tyler (see reference 23) at Pend Oreille, Idaho, 28 April 1957. The solar zenith angle was 33.4° For additional experimental details see Fig. 23. At the shallowest depth measured (4.24 m), the peak of the radiance distribution is at a slightly greater zenith angle than refracted rays from the sun (24.4°); see Fig. 29. At progressively greater depths the distribution becomes less sharply peaked and the maximum moves toward zero zenith angle. The radiance distribution is nearly in its asymptotic form at 66.1 m, the greatest depth at which data were taken. Corresponding trends appear in similar plots of data obtained by Sasaki in ocean water near Japan (see reference 26) and in Gullmar fjord by Jerlov and Fukuda (see reference 25).

by his colleagues,[29] which measures simultaneous vertical profiles of scalar irradiance $h(z)$ and attenuation coefficient $\alpha(z)$ in routine oceanographic surveys.

Shapes of the Underwater Radiance Distribution

The shapes of a typical family of underwater radiance distributions in the plane of the sun at progressively greater depths are shown by Fig. 26, which includes the same data plotted in Fig. 23. At shallow depths the distribution is sharply peaked, approximately in the direction of the refracted rays from the sun. At increasingly greater depths the distribution becomes less sharply peaked and the maximum moves progressively toward the zenith. The change in curve shape is better illustrated by Fig. 27, wherein the upper four curves of Fig. 26 have been superimposed at their respective maxima.

The lower two curves in Fig. 26 do not appear in Fig. 27 because their shape does not differ from that of the 41.3-m curve within the precision of the data. It may be noted, therefore, that the form of the radiance distribution

[29] R. W. Austin, Scripps Inst. Oceanog. Ref. **59-9**, (1959).

tribution changes throughout only the first 41 m of depth (about 16 attenuation lengths or *optical depths*). At that depth, however, the shift of the maximum toward the zenith is incomplete and continues to change rapidly as depth is progressively increased. Figure 28 shows how the maximum of the underwater daylight distribution shifts toward the zenith with increasing depth; it suggests, by extrapolation, that a depth of 20 attenuation lengths (100 m) or more is required in order for the true asymptotic radiance distribution to be reached.

Irradiance Profiles

When the underwater radiance distribution has its asymptotic form, the irradiance incident on a plane oriented in any direction will decrease exponentially with depth at the same rate as will the irradiance on planes oriented in any other directions. A family of semilogarithmic profiles of the irradiance on planes oriented in various directions is merely a group of parallel straight lines having a slope corresponding to $k(\infty)$, the limiting value of the attenuation function for scalar irradiance. In most ocean water the irradiance $H(z,-)$ on the upper surface of a horizontal plane at any depth z is approximately 50 times as great as the irradiance $H(z,+)$ on the lower surface of the same plane; the irradiance on planes oriented in all other directions at this depth lies between $H(z,-)$ and $H(z,+)$.

At lesser depths, where the underwater radiance distribution departs from its asymptotic form, the semilogarithmic irradiance profiles differ somewhat from parallelism and straightness. Such perturbations are, however, comparatively minor and for many purposes they are negligible. For example, some of the attenuation functions at a depth of 2.5 ft on an overcast day

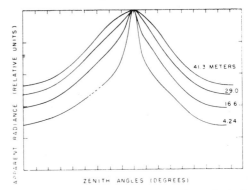

FIG. 27. In this figure the underwater radiance distribution curves for depths 4.24, 16.6, 29.0, and 41.3 m from Fig. 26 have been superimposed at their respective maxima in order to compare their shapes. The radiance curves for depths 53.7 and 66.1 m are not shown since, within the limits of experimental error, their shapes are identical with the curve for 41.3 m depth. Thus, the *shape* of the underwater radiance distribution has nearly completed its transformation to the asymptotic form at 41.3 m depth. The maximum of the curve has not, however, reached zero zenith angle at this depth and is, in fact, changing at maximum rate; see Fig. 28.

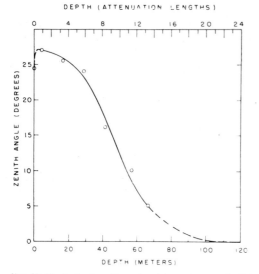

DEPTH (ATTENUATION LENGTHS)

ZENITH ANGLE (DEGREES)

DEPTH (METERS)

Fig. 28. Illustrating how the peaks of the underwater daylight radiance distributions shown in Fig. 26 shift toward zero zenith angle with increasing depth. At shallow depths in these data the peak occurs at a greater zenith angle than the direction (underwater) of rays from the sun. The extrapolated (dashed) portion of the curve suggests that a depth of more than 100 m is required to bring the peak to zero zenith angle; i.e., to complete the transformation of the light field to its asymptotic form.

(28 August 1959) at Diamond Island were $K(2.5,-) = 0.067$ ln/ft, $k(2.5) = 0.063$ ln ft, $K(2.5,+) = 0.051$ ln/ft, and $\alpha(2.5) = 0.18$ ln/ft.

Contrast Transmittance

Introduction. Underwater sighting ranges are always short compared with sighting ranges in clear air. Nearly all objects, therefore, subtend so large a visual angle when seen underwater that the exact size of the object is of almost no consequence. Except for very tiny objects or the fine details of larger ones, underwater sighting ranges depend almost entirely upon the contrast transmittance of the path of sight when ample daylight prevails. Along horizontal paths of sight dark objects (such as black-suited swimmers) approach detection threshold near the distance $4/\alpha(z)$ when viewed against a water background, although bright objects (including light sources) can be seen further.[30] For objects of sufficient angular size, horizontal daylight sighting ranges underwater are remarkably similar to horizontal daylight sighting ranges in the atmosphere if both are expressed in attenuation lengths. This quantitative similarity does not hold, however, when the path of sight is inclined either upward or downward because water, unlike air, absorbs light so strongly that all aspects of

[30] Along any underwater path of sight a remarkable proportion of the objects ordinarily encountered can be seen at limiting ranges between 4 and 5 times the distance $1/[\alpha(z) - K(z,\theta,\phi)\cos\theta]$, regardless of their size or the background against which they appear, provided ample daylight prevails [see Eqs. (14) and (15)].

daylight in the sea diminish rapidly with depth. Contrast reduction along inclined paths of sight through optically uniform water are treated after certain general principles have been discussed.

General case. A completely general phenomenological treatment of the reduction of apparent contrast by any scattering and absorbing medium has been given by the author and two of his colleagues in an earlier paper[31] concerned with the atmosphere; Eq. (1) through (10) of that paper and the discussions which accompany them apply also to the reduction of contrast along all underwater paths of sight, and the notation employed in reference 31 has been used throughout the present paper, except that z is used to denote depth (rather than altitude) and is positive from the sea surface downward. Although, in the interest of brevity, only one [Eq. (7)] of those equations is discussed here, they constitute the foundation for all of the relations which follow in this paper.

Equation (7) in reference 31 states that the ratio of the apparent contrast $C_r(z,\theta,\phi)$ of an object at distance r from an observer at depth z along a path of sight having zenith angle θ and azimuth ϕ to the inherent contrast $C_0(z_t,\theta,\phi)$ of a target at depth z_t is

$$C_r(z,\theta,\phi)/C_0(z_t,\theta,\phi) = T_r(z,\theta,\phi)\,_bN_0(z_t,\theta,\phi)/\,_bN_r(z,\theta,\phi), \quad (8)$$

where $T_r(z,\theta,\phi)$ is the beam transmittance of the path of sight for image-forming light and $_bN_0(z_t,\theta,\phi)/_bN_r(z,\theta,\phi)$ is the ratio of the apparent radiances of the background at the *terminals* of the path of sight. This equation is rigorously true despite any amount of nonuniformity in the water or in its lighting. Profiles of underwater radiance, such as those in Fig. 23, provide the two background radiance values required by Eq. (8) and the beam transmittance can be found from a profile of attenuation length by means of Eq. (16) in reference 31. It should be noted that the beam transmittance $T_r(z,\theta,\phi)$ must include the factor $[n(z)/n(z_t)]^2$ required by geometrical optics when the refractive index $n(z)$ of the medium at the observer differs from that at the target $n(z_t)$, as in the case of underwater observation through a flat face plate or a plane window.

Uniform water. If the underwater path of sight lies entirely within a single optically uniform stratum and if the profile of monochromatic apparent radiance (see Fig. 23) can be approximated by a straight line and represented by the differential equation

$$dN(z,\theta,\phi)/dr = -K(z,\theta,\phi)\cos\theta N(z,\theta,\phi), \quad (9)$$

where $r\cos\theta = z_t - z$, Eq. (10) of reference 31 can be replaced by differential equations of transfer for spectral field radiance

$$dN(z,\theta,\phi)/dr = N_*(z,\theta,\phi) - \alpha(z)N(z,\theta,\phi), \quad (10)$$

[31] S. Q. Duntley, A. R. Boileau, and R. W. Preisendorfer, J. Opt. Soc. Am. **47**, 499 (1957).

and for apparent spectral target radiance

$$d_t N(z,\theta,\phi)/dr = N_*(z,\theta,\phi) - \alpha(z)_t N(z,\theta,\phi). \quad (11)$$

Equations (9), (10), and (11) can be combined and integrated throughout the path of sight to produce the important relation

$$_t N_r(z,\theta,\phi) = _t N_0(z_t,\theta,\phi) \exp[-\alpha(z)r]$$
$$+ N(z_t,\theta,\phi) \exp[+K(z,\theta,\phi)r \cos\theta]$$
$$\times \{1 - \exp[-\alpha(z)r + K(z,\theta,\phi)r \cos\theta]\}, \quad (12)$$

where $_t N_r(z,\theta,\phi)$ is the apparent spectral radiance of the target and $_t N_0(z_t,\theta,\phi)$ is its inherent spectral radiance. In Eq. (12) the first term on the right represents

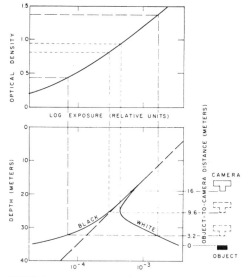

APPARENT RADIANCE (WATTS/Ω. SQ METER, mμ)

Fig. 29. Illustrating the effect of (vertical) object-to-camera distance on the apparent radiance (lower figure) and the photographic contrast (upper figure) of an object having both white and black areas submerged 35 m beneath the surface of deep, optically uniform water characterized by an attenuation length $(1/\alpha)$ of 3.2 m/ln, $(\alpha/K) = 2.7$, $H(z, +)/H(z, -) = 0.02$, and asymptotic radiance distribution. The prevailing spectral irradiance on the surface of the water is assumed to be 1 W/m², mμ.

As a downward-looking camera is lowered from the sea surface, the apparent radiance presented by the water decreases at the rate of $K = 0.116$ ln/m, as shown by the diagonal dashed line in the lower figure. At 19 m depth (i.e., an object-to-camera distance of 16 m or 5 attenuation lengths) the apparent radiances of the object differ but little from that of the surround. When the camera is 9.6 m (i.e., 3 attenuation lengths) above the target, the white area presents an apparent radiance significantly greater than the surround (diagonal dashed line) but the black area appears only slightly darker than the water background. Near this camera position the two terms in the right-hand member of Eq. (12) are equal, so that $dN(z,\pi,0)/dr = 0$; at greater camera depths the second term predominates. When the camera is 3.2 m or 1 attenuation length above the object, both the black and the white areas of the target differ markedly in apparent radiance from the surround (diagonal dashed line). The upper figure illustrates, by means of the characteristic curve of a negative material, the range of photographic densities corresponding with object-to-camera distances of 3.2 m (dashed lines) and 9.6 m (dotted lines).

residual image-forming light from the target and the second term represents radiance due to scattering of light in the sea throughout the path of sight, i.e., the *path radiance* $N_r*(z,\theta,\phi)$. A graphical illustration of Eq. (12) is provided by Fig. 29, which shows how black objects and white objects submerged in deep water appear to emerge gradually from the background as they are approached from above by a descending, downward-looking underwater observer or camera.

In Eq. (12), $\alpha(z)$ and $K(z,\theta,\phi)$ are considered to be constants throughout the path of sight. In uniform water this is true of $\alpha(z)$ but not of $K(z,\theta,\phi)$ unless the radiance distribution is asymptotic. Figure 24 illustrates how $K(z,\theta,\phi)$ changes with z and θ in the plane of the sun; corresponding figures can be constructed from Tyler's tables[23] to illustrate changes with ϕ. Such data should be used to ascertain the variation of $K(z,\theta,\phi)$ on the particular segment of the path of sight to be used; the degree of approximation represented by Eq. (12) [and by Eqs. (14), (15), and (16)] can then be estimated. Because underwater sighting ranges rarely exceed $2/K$, the effect of K variation is seldom appreciable, except near the surface of the sea. General equations, remarkably similar in form to Eqs. (12), (14), (15), and (16), have been written by Preisendorfer (private communication); these involve, for example,

$$\exp\left\{-\int_0^r [\alpha(z) - \cos\theta K(z,\theta,\phi)] dr'\right\}$$

instead of

$$\exp[-\alpha(z)r + \cos\theta K(z,\theta,\phi)r];$$

they are also applicable to nonuniform water and even to multi-media paths of sight.

Equation (12) also specifies the apparent radiance of any background against which a target may be seen; when used for this purpose the presubscript t (for target) should be changed to b (for background). Subtraction of the background form of Eq. (12) from Eq. (12) itself yields the relation

$$_t N_r(z,\theta,\phi) - _b N_r(z,\theta,\phi)$$
$$= [_t N_0(z_t,\theta,\phi) - _b N_0(z_t,\theta,\phi)] \exp[-\alpha(z)r]. \quad (13)$$

Equation (13) implies that along *any* underwater path of sight, radiance differences are transmitted with exponential attenuation at the same space rate as image-forming rays.

The two forms of Eq. (12) can be combined with the defining relations for inherent spectral contrast, $C_0(z_t,\theta,\phi)$, and apparent spectral contrast $C_r(z,\theta,\phi)$, which are, respectively,

$$C_0(z_t,\theta,\phi) = [_t N_0(z_t,\theta,\phi) - _b N_0(z_t,\theta,\phi)]/_b N_0(z_t,\theta,\phi),$$

and

$$C_r(z,\theta,\phi) = [_t N_r(z,\theta,\phi) - _b N_r(z,\theta,\phi)]/_b N_r(z,\theta,\phi).$$

When this is done, the ratio of inherent spectral contrast to the apparent spectral contrast is found to be

$$C_0(z_t,\theta,\phi)/C_r(z,\theta,\phi)$$
$$=1-[N(z_t,\theta,\phi)/{}_bN_0(z_t,\theta,\phi]$$
$$\times\{1-\exp[\alpha(z)r-K(z,\theta,\phi)r\cos\theta]\}. \quad (14)$$

If ${}_bN_0(z_t,\theta,\phi)=N(z_t,\theta,\phi)$, as in the special case of an object suspended in deep water, Eq. (14) reduces to

$$C_r(z,\theta,\phi)=C_0(z_t,\theta,\phi)$$
$$\times\exp[-\alpha(z)r+K(z,\theta,\phi)r\cos\theta]. \quad (15)$$

Whenever the underwater daylight radiance distribution has, effectively, its asymptotic form, the radiance attenuation function $K(z,\theta,\phi)$ is a constant, independent of z, θ, and ϕ. Equation (15) may then be written

$$C_r(z,\theta,\phi)/C_0(z_t,\theta,\phi)=\exp[-\alpha+K\cos\theta)]r. \quad (16)$$

The right-hand member of Eq. (16), sometimes called the *contrast reduction factor*, is independent of ϕ, the azimuth of the path of sight. This and other implications of Eq. (16) were discovered by the author in the

course of early experiments as illustrated, in part, by Figs. 30 and 31.

Horizontal paths of sight. Along horizontal paths of sight $\cos\theta=0$ in Eqs. (9), (12), (14), (15), and (16), which show that both the apparent radiance and the apparent contrast of objects seen horizontally underwater change with distance in a manner dependent on α but not on K. When $\cos\theta=0$, Eq. (10) indicates that some unique *equilibrium radiance* $N_q(z,\pi/2,\phi)$ must exist at each point such that the loss of radiance within the horizontal path segment is balanced by the gain, i.e.,

$$dN_q(z,\tfrac{1}{2}\pi,\phi)/dr=0=N_*(z,\tfrac{1}{2}\pi,\phi)-\alpha(z)N_q(z,\tfrac{1}{2}\pi,\phi). \quad (17)$$

Even in nonuniform water there is an equilibrium radiance for each element of horizontal path although this may differ from point to point. Inclined paths of sight do not have a true equilibrium radiance, as will be clear from Eq. (9), but they possess an exponential counterpart which is illustrated by the diagonal dashed line in Fig. 29.

A method[32] for measuring the attenuation coefficient

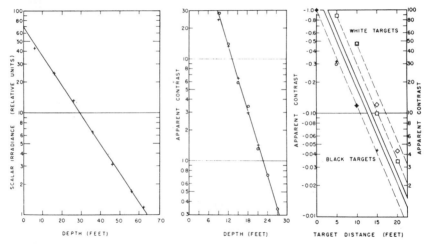

FIG. 30. Interrelated experiments from the September 1948 series at the Diamond Island Field Station: (*Left*) Semilogarithmic depth profile of scalar irradiance obtained by lowering a 6-in.-diameter, air-filled, hollow, translucent, opal glass sphere having a photovoltaic cell sealed in an opening at its bottom. The straightness of the curve indicates optical homogeneity of the water and a depth invariant attenuation coefficient $k(z)=0.066$ ln/ft. (*Center*) Semilogarithmic plot of the absolute apparent contrast of a horizontal, flat, white target lowered vertically beneath a telephotometer mounted in a small, hooded, glass-bottomed boat; calm water, clear sky, low sun. The long, straight portion of the curve illustrates Eq. (15) and its slope indicates that $\alpha(z)+K(z,\pi,0)=0.247$ ln/ft. Because the sun was low the radiance distribution was approximately asymptotic, so that $K(z,\pi,0)\approx k(z)=0.066$ ln/ft and, by subtraction $\alpha(z)=0.181$ ln/ft or the attenuation length $1/\alpha=5.5$ ft/ln. (*Right*) Two semilogarithmic plots of apparent contrast vs target distance along $60°$-downward-sloping paths of sight for black targets (lower portion) and white targets (upper portion) have been combined to demonstrate (1) that the apparent contrast is exponentially attenuated with target distance at the same space rate for both light targets and dark targets, (2) that this space rate is independent of azimuth, and (3) that Eq. (16) is valid. All four paths of sight have the same zenith angle, $\theta=150°$, but the azimuth angles relative to the plane of the sun are $\phi=0$ (circled points) and $\phi=45°$ (crosses), $\phi=95°$ (diamonds) and $\phi=135°$ (squares). The dashed straight lines are constructed parallel and, in accordance with Eq. (16), they have a slope $0.181+0.066\cos150°=0.214$ ln/ft. These lines were passed through the uppermost datum point of each series without regard to the lower points; the lines are provided solely to facilitate judgment of the slope and linearity of the data. Photographic underwater telephotometry; green light, calm water, clear sky, low sun.

32 S. Q. Duntley, J. Opt. Soc. Am. 37, 994(A) (1947) and U. S. Patent No. 2,661,650.

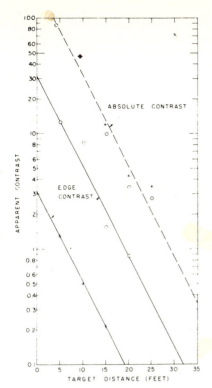

FIG. 31. Comparison of apparent absolute contrast with apparent edge contrast of white targets for two horizontal underwater paths of sight having azimuths relative to the direction of the sun of 95° (crosses) and 135° (circles), respectively. The three lines are parallel and correspond to an attenuation length $1/\alpha = 5.65$ ft/ln. The data are of 24 September 1948 at Diamond Island. Photographic telephotometry; green filter.

$\alpha(z)$ is suggested by Eq. (17) and the fact that in optically uniform water $_q N(z,\tfrac{1}{2}\pi,\phi) = N(z,\tfrac{1}{2}\pi,\phi)$; thus

$$\alpha(z) = N_*(z,\tfrac{1}{2}\pi,\phi)/N(z,\tfrac{1}{2}\pi,\phi). \qquad (18)$$

In Eq. (18), $N_*(z,\tfrac{1}{2}\pi,\phi)$ can be approximated by the apparent radiance of a very black object, such as an opening in a small black box, located at a unit distance which is small compared with the attenuation length, and $N(z,\tfrac{1}{2}\pi,\phi)$ is the apparent radiance of the unrestricted water background. This technique is especially convenient for documenting conditions in underwater photography by daylight. The value of $\alpha(z)$ so obtained agrees precisely with data obtained by (1) properly designed light beam transmissometers, (2) measurements of the apparent contrast of underwater objects observed along horizontal paths of sight, and (3) underwater telephotometry of the apparent radiance of the surface of a distant submerged frosted incandescent lamp or other diffusely emitting source.

Field experiments. Experimental explorations of the distribution of daylight in the sea and underwater image transmission phenomena were begun by the author in 1948 and are still in progress. Most of the physical principles discussed in this paper were discovered or generalized early in the course of these experiments. The data guided a collaborative development of the foregoing equations by Dr. Rudolph W. Preisendorfer and the author.[33-35]

Experiments were conducted concurrently in lakes and at sea almost from the beginning because optical principles can be explored better and more inexpensively in lakes whereas the magnitude of the optical constants of ocean waters can be measured only at sea. Most of the data used in this paper to illustrate principles were obtained at a field station established by the author in 1948 at Diamond Island in Lake Winnipesaukee, New Hampshire. Examples of data from the field station are provided in Fig. 30. These data, taken from the 1948 series, illustrate several important principles which are implied and summarized by Eq. (16). Figure 30 shows that the attenuation coefficients $k(z)$ and $\alpha(z)$ obtained by means of a depth profile of scalar irradiance and measurements of the apparent radiance of a white object lowered vertically (in the manner of a Secchi disk) can be used with Eq. (16) to predict the apparent contrast of any object, black or white, along various underwater paths of sight. Measurements of apparent contrast with highly refined photoelectric equipment have been made along many paths of sight and under many kinds of lighting conditions in the course of the field station experiments; all of these experiments support the validity of Eqs. (15) and (16).

The water-clarity meter pictured in Fig. 25 produces a profile of scalar irradiance similar to that shown in Fig. 30 and, therefore, a measure of $k(z)$; it also measures the attenuation coefficient $\alpha(z)$, providing, thereby, the necessary input information for using Eq. (16) to calculate contrast reduction, since $K = k(z)$.

Telephotometry of either black or white targets along any two paths of sight having different inclinations (i.e., zenith angle θ) yields two values of the contrast attenuation coefficient $(\alpha - K \cos\theta)$ from which α and K can be found. The use of a horizontal path for determining α, and a downward vertical path for determining $\alpha + K$, is often a convenient choice.

Absolute contrast. The water immediately surrounding a submerged white object sometimes appears to glow. This effect is caused by the intense small-angle forward scattering of light which is reflected by the target in directions adjacent to that of the observer. The effect is most noticeable when a strongly lighted white object is observed against a dark background. The apparent radiance of the scattered glow has been found to be attenuated at the same space rate as the target itself; this is shown by Fig. 31 wherein the semilogarithmic attenuation curves for apparent *absolute contrast* and apparent *edge contrast* are parallel. Apparent absolute contrast is relative to the apparent background

[33] S. Q. Duntley, Proc. Armed Forces–Natl. Research Council Vision Committee **23**, 123 (1949); **27**, 57 (1950); **28**, 60 (1951).

[34] S. Q. Duntley and R. W. Preisendorfer, MIT Rept. N5ori 07864 (1952).

[35] R. W. Preisendorfer, Scripps Inst. Oceanog. Ref. **58-42** (1957).

radiance that would be observed if the target were absent; apparent edge contrast is relative to the apparent background radiance which appears immediately adjacent to the target. Ordinarily, few underwater objects are white enough to cause the two types of contrast to differ significantly. When the glow is prominent, absolute contrast is usually the more meaningful measure of object detectability, but a full treatment of this topic can be made only in context with details concerning the characteristics of the detector (eye, camera, etc.), a matter beyond the scope of this paper.

Absorption

If radiant power in the sea is to be useful for heating or for photosynthesis it must be absorbed. The monochromatic radiant power absorbed per unit of volume at any depth depends upon the amount of power received by the volume element and the magnitude of the absorption coefficient; i.e., upon the product of the scalar irradiance and the volume absorption coefficient.[36] A more frequently useful relation[37] has been evolved as follows: The net inward flow of radiant power to any element of volume dz in any horizontal lamina of thickness dz at depth z in the sea is

$$\frac{dP(z)}{dz} = -\frac{d}{dz}\{H(z,-)-H(z,+)\}$$
$$= \frac{d}{dz}\left\{H(z,-)\left[1-\frac{H(z,+)}{H(z,-)}\right]\right\}. \quad (19)$$

The ratio $H(z,+)/H(z,-)$, sometimes called the *reflection function* of water, has been found by experiment to be virtually independent of depth and to have a value of 0.02 ± 0.01 for most natural waters unless large quantities of suspended matter are present; the reflection function is rigorously independent of depth when the underwater daylight radiance distribution has its asymptotic form in optically uniform water. To the extent to which 2% effects are negligible, Eq. (19) becomes

$$dP(z)/dv \approx H(z,-)K(z,-), \quad (20)$$

since, by definition, $K(z,-) = -[dH(z,-)/dz]/H(z,-)$. Thus, the radiant power absorbed per unit of volume at any depth in the sea can be measured simply by lowering an upward-facing, diffusely collecting, flat photocell and determining the product of the magnitude and slope of the resulting profile of downwelling irradiance, as illustrated by Fig. 32.

Alternatively, the quantity $\{H(z,-)-H(z,+)\}$ can be measured directly by lowering an assembly of two diffusely collecting, flat photocells mounted back to back so that one faces upward and the other downward. Such an assembly, sometimes called a *janus cell*, can

---

[36] R. W. Preisendorfer, Scripps Inst. Oceanog. Ref. **58-41**, (1957).
[37] S. Q. Duntley, Natl. Acad. Sci./Natl. Research Council Publ. **473**, 85 (1956).

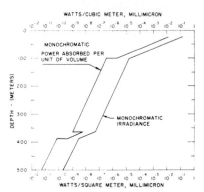

FIG. 32. Superimposed semilogarithmic plots of monochromatic downwelling irradiance vs depth and monochromatic radiant power absorbed per unit of volume vs depth illustrate the (approximate) relation between these quantities expressed by Eq. (20). Monochromatic downwelling irradiance is the total monochromatic radiant power per unit of area received by the upper surface of a horizontal plane at arbitrary depth z. The product of this irradiance and its depth attenuation function (slope of its depth profile) is, within about 2%, equal to the monochromatic power absorbed per unit of volume. Thus, at a depth of 50 m in Fig. 32, $H(50,-)=6.3\times10^{-3}$ W/(m², mμ), $K(50,-)=0.114$ ln/m, and $dP(50)/dz \approx (6.3\times10^{-3})(0.114) = 7.2\times10^{-4}$ W/(m³, mμ). Neither of the curves in this figure represent specific experimental data, but the irradiance profile is typical of the Pacific Ocean off California. The presence of a deep scattering layer is shown below 350 m.

be used to measure $dP(z)/dv$ by means of Eq. (19) in turbid waters for which $\{1-[H(z,+)/H(z,-)]\}$ is not negligible.

CONCLUSION

Although no research program is ever fully completed and the author hopes to participate in studies of light in the sea for many years to come, the investigations which, with many colleagues, have been made thus far, coupled with the findings of other workers all over the world, have produced a sufficient quantitative understanding of the optical properties of ocean water and the behavior of underwater light to provide scientific guidance and optical engineering methods for those persons whose interests or occupations involve light in the sea.

ACKNOWLEDGMENTS

The long research program, spanning two decades, from which this paper is drawn has involved too many persons to permit complete acknowledgment here. Special mention should be made, however, of the important technical contributions of Dr. David L. MacAdam, Willard P. Greenwood, Capt. Dayton R. E. Brown, Professor George E. Russell, John Frankovitch, Frederick C. Spooner, Robert W. Sandberg, Walter Rutkowski, Robert J. Uhl, Frances Richey, Dr. Rudolph W. Preisendorfer, Roswell W. Austin, Almerian R. Boileau, John E. Tyler, Justin J. Rennilson, William Hadley Richardson, Dr. William H. Culver, Theodore J. Petzold, Charles W. Saunders, Jr., Sidney Lindroth, Alden D. J. Hooton, Roger A. Howerton, and Clarence Fred Pinkham who, since 1949, has superintended nearly all of the field operations.

Part II

RADIATIVE TRANSFER THEORY

Editor's Comments
on Papers 3 and 4

3   PREISENDORFER
Application of Radiative Transfer Theory to Light Measurements in the Sea

4   MOREL
Excerpt from *Diffusion de la Lumière par les Eaux de Mer. Résultats Expérimentaux et Approche Théorique*

The early research measurements of light in the sea suffered from serious deficiencies. Due to difficulties associated with optical design and absolute calibration of instruments, the results obtained by one laboratory could not be accurately related to the results from any other laboratory. Instrument components, especially the detectors and optical filters, were not adequate to the task imposed upon them by the selective spectral absorbing properties of deep ocean water.

Although by 1950 there was, in the literature, abundant general theory regarding radiative transfer in scattering-absorbing media, these theories were not in a form readily applicable to radiative transfer in the sea.

Some of the early contributors to the development of a general theory of radiative transfer in scattering-absorbing media were: L. C. Wiener (1900) who clearly formulated the idea of scattering-order decomposition of the light field; A. Schuster (1905) who formulated the concept of self-illumination of a scattering-absorbing medium and was the first to write down the two-flow equations; K. Schwarzschild (1906) and L. V. King (1913) who developed the integral and integrodifferential equations of radiative transfer; V. A. Ambarzumian (1943) who formulated the invariance principle, which was elaborated and modernized by S. Chandrasekhar (1950) and R. Bellman and R. Kalaba (1956); W. Hopf (1934) who gave mathematical proofs showing the existence of solutions to the equations of Schwarzschild and King; A. Gershun (1936) who treated the fundamental geometry of the photometric (radiometric) quantities and illustrated methods for their measurement; and J. Lehner and G. M. Wing (1956) who clarified the nature of the spectrum of the general integrodifferential equation and hence provided for its solution.

The Russian scientist, A. Gershun, was perhaps the first to fully and systematically develop the geometrical concepts involved in radiometry, and to suggest practical methods for their measurement. In their translation of Gershun's 1936 paper, Parry Moon and Gregory Timoshenko have adhered to the specialized photometric terminology (rather than to radiometric terminology). Photometric terminology will not be used in all sections of this book. However, it should be remembered that photometric terms do no more than specify the spectral composition of the energy that is under discussion (or is being measured), and, as Moon and Timoshenko have themselves pointed out, the concepts or theorems set forth by Gershun in photometric terminology will be exactly the same for any other chosen band of wavelengths.

R. W. Preisendorfer (1960), Paper 3, has incorporated the geometrical concepts introduced by Gershun into a general theory of monochromatic radiative transfer applicable to ocean and lake water. Preisendorfer's theory makes use of directly measurable radiometric quantities. His theory applies to water sample volumes that are much larger than the scattering particles they contain and the circulation of these particles through the sample volume is therefore not a detectable feature. At the same time, the water sample volumes employed for measurement are generally small compared to many ocean features that are of interest to physical oceanographers. These two factors determine the validity of applying this theory to light phenomena in the sea, and to the interpretation of the measurements. An important but elementary feature of Preisendorfer's work is his formal separation of the measurable optical properties of ocean water into two categories: the "inherent" optical properties, having operational values that do not change with variations in the natural radiance distribution (from sun and sky), and the "apparent" optical properties, having operational values that do vary systematically with the natural radiance distribution.

This formal separation is in recognition of the experimental fact that the transmission of radiant flux through a scattering-absorbing medium will vary systematically with the geometrical distribution of the flux that is being used for the measurement. Although this dependence of optical transmission on the geometrical distribution of the incidence flux was a well known phenomenon in photography in the 1920s (see, for example, Tuttle, 1926), it was not recognized as an oceanographic phenomenon until many years later.

Most recently Dr. Preisendorfer (1976) has published the first five volumes of *Hydrologic Optics,* a six-volume work that develops the theory of radiative transfer in detail and demonstrates various applications of this theory to specific problems in oceanography. The principal optical properties discussed by Preisendorfer are listed in table 1, which also gives the units and commonly used symbols. Figure 1 indicates the interrelationships between the major optical properties defined by Preisendorfer. Figure 1 shows, in heavy boxes, the three basic measurements required for a complete optical documentation of an ocean water type. These are: the volume scattering function, $\sigma(\theta)$; the distribution of radiance around a point at a fixed depth, $N(Z, \theta, \varphi)$; and the beam transmittance, T.

The exact documentation of a water type requires a homogeneous

Table 1   Concepts and symbols in optical oceanography

	Units	JOSA Radiometric Symbol Usage	Symbol Adopted by IAPSO*
Volume scattering function	$m^{-1}\Omega^{-1}$	$\sigma(\Theta)$	$\beta(\Theta)$
Backward scattering coefficient	m^{-1}	b	b_b
Forward scattering coefficient	m^{-1}	f	b_f
Total scattering coefficient	m^{-1}	s	b
Radiance at depth z in direction (θ,φ)	watts $m^{-2}\Omega^{-1}$	$N(z,\theta,\varphi)$	L
Scalar irradiance (Downwelling (−))	watts m^{-2}	$h(z,-)$	—
Scalar irradiance (Upwelling (+))	watts m^{-2}	$h(z,+)$	—
Total scalar irradiance	watts m^{-2}	$h(z)$	E_o
Irradiance on a horizontal plan from the upper hemisphere[a]	watts m^{-2}	$H(z,-)$	E_d
Irradiance on a horizontal plane from the lower hemisphere[a]	watts m^{-2}	$H(z,+)$	E_u
Distribution functions	(ratio)	$D(z,\pm)$	—
Reflectance function (Irradiance Reflectance)[b]	(ratio)	$R(z,-)$ or (R)	R
Diffuse attenuation coefficient for irradiance from the upper hemisphere	m^{-1}	$K(z,-)$	—
Diffuse attenuation coefficient for irradiance from the lower hemisphere	m^{-1}	$K(z,+)$	—
Absorption coefficient	m^{-1}	a	a
Ratio of the radiance at the end of a path, r, to the input radiance	(ratio)	N_r/N_o	—
Beam transmittance (for specified path length)	(ratio)	T	T
Total attenuation coefficient	m^{-1}	α	c

*International Association for Physical Sciences of the Ocean
[a]Called "downward" and "upward" irradiance by IAPSO
[b]Called "Irradiance Ratio" by IAPSO

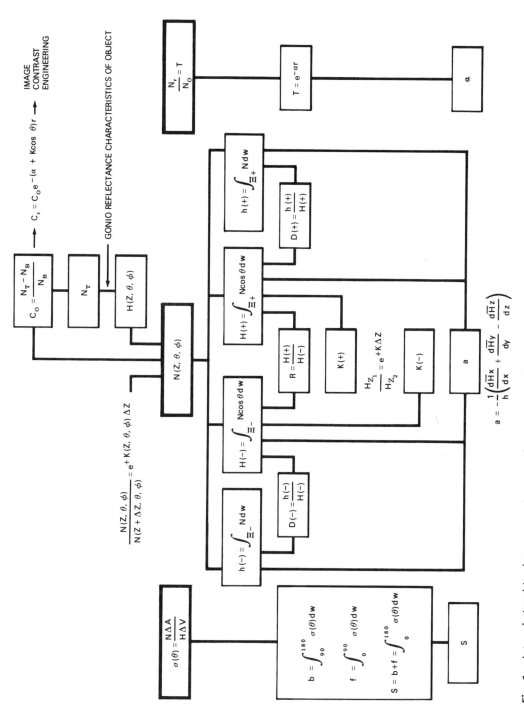

Fig. 1. Interrelationships between major optical properties. *Originally published in Radiative Energy Transfer, R. Goulard, S. M. Scala, and R. N. Thomas, eds., (London, 1968), p. 340. Copyright © 1968 by Pergamon Press, Ltd.*

hydrosol which would therefore exhibit a single value for *T* for each wavelength, a single set of values of σ as a function of the scattering direction θ for each wavelength, and sets of monochromatic values of radiance distribution at a selection of observation angles, θ and φ, obtained at a series of selected depths. The radiance distribution data should be measured under critically stable, natural lighting conditions at the surface of the water. Polarization effects must be eliminated from all measurements, unless the objective of the measurements is to study polarization. Presiendorfer's theory has the immense advantage of detailed continuity and completeness, starting as it does with the fundamental concepts of radiative transfer and continuing on to the basic concepts of instrument design.

The scattering of light by sea water has been extensively studied both theoretically and experimentally by André Morel. The result of his research and the relationship of his work to the work of others who have contributed to this important aspect of radiative transfer are brought together in Paper 4. The first chapter of this work is a review of experimental work on the volume scattering function of sea water. A distinction is made between the molecular scattering of the water and the scattering of the particles in suspension. Magnitude variations in the volume scattering function are closely associated with variations in the concentration of the particles in suspension. If the molecular scattering of the water is subtracted from the total scattering, it is found that the volume scattering function due to the particles alone is only slightly variable when calculated on a relative basis. However, the particle scattering coefficients vary with wavelength approximately as λ^{-1}.

The second chapter reviews the theories of Rayleigh, Rayleigh-Gans, and Mie as they apply to the marine particles. Numerical results, obtained using the Mie theory, are given. These results are for spherical particles having index of refraction varying from 1.02 to 1.15 and with a size parameter ($\alpha = 2\pi(r/\lambda)$) ranging from 0.2 to 200. The computed quantities are Q, the efficiency factor for scattering; and $i_1(\Theta)$, and $i_2(\Theta)$, the intensity functions for the two polarized components. The scattering angle, Θ, varies from 0° to 180° in 2° increments.

These numerical results are combined to compute the scattering properties of polydispersed systems in Morel's section 3, "Diffusion par un Ensemble de Particules Polydisperse," which is included in Paper 4. At the end of the paper, the author supplies useful definitions and discusses some problems, including the adaptation of the calculus to the computer. A list of eighty-eight references to the literature is also given.

REFERENCES

Ambarzumian, V. A. 1943. Diffuse reflection of light by a foggy medium. *C. R. (Dokl.) Acad. Sci. USSR.* **38:**229.

Bellman, R., and R. Kalaba. 1956. On the principle of invariant imbedding and propagation through inhomogeneous media. *Natl. Acad. Sci. (U.S.A.) Proc.* **42:**629.

Busbridge, I. W. 1960. *The mathematics of radiative transfer.* Cambridge Univ. Press, London.

Chandrasekhar, S. 1950. *Radiative transfer.* Oxford Univ. Press, London.

Gershun, A. 1963. *The Light Field.* Svetovoe pole, Moscow.

Hopf, W. 1934. *Mathematical problems of radiative equilibrium.* Cambridge Univ. Press, London.

King, L. V. 1913. On the scattering and absorption of light in gaseous media, with applications to the intensity of sky radiation. *Philos. Trans. R. Soc. London* **212A:**375.

Lehner, J., and G. M. Wing. 1956. Solution of the linearized boltzmann transport equation for the slab geometry. *Duke Math. J.* **23:**125.

Moon, P., and G. Timoshenko. 1939. The light field. *J. Math. Phys.* **18:**2.

Preisendorfer, R. W. 1976. *Hydrologic Optics.* U.S. Dept. Commerce, U.S. Govt. Printing Office, Washington, D.C.

Schuster, A. 1905. Radiation through a foggy atmosphere. *Astrophys. J.* **21:**1.

Schwarzschild, K. 1906. Uber das gleichgewicht der sonnenatmosphäre. *Ges. Wiss. (Göttingen) Math.-Phys. Klasse Nachr.,* p. 41.

Tuttle, C. 1926. The relation between diffuse and specular density. *Opt. Soc. Am. J. (and Rev. Sci. Instr.)* **12:**483A, 559.

Wiener, L. C. 1900. Die helligkeit des klaren himmels und die beleuchtung durch sonne, himmel und rückstrahlung. *Dtsch. Akad. Naturforsch. Nova Acta Leopold.* **73:**1.

3

Reprinted from *Symp. Radiant Energy in the Sea, Helsinki, 1960,* I.U.G.G. Monogr. No. 10, pp. 11–29

APPLICATION OF RADIATIVE TRANSFER THEORY TO LIGHT MEASUREMENTS IN THE SEA

by

R. W. Preisendorfer

Introduction

The equation of transfer for radiance N:

$$\frac{n^2}{v} \frac{D[N/n^2]}{Dt} = -\alpha N + N* + N\eta \ , \tag{1}$$

where

$$N* = \int_{\Xi} N\sigma d\Omega \ ,$$

forms the heart of modern radiative transfer theory. For many years the equation of transfer in various simpler forms has been an extremely powerful tool in the hands of the theorists working in the fields of astrophysical and geophysical optics. The purpose of this discussion is to develop the thesis that the general equation of transfer may be used with equal incisiveness and to great practical advantage by the practicing experimenter to help him understand and collate with precision the results of his radiometric measurements of the light fields in the sea. In particular, I shall show how the equation may serve as a basis for the systematic classification of the principal optical properties of the sea, and of how the beginning of an exact theory of these properties may be achieved.

To understand the basic motivation for the formulations presented below, it is necessary first to understand the natural and fundamentally different ways in which the equation of transfer is viewed by theorists and experimenters: The theorist on the one hand, sees the equation of transfer as a set of mathematical operations on the radiance function. By means of the equation and various solution procedures, he

46

strives to determine the function N throughout a subregion of the optical medium of interest once he is given the values of N on its boundary along with the values of α, σ and other pertinent information throughout the region. The experimenter, on the other hand, sees the equation as a symbolic summary of certain basic operations with actual light measuring instruments. By means of the equation in various rearranged forms, he strives to determine the values of the functions α and σ in a given subregion of the medium once he has measured the function N in that region.

Since the equation of transfer in the experimenter's viewpoint is a symbolic representation of radiometric phenomena observed on a macroscopic scale, the resultant classification of optical properties will thus be phenomenological in nature (relative to such classifications as may be developed by means of, say, the maxwellian or quantum theories).

As the phenomenological classification proceeds, we shall see that the optical properties of the sea can quite naturally be divided into two mutually exclusive and exhaustive classes: The class consisting of the <u>inherent optical properties</u>, and the class consisting of the <u>apparent optical properties</u>. Now from the viewpoint of either the theorist or experimenter, an optical property of a medium is simply a function (defined by a certain sequence of realizable operations) which assigns to a given point in the medium a number representing the magnitude of the property at that point. With this in mind the two classes may be broadly defined as follows: An optical property is <u>inherent</u> if its operational value at a given point in a given medium is invariant under all changes of the radiance distribution at that point. An optical property is <u>apparent</u> if its operational value at a given point in a given medium is not invariant under all changes of the radiance distribution at that point. A practical test for these properties is the following: Suppose that a small element of volume of the medium about the given point were in imagination excised from the medium and placed in a new radiometric environment. Then apply the operational definition of the optical property to the element of volume in its new setting. If the operation remains meaningful and its value does not change, then the concept belongs to the class of inherent optical properties. If the operation becomes meaningless or the value of the function changes, then the concept belongs to the class of apparent optical properties.

By applying the preceding criterion to various cases, it becomes evident that the apparent optical properties depend in a rather complicated way on the inherent optical properties and on the ephemeral light field throughout the medium. This is the case for such well-known apparent optical properties as the logarithmic derivatives, with respect to depth, of upwelling and downwelling irradiances, and the reflectance functions of the sea. However, these concepts and the others classified below are worthy of the appellation "optical property" because: (i) their gross behavior with depth exhibits striking regularities, which are reproducible in medium after medium; (ii) it is possible to formulate exact mathematical interrelationships between these concepts and the inherent optical properties which hold in all practical cases; and finally (iii) the use of apparent optical properties reduces to a practical level the solution of underwater visibility problems and pertinent problems of marine biology. By experimentally determining the apparent optical properties of real media one is in effect solving on practical level certain particularly difficult problems in the theoretical-analytical aspects of the radiative transfer theory.

INHERENT OPTICAL PROPERTIES

Preliminary Definitions

The equation of transfer (1) forms the basis of all the deductions of the present discussion: From the equation or the various terms appearing in it we deduce the operational definitions of the inherent and apparent optical properties and some of their principal interrelations. It is, therefore, essential to make clear the definitions of the various terms appearing in the equation; this will now be done.

The radiometric quantity which is the most basic from the viewpoint of either the theorist or experimenter is that of <u>radiance</u>: The amount of radiant flux of a given wave-length crossing a unit area within a unit solid angle about the normal to the plane of the area. This concept is denoted by the symbol N which stands for the function with the following property: Let $\underline{x}$ be a vector denoting position of a point in an optical medium and let ξ denote a unit vector at $\underline{x}$. Let radiant flux of wave-length λ exist at x at time t. Then $N(\underline{x}, \underline{\xi}, t, \lambda)$ denotes the radiance of wave-length λ at $\underline{x}$ flowing in the direction $\underline{\xi}$ at time t. Here and below, N will be associated with an arbitrary fixed wave-length λ. Thus the symbol λ may, without loss of generality, be dropped from the notation.

To see the physical significance of the equation of transfer's various terms, we follow in imagination a packet of photons as it traverses a path through the neighborhood of a point in a scattering-absorbing medium. First of all, from geometrical optics we know that the quotient N/n^2, where n is the <u>index of refraction</u>, is invariant along a path through a region which exhibits no scattering, absorption, or sources of radiant flux. In such a case the equation of transfer is simply

$$\frac{1}{v}\frac{D[N/n^2]}{Dt} = 0 , \tag{2}$$

where v is the <u>speed of light</u> at the point instantaneously occupied by the packet. If the path were now directed through a region in which, there was pure absorption -that is, no scattering mechanisms present, and no sources of flux, then (2) becomes

$$\frac{1}{v}\frac{D[N/n^2]}{Dt} = -a\frac{N}{n^2} , \tag{3}$$

where a is the <u>volume absorption function</u>.
In reality, scattering mechanisms are coexistent with absorbing mechanisms. In this case there is, in addition to the absorption-loss term $a(N/n^2)$, a scattering-loss term $s(N/n^2)$ in the equation:

$$\frac{1}{v}\frac{D[N/n^2]}{Dt} = -(a+s)\frac{N}{n^2} , \tag{4}$$

where s is the <u>volume total scattering function</u>. However, the scattering mechanism simultaneously increases the population of the streaming packet of photons by having photons scattered into the direction of its travel. Hence, the equation, when taking this into account, becomes:

$$\frac{1}{v}\frac{D[N/n^2]}{Dt} = -\alpha\frac{N}{n^2} + \frac{N_*}{n^2} , \tag{5}$$

where we have set

$$\alpha = a + s \tag{6}$$

which is the <u>volume attenuation function</u>,
and where

$$N_*(\underline{x},\xi,t) = \int_{\Xi} N(\underline{x},\underline{\xi}',t)\, \sigma(\underline{x},\underline{\xi}';\underline{\xi},t)\, d\Omega(\underline{\xi}') \tag{7}$$

defines the <u>path function</u> N_* ; and introduces the <u>volume scattering function</u> σ. The symbol Ω stands for the solid angle measure, and Ξ is the set of all unit vectors (the unit sphere) in space. Finally, if sources are present, we add to (5) the <u>emission (source) function</u> N_η, to obtain (1). Both $N_*(\underline{x}, \xi, t)$ and $N_\eta(\underline{x}, \xi, t)$ have units of radiance per unit length, the unit of length being in the direction of ξ. The functions $\gamma, \alpha, \sigma, a$, and s have dimensions of inverse length. For a rigorous derivation of the equation of transfer from first principles, see reference (1.)

The Steady-State, Emission-Free Equation of Transfer

Light fields in the sea are in the steady state or, more precisely, in the quasi-steady state. Furthermore, the values of the emission function, except for some cases which are for the present purposes only of academic interest, are all zero. Finally, the index of refraction function n is essentially constant <u>within</u> the body of the sea: the domain of present interest. Thus, I shall at present adopt the form of the equation of transfer which results from the introduction of these conditions into (1); the result is:

$$\frac{dN}{dr} = -\alpha N + N_* \; ,$$

(8)

where d/dr is the steady state Lagrangian derivative:

$$\frac{d}{dr} = \frac{dx}{dr}\frac{\partial}{\partial x} + \frac{dy}{dr}\frac{\partial}{\partial y} + \frac{dz}{dr}\frac{\partial}{\partial z} \; ,$$

i.e.

$$\frac{d}{dr} = \underline{\xi} \cdot \nabla \; ,$$

(9)

where

$$\nabla = \left(\frac{\partial}{\partial x} , \frac{\partial}{\partial y} , \frac{\partial}{\partial z} \right) , \qquad \text{and} \qquad \underline{\xi} = \left(\frac{dx}{dr} , \frac{dy}{dr} , \frac{dz}{dr} \right) ,$$

(10)

in the usual cartesian coordinate system for space.

The physical significance of d/dr is simply that of a rate of change with distance r along a path with direction $\underline{\xi}$ at the point of interest. Equation (8) will form the basis of all the deductions below. It may be of interest to note that all of the results below may be generalized to the transient emission case. However, it is quite clear that such a case transcends present day interests, and will thus be reserved for subsequent study. However, the case of polarized radiance is of fundamental and immediate interest. The classification of the optical properties outlined below is easily extended to this case, but since this would introduce certain inessential complications, we omit it from the present discussion.

Enumeration of the Inherent Optical Properties

The <u>inherent optical properties</u> are:

α :   the volume attenuation function
σ :   the volume scattering function
a :   the volume absorption function
s :   the volume total scattering function

These properties are <u>inherent</u> in the sense described in the introduction. Their inter-relations are very simple:

$$s(\underline{x}) = \int_{\Xi} \sigma(\underline{x}, \underline{\xi}, \underline{\xi}') \, d\Omega(\underline{\xi}')$$

(11)

$$\alpha(\underline{x}) = a(\underline{x}) + s(\underline{x})$$

(12)

at each point $\underline{x}$ of the optical medium.

Thus either the pair (α, σ), or (a, σ) may be considered as the fundamental pair of inherent optical properties. That is, there are essentially two fundamental inherent optical properties; which two are considered as the fundamental pair is principally a matter of personal taste. As we shall see below, the equation of transfer provides operational definitions of comparable simplicity for either choice of α or a. Before leaving the enumeration of the inherent optical properties, we append two auxiliary inherent optical properties of comparatively minor importance:

$$f(\underline{x}, \underline{n}) = \int_{\underline{\xi} \cdot \underline{n} \geq 0} \sigma(\underline{x}, \underline{\xi}, \underline{\xi}') \, d\Omega(\underline{\xi}')$$

(13)

and

$$b(\underline{x}, \underline{n}) = \int_{\underline{\xi} \cdot \underline{n} \leq 0} \sigma(\underline{x}, \underline{\xi}, \underline{\xi}') \, d\Omega(\underline{\xi}')$$

(14)

which define the volume forward and volume backward scattering functions f and b respectively at $\underline{x}$ for the direction $\underline{n}$. If the medium is isotropic, then f and b are independent of $\underline{n}$. Furthermore,

$$s(\underline{x}) = f(\underline{x}) + b(\underline{x}) \tag{15}$$

We shall, henceforth, assume the sea to be an isotropic medium, so that $\underline{n}$ in (13), (14) may be dropped from the notation. This was already done in (11), (12), and (15) above.

Operational Definitions of the Inherent Optical Properties

The inherent optical properties will now be given operational definitions, i.e., definitions which show how the properties may be obtained by actual operations in natural or artificial light fields with a radiance meter and related radiometric equipment. Some of these operational definitions (for α and σ) are well-known. However, we are concerned, as stated above, with the attempt to unify and classify the optical properties by means of physically realizable operations with light-measuring apparatus guided by the equation of transfer. While the procedures described below will be idealized, it should be emphasized that they outline realizable operations which may be followed in obtaining their numerical values. I shall leave to the experimenters the necessary precautions that must be taken in actual practice in order to avoid the obvious pitfalls accompanying non-ideal instruments, perturbations of light fields by measuring equipment, etc.

An examination of the equation of transfer (8) shows that, in order to solve it for the volume attenuation function, one must generally know dN/dr, N_* and N:

$$\alpha = \frac{N_*}{N} - \frac{1}{N} \frac{dN}{dr} \tag{16}$$

However, by judiciously choosing a natural light field, or by generating an appropriate artificial light field, (16) may be simplified by causing, in turn, each of the two terms to be zero, thus yielding two basic methods of defining α. Recall that the function N_* represents radiance generated per unit length by the scattering of radiant flux into the direction of motion of the photons. Hence, if N_* can be made zero -or essentially negligible compared with the term $\frac{1}{N} \frac{dN}{dr}$, (16) reduces to

$$\alpha = -\frac{1}{N} \frac{dN}{dr} \tag{17}$$

This is the basic operational definition of α in local (differential) form. Formula (17) is the basis of the well-known beam-transmittance method of measuring α, wherein a radiance meter looks across an optically short path of length r within the medium at a controlled known source of flux of inherent radiance N_0. If N_r is the radiance measured at distance r, and α is known to be constant along the path, then (17) implies

$$\alpha = -\frac{1}{r} \ln\left(\frac{N_r}{N_0}\right) \tag{18}$$

which is the operational definition of α in global form. An analysis of the perturbations of the light field engendered by this method may be found in reference 2.

A less general, but quite interesting alternative operational definition of α is based on the generally valid fact that in the sea, the radiance along a horizontal path is invariant with respect to distance r along the path. It follows that, in this instance, (16) reduces to

$$\alpha = \frac{N_*}{N} \tag{19}$$

Now suppose that a radiance meter is built in such a way as to look across a horizontal path of small unit length at a black target (i.e., a target of zero inherent radiance in the wave-length in question). Then, this device will in principle record N_*. Another radiance meter, simultaneously records the actual horizontal radiance N at the same depth and in the same direction. Then (19) yields α by combining these readings as shown. This procedure was originated by Duntley in his studies of atmospheric optics and is quite successful in that context. In view of the excellent techniques now available by means of the more general definitions (17), (18), the definition summarized by (19) is perhaps only of academic interest for the case of light measurements in the sea.

The operational definition of the volume scattering function σ is based not on the entire equation of transfer, but only its path function term (7) now written for the steady state case. Suppose that the radiance function in the integrand were zero for all directions ξ' in Ξ except over a subset Ξ_0 of small solid angle measure $\Delta\Omega_0$ about ξ_0 a representative direction for Ξ_0. Furthermore, suppose that N over Ξ_0 is uniform of magnitude $N_0(\underline{x}, \xi_0)$. Then the corresponding path function value ΔN_* $(\underline{x}, \xi)$ at $\underline{x}$ in an arbitrary direction ξ is related to $\Delta\Omega_0$ and $N_0(\underline{x}, \xi_0)$ by the following special case of (7):

$$\Delta N_*(\underline{x}, \xi) = N_0(\underline{x}, \xi_0)\, \sigma(\underline{x}, \xi_0, \xi)\, \Delta\Omega_0$$

From this follows the basic operational definition of σ at $\underline{x}$, for the pair of directions $\xi_0\,\xi$:

$$\sigma(\underline{x}, \xi_0, \xi) = \frac{\Delta N_*(\underline{x}, \xi)}{N_0(\underline{x}, \xi_0)\, \Delta\Omega_0} \tag{20}$$

Recall that $\Delta N_*(\underline{x}, \xi)$ is the radiance generated by flux at $\underline{x}$ moving in the direction ξ_0 within a small solid angle about ξ_0 which is subsequently scattered in the direction ξ.

An alternate form of (20) is obtained if one notes first of all that $N_0(\underline{x}, \xi_0)\Delta\Omega_0$ is simply the irradiance $\Delta H_0(\underline{x}, \xi_0)$ at $\underline{x}$ generated by flux in the direction ξ_0, and secondly that $\Delta N_*(\underline{x}, \xi)\,\Delta V(\underline{x})$ is the radiant intensity $\Delta^2 J_*(\underline{x}, \xi)$ at $\underline{x}$ in the direction ξ generated by an element of volume of scattering matter at $\underline{x}$. Thus (20) may be written as:

$$\sigma(\underline{x}, \xi_0, \xi) = \frac{\Delta^2 J_*(\underline{x}, \xi)}{\Delta H_0(\underline{x}, \xi_0)\, \Delta V(\underline{x})} \tag{21}$$

where $\Delta V(\underline{x})$ is the actual volume irradiated.

This form of the definition, in the last analysis, is less operationally meaningful than (20), since the concept of radiant intensity is definable operationally only in terms of radiances and measurable areas, thus returning to (20). For further details concerning (20), see reference 3.

A straight-forward determination of the inherent optical property s follows from knowledge of σ and use of (11). However, the basic operational definition of the volume total scattering function s follows from (7) by the simple expedient of integrating (7) over Ξ. The result is

$$h_*(\underline{x}) \equiv \int_{\Xi} N_*(\underline{x}, \xi)\, d\Omega(\xi) = h(\underline{x})\, s(\underline{x}) \tag{22}$$

where

$$h(\underline{x}) = \int_{\Xi} N(\underline{x}, \xi)\, d\Omega(\xi) \tag{23}$$

is the scalar irradiance at $\underline{x}$. $h(\underline{x})$ is related to the radiant density (radiant energy per unit volume) $u(\underline{x})$ at $\underline{x}$ by: $h(\underline{x}) = u(\underline{x})\, v(\underline{x})$, where $v(\underline{x})$ is the speed of light at $\underline{x}$. Thus, from (22), we have:

$$s(\underline{x}) = \frac{h_*(\underline{x})}{h(\underline{x})} \tag{24}$$

In order to determine s(x) in accordance with (24) we need only integrate, either manually or automatically, the unit path radiance readings N_* over Ξ at $\underline{x}$. The scalar irradiance at $\underline{x}$ is measurable by suitably designed collecting spheres (see, e.g., reference 3). Observe the interesting similarity of structure between the operational definitions (19) and (24). For further details about (24) and other measurement procedures, see reference 4.

A straight-forward determination of the inherent optical property a from knowledge of s and α is made by means of (12). However, the basic operational definition of the volume absorption function a follows immediately from (8) in the form:

$$\xi \cdot \nabla N = -\alpha N + N_* \tag{25}$$

after defining the concept of <u>vector irradiance</u> $\underline{H}$:

$$\underline{H}(\underline{x}) = \int_{\Xi} \underline{\xi} N(\underline{x}, \underline{\xi}) \, d\Omega(\underline{\xi}) \tag{26}$$

which is the radiometric counterpart to Gershun's light vector $\underline{D}$. Integrating (25) over Ξ, we have :

$$\nabla \cdot \underline{H}(\underline{x}) = -\alpha(\underline{x}) h(\underline{x}) + s(\underline{x}) h(\underline{x})$$

by using (22) and (26). This, in view of (12), reduces to :

$$\nabla \cdot \underline{H}(\underline{x}) = -a(\underline{x}) h(\underline{x}) \tag{27}$$

which yields the basic operational definition of a(<u>x</u>) (<u>local form</u>) :

$$a(\underline{x}) = -\frac{\nabla \cdot \underline{H}(\underline{x})}{h(\underline{x})} \tag{28}$$

For further details concerning this definition, see reference 5. To illustrate the use of (28) for the case of the sea, we once again make use of the generally valid fact that the light field in the sea is stratified, i.e., that it depends spatially only on depth z below the surface. Suppose then that z represents depth measured with positive sense downward, and that $\underline{n}$ represents the unit outward normal to the surface of the sea (Fig. 1).

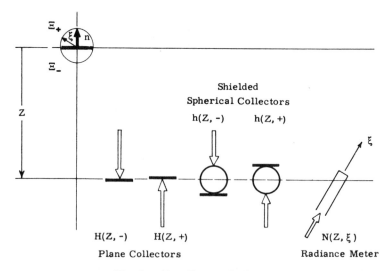

<u>Fig. 1</u> . Two-flow analysis of light field.

Then the divergence of the irradiance vector $\underline{H}(\underline{x})$ is

$$\nabla \cdot \underline{H}(\underline{x}) = \frac{d}{dz} \left[H(z,-) - H(z,+) \right]$$

where $H(z,+)$ is the upwelling irradiance at depth z across a horizontal surface, and $H(z,-)$ is the downwelling irradiance at that depth across the same surface. Then (28) becomes

$$a(z) = \frac{1}{h(z)} \frac{d\bar{H}(z,+)}{dz} \tag{29}$$

Here I have set

$$\bar{H}(z,+) = H(z,+) - H(z,-) \tag{30}$$

Each term on the right side of (29) is directly measurable. The irradiances $H(z, \pm)$ are related to the radiance distribution at depth z by the formulas:

$$H(z,\pm) = \int_{\Xi_\pm} |\underline{\xi} \cdot \underline{n}| N(z,\underline{\xi}) \, d\Omega(\underline{\xi}) \tag{31}$$

where Ξ_+ is the subset of Ξ consisting of all $\underline{\xi}$ such that $\underline{\xi} \cdot n \geqslant 0$; and Ξ_- consists of all $\underline{\xi}$ such that $\underline{\xi} \cdot n \leqslant 0$.

A global form of the definition of the volume absorption function is obtained by assuming the function a to be constant over an arbitrary sub-region X of the medium and then applying the divergence theorem to (27). The result is:

$$a = \frac{\bar{P}(S,-)}{vU(X)} \tag{32}$$

where a   is the (constant) value of the volume absorption function throughout X, $U(X)$ is the radiant energy content of X, $\bar{P}(S,-)$ is the net inward flux across the simple closed surface S bounding X, and v is the speed of light throughout X. This form may be of some practical interest in laboratory procedures for determining a. The nume- rator of (32) may be obtained by traversing the boundary of X with flat plate (irradiance) collectors. The denominator is obtained by probing the interior of X with a spherical (scalar irradiance) collector and integrating the values over X. By suitably adjusting the light field within X to attain maximum (e.g., spherical) symmetry, the number of probing operations can be reduced to a minimum. It should be clear that this method is independent of the structure of the light field within X.

This completes the present discussion of inherent optical properties. We now go on to discuss the second main group of optical properties of the sea.

APPARENT OPTICAL PROPERTIES

Introduction and Preliminary Definitions

The total set of apparent optical properties of natural waters consists at present of twenty quantities. The properties are apparent in the sense explained in the introduction. However, for the reasons also touched on in the introduction, these properties have been found to possess striking regularities and to be extremely useful in the experimental study of the light fields in natural hydrosols. Over recent years, at least five of them (the K-functions for irradiance and the reflectance functions) have found a permanent place in the experimentalist's stock in trade in his study of the optics of the sea.

Historically, the apparent optical properties (with the exception of $K(z, \underline{\xi})$) are traceable (at least in their most rudimentary form) to the classical two-flow analysis of the light field in plane-parallel media. In this context they were truly "optical properties" and fixed in magnitude throughout the media. Thus, the present day apparent optical properties are outgrowths of man's conceptual decomposition of the light fields in such media into two counter-moving streams: An "upwelling" and a "downwelling" stream. An outline of the salient history of this concept and its evolution over the past fifty-five years may be found in references 6 and 7. I might venture to say that a complete understanding of the classical theory is indispensible for a thorough appreciation of the role of the apparent optical properties in the general modern theory given below.

It has been found possible to characterize nine of the basic optical properties by means of a basic quartet of directly observable irradiances. These irradiances require less effort to be determined than the radiance distribution and of course contain less information than the latter. However, the quartet supplies enough information about the depth dependence of the light field, its directional structure, and overall flux transmitting and reflecting properties of the medium to allow many practical problems to be solved with satisfactory precision and completeness. These facts along with their regularity features hold the secret of the utility of the apparent optical properties.

The basic irradiance quartet is:

$$H(z,+) \qquad h(z,+)$$
$$\tag{33}$$
$$H(z,-) \qquad h(z,-)$$

where the $H(z, \pm)$ have already been defined in (31), and

$$h(z,\pm) = \int_{\Xi\pm} N(z,\xi)\,d\Omega(\xi). \tag{34}$$

Observe that the notation reflects our standing assumption about the optical structure of the sea: It is horizontally stratified; the character of the stratification is, however, arbitrary. The quantities $h(z,\pm)$ are the <u>upwelling</u> (+) and <u>downwelling</u> (-) <u>scalar irradiances</u> at depth z (Fig. 1). They are related to the scalar irradiance h(z) (Eq.(23)) by:

$$h(z,+) + h(z,-) = h(z)$$

Methods of measurement of $h(z,\pm)$ other than by means of integrated N-values are discussed in reference 3. The quantities $H(z,\pm)$ are measured by horizontal flat plate collectors or from N by integration.

Operational Definitions of the Apparent Optical Properties

The apparent optical properties along with their operational definitions are:

$$R(z,\pm) = \frac{H(z,\mp)}{H(z,\pm)} \tag{35}$$

$$K(z,\xi) = \frac{-1}{N(z,\xi)}\,\frac{dN(z,\xi)}{dz} \tag{36}$$

$$K(z,\pm) = \frac{-1}{H(z,\pm)}\,\frac{dH(z,\pm)}{dz} \tag{37}$$

$$k(z,\pm) = \frac{-1}{h(z,\pm)}\,\frac{dh(z,\pm)}{dz} \tag{38}$$

$$k(z) = \frac{-1}{h(z)}\,\frac{dh(z)}{dz} \tag{39}$$

$$D(z,\pm) = \frac{h(z,\pm)}{H(z,\pm)} \tag{40}$$

In these formulas read all upper signs together and all lower signs together. For example:

$$R(z,-) = \frac{H(z,+)}{H(z,-)}$$

is the reflectance of the sea at depth z, with respect to the downwelling flux at that level. Since

$$R(z,+) = \frac{1}{R(z,-)} \, ,$$

it is sufficient to discuss only R(z, -).

The quantity $K(z, \xi)$ is the diffuse attenuation function for radiance.

The quantities $K(z,\pm)$ are the values of the diffuse attenuation functions for irradiance at depth z. Knowledge of these functions over a depth interval from z_1 to z_2, $0 \leqslant z_1 \leqslant z_2$ is tantamount to knowledge of H(z, $\pm$) over the same interval; for knowing H(z, $\pm$) over this interval yields K(z, $\pm$) by means of (37). On the other hand, (37) implies :

$$H(z_2, \pm) = H(z_1, \pm) \exp\left\{-\int_{z_1}^{z_2} K(z', \pm) \, dz'\right\} \, , \tag{41}$$

Observe that if $K(z, \pm) = K$ a given value over $[z_1, z_2]$ then from (41),

$$H(z_2, \pm) = H(z_1, \pm) \exp\left\{-K(z_2 - z_1)\right\} \, ,$$

which shows that $K(z, \pm)$ are the exact generalizations of the classical diffuse attenuation function.

The $k(z, \pm)$ are the values of the diffuse attenuation functions for upwelling (+), downwelling (-) scalar irradiance. The subtle distinction required between the little k's and the big K's is a direct consequence of the increasing demands for accuracy by modern experimentalists in their basic determinations of the structure of light fields in natural hydrosols. $k(z)$ is the value of the diffuse attenuation function for scalar irradiance. The interrelations among these K-functions will be illustrated below in the theoretical section.

Finally, $D(z, \pm)$ is the value of the distribution function for upwelling (+) downwelling (-) flux, at depth z. These numbers are indicators of the directional structure of the radiance distribution at depth z. For example, if N were collimated within Ξ_- and the collimated beam was in the direction defined by ξ , then

$$D(z,-) = \frac{1}{|\xi \cdot n|}$$

If, on the other hand, N were uniform over Ξ_- , then

$$D(z,-) = 2$$

The distribution functions, besides being simple conceptual indicators of the structure of the radiance distribution, are indispensible analytical links between the various optical properties, as shall be seen in detail below.

Examples of Values of D, K, a, R

Table I exhibits values of some of the apparent optical properties. Depths and units are in meters. Data were associated with wave-length 480 mμ and were derived from radiance information summarized in reference 8. The optical medium (Lake Pend Oreille, Idaho, U.S.A.) was found to be essentially homogeneous with $\alpha = 0.402/$meter. The sky was clear and sunny with sun at about 40° from zenith.
The values a(z) were obtained by means of (29).

Table I. Examples of the Values of D(Z, ±), K(Z, ±), a(Z), R(Z, -)

Z (Meters)	D(Z, -)	D(Z, +)	K(Z, -)	K(Z, +)	a (Z)	R(Z, -)
4.24	1.247	2.704				0.0215
7.33			0.129	0.126		
10.42	1.288	2.727	0.153	0.150	0.115	0.0184
13.50			0.178	0.174		
16.58	1.291	2.778	0.174	0.172	0.118	0.0204
22.77			0.171	0.170		
28.96	1.313	2.781	0.169	0.169	0.117	0.0227
35.13			0.167	0.167		
41.30	1.315	2.757	0.165	0.165	0.117	0.0235
47.50			0.162	0.163		
53.71	1.307	2.763	0.158	0.158	0.112	0.0234
59.90			0.154	0.154		

Hybrid Optical Properties

The classification of the optical properties of the sea would not be complete without some provision for describing the direct scattering and absorbing counterparts to α , a, s, f and b which belong exclusively to the upwelling and downwelling streams. These descriptions are supplied by the set of hybrid optical properties which form a subset of the class of apparent optical properties. They receive their name from the fact that they escape any simple operational definitions of the kind described so far, and because their natural definitions are achieved by mating inherent optical properties with apparent optical properties by means of certain well defined analytical bonds. Thus, the volume attenuation functions for the two streams are:

$$\alpha(z,\pm) = \alpha(z) \, D(z,\pm); \tag{42}$$

the volume absorption functions for the two streams are:

$$a(z,\pm) = a(z) \, D(z,\pm); \tag{43}$$

the volume total scattering functions for the two streams are :

$$s(z,\pm) = s(z) \, D(z,\pm); \tag{44}$$

and, finally, the forward scattering functions for each stream are:

$$f(z,\pm) = \frac{1}{H(z,\pm)} \int_{\Xi \pm} \left[\int_{\Xi \pm} N(z,\xi') \sigma(z,\xi',\xi) \, d\Omega(\xi') \right] d\Omega(\xi) \tag{45}$$

and the backward scattering functions for each stream are:

$$b(z,\pm) = \frac{1}{H(z,\pm)} \int_{\Xi \mp} \left[\int_{\Xi \pm} N(z,\xi') \sigma(z,\xi',\xi) \, d\Omega(\xi') \right] d\Omega(\xi) \tag{46}$$

Clearly,

$$s(z,\pm) = f(z,\pm) + b(z,\pm) , \tag{47}$$

where we continue to read upper signs together for the upwelling stream and lower signs together for the downwelling stream:
and furthermore

$$\alpha(z,\pm) = a(z,\pm) + s(z,\pm) , \tag{48}$$

56

in complete analogy to (12) and (15).

The list of optical properties as it stands at present is summarized in Table II below:

CLASSIFICATION OF OPTICAL PROPERTIES OF THE SEA

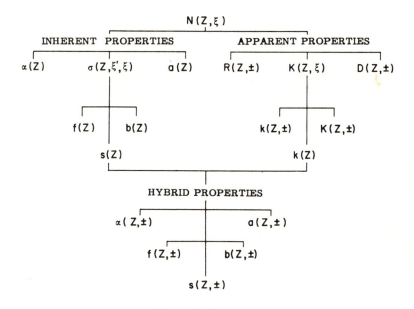

GENERAL THEORY OF OPTICAL PROPERTIES

I shall now briefly outline a theory, based on the exact equation of transfer, which ties together the various inherent and apparent optical properties discussed above. The theory in its present form is not intended to be exhaustive; rather, it is intended to illustrate the basic thesis of the present discussion, namely: That the equation of transfer can be turned into a useful, incisive tool in the hands of the experimental investigator who, having the light field of the sea at his finger-tips, so to speak, is concerned with the experimental documentation of the salient optical properties of the sea. Other than the generally valid assumption of the predominantly horizontal stratification of the light field in natural waters, the theory incorporates no further assumptions about such intricate features as the depth and directional dependence of the submarine radiance distribution, nor of the nature of the stratification of the optical structure of the sea. In short, the theory proposed below is generally applicable to any arbitrary stratified plane-parallel optical medium, of arbitrary depth, with arbitrary reflecting upper and lower boundaries, and with arbitrary incident radiance distribution at its upper boundary.

The theory of the optical properties of the sea is perhaps best expounded in the limited space at hand by attempting to answer some of the more fundamental questions that may be raised by the practicing optical oceanographer. The paragraphs below, it is hoped, will supply answers to some of these questions, and show by example how further questions of similar natures may be resolved once they are formulated.

The General Transport Equations for Irradiance

The core of the theory of optical properties resides in a pair of general equations which govern the depth behavior of the upwelling and downwelling irradiance $H(z,\pm)$. The starting point of their derivation is the equation of transfer in the form:

$$\xi \cdot \nabla N(z, \xi) = -\alpha(z) N(z, \xi) + N*(z, \xi) \tag{49}$$

If each term of (49) is integrated over Ξ_-, we shall obtain the equation governing $H(z, -)$ as follows:

First,

$$\int_{\Xi_-} \xi \cdot \nabla N(z, \xi) \, d\Omega(\xi) = \nabla \cdot \int_{\Xi_-} \xi N(z, \xi) \, d\Omega(\xi) = \frac{dH(z, -)}{dz}$$

since the sea is stratified.

Second,

$$\int_{\Xi_-} \alpha(z) N(z, \xi) \, d\Omega(\xi) = \alpha(z) h(z,-) = \alpha(z,-) H(z,-)$$

which follows from (34) and (40).

Finally,

$$\int_{\Xi_-} N*(z, \xi) \, d\Omega(\xi) = \int_{\Xi_-} \left[\int_{\Xi} N(z, \xi') \sigma(z, \xi'; \xi) \, d\Omega(\xi') \right] d\Omega(\xi)$$

$$= \int_{\Xi_-} \left[\left(\int_{\Xi_+} + \int_{\Xi_-} \right) N(z, \xi') \sigma(z, \xi'; \xi) \, d\Omega(\xi') \right] d\Omega(\xi)$$

$$= b(z, +) H(z, +) + f(z, -) H(z, -) \ ,$$

where the first equality follows from the definition of N_*, the second from the definition of Ξ, and the third from the interchange of the order of integrations, and use of (45), (46).

Assembling these results, we have, making use of (47) and (48):

$$\frac{dH(z, -)}{dz} = -\left[a(z, -) + b(z, -) \right] H(z, -) + b(z, +) H(z, +) \ ; \tag{50}$$

and in a similar manner we deduce:

$$-\frac{dH(z, +)}{dz} = -\left[a(z, +) + b(z, +) \right] H(z, +) + b(z, -) H(z, -) \ , \tag{51}$$

which are the desired, exact, transport equations for $H(z, \pm)$. For a more general derivation within the time dependent, emission context, see reference 6. For the relationship between (50), (51) and the classical Schuster two-flow equations, see references 6 and 7. The present equations are simultaneously more exact, more general, and operationally more meaningful than the classical two-flow equations; nevertheless, they are identical in spirit, and the historical indebtedness of the present set with respect to the classical set is most certainly acknowledged. The remainder of the discussion will be devoted to a brief survey of some of the fundamental interrelations between optical properties which may be deduced from the transport equations for $H(z, \pm)$.

Formulas for K (z, ±) and R (z, ±) in Terms of Hybrid Optical Properties

Dividing through (50) by H(z, -) and applying the operational definitions of $K(z, -)$ and $R(z, -)$ we have:

$$K(z,-) = a(z,-) + b(z,-) - b(z,+) R(z,-) , \qquad (52)$$

Similarly, from (51):

$$-K(z,+) = a(z,+) + b(z,+) - b(z,-) R(z,+) \qquad (53)$$

These equations show how the experimental K-functions are linked to the absorption and scattering properties of the optical medium. From these, we immediately deduce the exact counterparts to some well-known classical equations for R(z, -):

$$R(z,-) = \frac{a(z,-) + b(z,-) - K(z,-)}{b(z,+)} \qquad (54)$$

and

$$R(z,-) = \frac{b(z,-)}{a(z,+) + b(z,+) + K(z,+)} \qquad (55)$$

By making use of (47) and (48) these may be cast into the alternate forms:

$$R(z,-) = \frac{\alpha(z,-) - f(z,-) - K(z,-)}{b(z,+)} \qquad (56)$$

$$R(z,-) = \frac{b(z,-)}{\alpha(z,+) - f(z,+) + K(z,+)} \qquad (57)$$

The Basic Reflectance Relation

The preceding formulas are simple, immediate consequences of the transport equations for H(z, ±). A more fundamental relation exists and is singled out for special attention because it relates the six direct observables of principal interest : The two K-functions for irradiance, the two distribution functions, the volume absorption function and the reflectance function.

This relation is:

$$R(z,-) = \frac{K(z,-) - a(z,-)}{K(z,+) + a(z,+)} \qquad (58)$$

and follows from (52), (53) by performing the operations on them indicated in (58), and simplifying the result.

The Characteristic Equation for the Pair K(z, ±)

The classical theory gives a convenient expression for k (the classical diffuse absorption coefficient of which, as we have seen, K(z, ±) are the immediate generalizations) in terms of the volume absorption coefficient a and the backward scattering coefficient b* for diffuse flux. This classical relation is:

$$k = 2\left[a(a + b^{*})\right]^{1/2} \qquad (59)$$

The modern exact theory yields the following counterpart to (59):

$$1 = \frac{b(z,-)}{K(z,-) - a(z,-)} - \frac{b(z,+)}{K(z,+) + a(z,+)} \tag{60}$$

which is the <u>characteristic equation</u> for the pair $K(z, \pm)$. It may be deduced from (52) and (58).

The characteristic equation reduces to (59) by setting

$$b(z,-) = b(z,+) = b^* \ , \quad a(z,-) = a(z,+) = 2a \ , \ \text{and} \ K(z,+) = K(z,-) = k$$

The Basic Inequalities

What relation does the magnitude of, say, $K(z, -)$ have to $\alpha(z)$ and $a(z)$? Can one rigorously show that the well-known "fact" $R(z, -) \leqslant 1$ is really true, and under what conceivable conditions can one have $R(z, -) > 1$? Is there some invariable relation between the relative depth rates of change of $H(z, +)$ and $H(z, -)$ which is sufficiently sensitive to provide a check of empirical determinations of these quantities?

To answer these questions and to prepare the ground work for answering more of this general kind, consider once again equation (50) now written in the form

$$K(z,-) = \alpha(z,-) - \frac{1}{H(z,-)} \int_{\Xi_-} N_*(z,\xi) \, d\Omega(\xi) . \tag{61}$$

Since the substracted number on the right is never negative, we deduce that

$$K(z,-) \leq \alpha(z,-)$$

for all z.

Furthermore, whenever $K(z,+) \geq 0$ we have from (58)

$$a(z,-) \leq K(z,-) .$$

Combining these two inequalities, we answer the first question in the form.

$$a(z) \leq \frac{K(z,-)}{D(z,-)} \leq \alpha(z) \tag{62}$$

These are the sharpest possible inequalities, since examples can be found for which equality holds on either side. A similar inequality holds for the upwelling stream:

$$a(z) \leq \frac{-K(z,+)}{D(z,+)} \leq \alpha(z) \tag{63}$$

For a detailed discussion of these inequalities and those below, see reference 7.

To answer the second question, we return to (29) which, upon integration over a depth interval $\left[z_1, z_2\right]$ yields:

$$\bar{H}(z_1,-) + \bar{H}(z_2,+) = \int_{z_1}^{z_2} a(z) h(z) \, dz \geq 0 \ ,$$

where

$$\bar{H}(z, \pm) = H(z, \pm) - H(z, \mp) \ ;$$

hence

$$\bar{H}(z_2, -) \leq \bar{H}(z_1, -) \tag{64}$$

from which we conclude that in source-free media, the net downward irradiance never increases. If the medium is finitely deep with a bottom of reflectance r, $0 \leqslant r \leqslant 1$, then (64) clearly implies

$$0 \leq R(z, -) \leq 1$$

for all depths z. If the sea is optically infinitely deep, then again, since $\bar{H}(\infty, +)=0$,

$$\bar{H}(z, -) = \int_z^\infty a(z') h(z') \, dz' \geq 0$$

so that:

$$0 \leq H(z, +) \leq H(z, -) \qquad \text{i.e.} \qquad 0 \leq R(z, -) \leq 1 \tag{65}$$

once again for all depths z.

To answer the third question, one may start with (58) to find

$$K(z, -) - K(z, +) R(z, -) = a(z, -) + a(z, +) R(z, -) \geq 0$$

whence:

$$K(z, -) \geq K(z, +) R(z, -) \tag{66}$$

or, equivalently

$$\frac{dH(z, -)}{dz} \leq \frac{dH(z, +)}{dz} \tag{67}$$

for all depths in an arbitrarily stratified source-free sea. These inequalities should by now show the versatility of the exact two-flow theory, so we may go on to consider further questions.

The Relation Between the Depth Behavior of R(z, -) and the K-Functions

It is of interest to characterize the variation of the reflectance function in terms of the experimental K-functions. The desired formula follows immediately from the definition of R(z, -) upon differentiation:

$$\frac{dR(z, -)}{dz} = R(z, -) \left[K(z, -) - K(z, +) \right] \tag{68}$$

From this we see that the constancy of R(z, -) over an interval is equivalent to the equality of K(z, +) and K(z, -) over that interval. If, on the other hand, K(z, -) exceeds K(z, +) at some depth z then R(z, -) is increasing at depth z; and so on.

The Basic Relations Among the Various K-Functions

The K-functions for the irradiances and the K- functions for the scalar irradiances are related by the formulas:

$$k(z, \pm) = K(z, \pm) - \frac{1}{D(z, \pm)} \frac{dD(z, \pm)}{dz} \tag{69}$$

Thus, the little k's and the big K's coincide at a certain level when and only when the directional structure of the radiance distribution does not change in a vertical neighborhood of that level.

For the purposes of unifying all of the K-functions, we may use $K(z, \xi)$, perhaps the most fundamental of them all. This is the K-function associated with the radiance function:

$$K(z, \xi) = \frac{-1}{N(z, \xi)} \frac{dN(z, \xi)}{dz} \tag{70}$$

From this and the preceding definitions, it may be deduced that:

$$K(z, \pm) = \frac{\int_{\Xi_\pm} N(z, \xi) K(z, \xi) \, \xi \cdot n \, d\Omega(\xi)}{\int_{\Xi_\pm} N(z, \xi) \, \xi \cdot n \, d\Omega(\xi)} \tag{71}$$

$$k(z, \pm) = \frac{\int_{\Xi_\pm} N(z, \xi) K(z, \xi) \, d\Omega(\xi)}{\int_{\Xi_\pm} N(z, \xi) \, d\Omega(\xi)} \tag{72}$$

$$k(z) = \frac{\int_{\Xi} N(z, \xi) K(z, \xi) \, d\Omega(\xi)}{\int_{\Xi} N(z, \xi) \, d\Omega(\xi)} \tag{73}$$

These representations will form a basis for the important conclusions about the asymptotic values of the K-functions in optically deep seas, which appear in the closing paragraph below.

The Covariation of K(z, -) and D(z, -)

If measurements of K(z, -) are made within a small depth interval about a fixed depth in the sea over a period of time during which only the external lighting conditions change, how does this change manifest itself in the recorded values of K(z, -)? Alternatively, how does K(z, -) depend on the shape of the radiance distribution at depth z, assuming the inherent optical properties do not change?

An answer to this question may be made by appealing once again to (61). Let

$$h*(z, -) = \int_{\Xi_-} N*(z, \xi) \, d\Omega(\xi) \tag{74}$$

The physical significance of this concept is readily discerned: $h_*(z, -)$ is the downwelling scalar irradiance generated by radiant flux scattered by a unit volume at depth z. With this definition (61) becomes

$$K(z, -) = \left[\alpha(z) - \frac{h*(z, -)}{h(z, -)} \right] D(z, -) \tag{75}$$

On the basis of this formula it may be shown that in all forward scattering media, of which the sea is an outstanding example, K(z, -) varies directly (but not necessarily linearly) with D(z, -). Thus, if D(z, -) increases, remains the same, decreases over a period of time, the inherent optical properties remaining unchanged, then K(z, -) likewise increases; remains the same, decreases over the same time period. For further details on this and related matters, along with some useful rules of thumb deduced from (75), see reference 9.

The Asymptotic Light Field

I have mentioned several times throughout this discussion that the apparent optical properties of the sea, despite their dependence on highly variable lighting conditions, nevertheless, exhibit certain striking regularities in both time and space. In this, the concluding part of the discussion, I wish to bring to your attention perhaps one of the most remarkable of all these regularities and its manifold consequences. This concerns the asymptotic (or limiting) behavior of the light field as great depths are approached in deep natural waters. The experimental evidence for this phenomenon is by now well established. The pioneering experiments by Whitney are perhaps the best known and best documented early instances in this matter (reference 10).

I wish now to formulate this phenomenon in terms of the apparent optical properties defined above and to draw out some of its practical consequences. For further consequences see reference 11. Proofs of the asymptotic radiance theorem will be omitted; they may be found in references 12 (the general case) and 13 (the case of the sea).

Briefly and simply the asymptotic radiance theorem states that, regardless of the external lighting conditions, the directional (ξ) and depth dependences (z) of the radiance distribution $N(z, \xi)$ multiplicatively uncouple at great depths in all homogeneous, or eventually homogeneous waters. That is, at great depths we may write:

$$N(z, \xi) = N(z_0, \xi) e^{-k_\infty (z - z_0)}$$ (76)

where z_0 is some depth below which

$$K(z, \xi) = k_\infty$$ (77)

Here k_∞ is a fixed magnitude independent of direction ξ and which depends only on the inherent optical properties of the sea. Thus the _shape_ of the radiance distribution eventually settles down to some fixed form which is independent of the external lighting conditions and dependent only on α and σ. More precisely, if

$$\lim_{z \to \infty} \sigma(z, \xi, \xi') \equiv \sigma(\xi, \xi')$$
exists

and

$$\lim_{z \to \infty} \alpha(z) \equiv \alpha$$
exists

in an infinitely deep sea, then

$$\lim_{z \to \infty} K(z, \xi) = \lim_{z \to \infty} k(z) = k_\infty$$ (78)

exists for each ξ in Ξ. This is the content of the _asymptotic radiance theorem_ in terms of inherent and apparent optical properties. The two versions of the theorem (76) and (78) are equivalent. From this we may deduce manifold consequences. For example:

(i) $$\lim_{z \to \infty} K(z, \pm) = \lim_{z \to \infty} k(z, \pm) = k_\infty$$ (79)

which follows from (78) and (71), (72).

(ii) $$\lim_{z \to \infty} D(z, \pm) \equiv D(\pm)$$ (80)
exists

which follows from (78) using (76) and (40)

(iii) $$\lim_{z \to \infty} R(z, -) = \frac{k_\infty - D(-) a}{k_\infty + D(+) a}$$ (81)

which follows from (58), (78), (79) and (80).

From these few examples we see the vast simplifications, implied by the asymptotic radiance theorem, which are in store for the experimental study of light fields at great depths in essentially homogeneous media.

More generally, from these examples and the arguments presented above it is hoped that the main point of the present discussion has been made: By taking a strictly operational point of view of the general equation of transfer, it may be used without once solving it, or subjecting it to vitiating assumptions about the structure of of either the medium or the light field within the medium to erect a systematic classification of the optical properties of the sea and the attendant theory for these properties.

References

1. Preisendorfer, R.W., A Mathematical Foundation for Radiative Transfer Theory, J. Math. and Mech., 6, 685-730 (1957).
2. Preisendorfer, R.W., A General Theory of Perturbed Light Fields, with Applications to Forward Scattering Effects in Beam Transmittance Measurements. Scripps Institution of Oceanography, La Jolla, California, S10 Ref. 58-37 (1958).
3. Preisendorfer, R.W., and J.E. Tyler, The Measurement of Light in Natural Waters: Radiometric Concepts and Optical Properties, Scripps Institution of Oceanography, La Jolla, California, S10 Ref. 58-69 (1958).
4. Preisendorfer, R.W., On the Direct Measurement of the Total Scattering Function, Scripps Institution of Oceanography, La Jolla, California, S10 Ref. 59-41 (1959).
5. Preisendorfer, R.W., The Divergence of the Light Field in Optical Media, Scripps Institution of Oceanography, La Jolla, California, S10 Ref. 58-41 (1957).
6. Preisendorfer, R.W., Unified Irradiance Equations, Scripps Institution of Oceanography, La Jolla, California, S10 Ref. 58-43 (1957).
7. Preisendorfer, R.W., Directly Observable Quantities for Light Fields in Natural Hydrosols, Scripps Institution of Oceanography, La Jolla, California, S10 Ref. 58-46 (1958).
8. Tyler, J.E., Radiance Distribution as a Function of Depth in the Submarine Environment, Scripps Institution of Oceanography, La Jolla. California, S10 Ref. 58-25 (1958).
9. Preisendorfer, R.W., The Covariation of the Diffuse Attenuation and Distribution Functions in Plane Parallel Media, Scripps Institution of Oceanography, La Jolla, California, S10 Ref. 59-52 (1959).
10. Whitney, L.V. , The Angular Distribution of Characteristic Diffuse Light in Natural Waters, J. Mar. Res. 4, 122-131 (1941).
11. Preisendorfer, R.W., Some Practical Consequences of the Asymptotic Radiance Hypothesis, Scripps Institution of Oceanography, La Jolla, California, S10 Ref. 58-60 (1958).
12. Preisendorfer, R.W., A Proof of the Asymptotic Radiance Hypothesis, Scripps Institution of Oceanography, La Jolla, California, S10 Ref. 58-57 (1958).
13. Preisendorfer, R.W., A Theoretical Proof of the Existence of Characteristic Diffuse Light in Natural Waters, J. Mar. Res., 18, 1-9 (1959).

DIFFUSION DE LA LUMIERE PAR LES EAUX DE MER.
RESULTATS EXPERIMENTAUX ET APPROCHE THEORIQUE

A. Morel

[Editor's Note: In the original, material precedes this excerpt.]

3 - DIFFUSION PAR UN ENSEMBLE DE PARTICULES POLYDISPERSE

Le vocable "ensemble polydispersé" est pris au sens où on l'entend généralement. Il désigne un ensemble de particules de même forme (sphérique en l'occurence) et de même nature (donc de même indice) qui ne diffèrent entre elles que par leur taille. Leur nombre varie avec la taille selon une loi de distribution.

On indiquera d'abord (§ 3.1) les formules permettant le calcul des propriétés diffusantes d'un tel ensemble. Puis on examinera les prévisions qu'on peut faire concernant le résultat des calculs (§ 3.2) dans le cas où on adopte une distribution selon une loi puissance. Ceci conduit à s'interroger sur la signification des calculs, compte tenu du fait que ceux ci doivent être exécutés en posant des limites mathématiques dont la réalité physique n'est pas évidente. Enfin, et par comparaison avec les lois puissance, le cas d'autres distributions est envisagé.

3.1 - CALCUL DES PROPRIETES DIFFUSANTES

La population de particules est caractérisée par une fonction de distribution établie vis à vis d'un paramètre géométrique caractérisant la taille : dans ce qui suit on utilisera le paramètre α, c'est à dire la taille relative $2 \pi r / \lambda$ qui intervient dans les calculs de diffusion et en conserve la généralité (pour les applications, les rayons des particules supposées sphériques étant fixés, le changement de longueur d'onde se traduit par une variation inversement proportionnelle de α). La fonction de distribution $F(\alpha)$ correspond à une fréquence ou probabilité d'apparition : si la totalité de la population comporte N particules dont les tailles vont de zéro à l'infini (ou d'une taille minimale α_m à une taille maximale α_M), la quantité :

$$\frac{1}{N} \int_{\alpha_1}^{\alpha_2} F(\alpha) \, d\alpha$$

est la probabilité relative d'apparition des particules dont le paramètre de taille est compris entre α_1 et α_2 ; la fonction $F(\alpha)$ est supposée continue et intégrable dans l'intervalle $0, \infty$ ou α_m, α_M.

L'additivité des intensités diffusées par des particules réparties au hasard rend, d'un point de vue formel, le calcul simple ; il est entendu que les intégrations prévues ci dessous sont pratiquement remplacées par des sommations, l'incrémentation en taille $d\alpha$ étant celle que commandent les calculs préparatoires d'indicatrices individuelles (cf. § 2.2, 2e partie).

3.1.1 - Indicatrice de diffusion pour un nombre donné de particules.

Pour chaque angle, les fonctions d'intensité $i_1(\theta)$ et $i_2(\theta)$ pour l'ensemble des particules sont données par l'intégrale portant sur les produits $i_1(\theta,\alpha) F(\alpha)$ et $i_2(\theta,\alpha) F(\alpha)$. En normalisant par l'intégrale de la fonction $F(\alpha)$ entre les mêmes limites, le calcul est rendu indépendant du nombre total de particules, sinon il dépendrait à la fois de la distribution et des limites adoptées. On en vient donc à calculer :

$$i_1(\theta) = \frac{\int_{\alpha_m}^{\alpha_M} F(\alpha) \, i_1(\theta,\alpha) \, d\alpha}{\int_{\alpha_m}^{\alpha_M} F(\alpha) \, d\alpha} \quad , \tag{3.1}$$

et $i_2(\theta)$ de la même façon. La fonction d'intensité totale $i_T(\theta) = (1/2)(i_1(\theta) + i_2(\theta))$ et le taux de polarisation $(i_1(\theta) - i_2(\theta)) / (i_1(\theta) + i_2(\theta))$ en sont déduits. Pour l'application, si N est le nombre total de particules dans l'unité de volume, le coefficient angulaire de diffusion est obtenu simplement en calculant (cf. annexe I "définitions") :

$$\beta(\theta) = N \frac{\lambda^2}{4 \pi^2} i_T(\theta) \quad .$$

L'adoption pour λ d'une valeur donnée se répercute sur les limites α_m et α_M à utiliser.

3.1.2 - Facteur d'efficacité moyen.

Pour un ensemble de particules, il est défini comme le rapport de la somme des sections efficaces à la somme des aires des sections géométriques. Pour les particules sphériques considérées, et en utilisant le paramètre α, on a :

$$\overline{Q} = \frac{\displaystyle\int_{\alpha_m}^{\alpha_M} F(\alpha)\ \alpha^2\ Q\ (\alpha)\ d\alpha}{\displaystyle\int_{\alpha_m}^{\alpha_M} F(\alpha)\ \alpha^2\ d\alpha} \quad . \tag{3.2}$$

On peut également utiliser le paramètre $\rho = 2\ \alpha(n-1)$ qui combine à la fois taille et indice relatifs. Ceci est avantageux lorsque l'indice est assez proche de 1 car alors la formule de Van de Hulst (2.29) est applicable ; son expression est suffisamment simple pour que l'intégrale au numérateur de l'équation 3.2 puisse être calculée numériquement avec un pas de ρ très petit.

3.1.3 - Indicatrice normalisée.

Pour comparer les résultats obtenus en faisant varier la loi de distribution et ses limites, il est finalement moins pratique de fixer le nombre de particules (comme le fait l'équation 3.1) que de fixer, au contraire, le coefficient total de diffusion. Ce cas correspond à l'examen de collections de particules dont le nombre et la répartition sont variables, mais qui présentent le même effet global de diffusion, en l'occurence une diffusion unitaire ; les coefficients de diffusion normalisés (cf. Annexe I, "définitions") sont calculés par :

$$\overline{\beta}_1\ (\theta) = \frac{1}{\pi}\ \frac{\displaystyle\int_{\alpha_m}^{\alpha_M} F(\alpha)\ i_1\ (\theta,\alpha)\ d\alpha}{\displaystyle\int_{\alpha_m}^{\alpha_M} F(\alpha)\ Q\ (\alpha)\ \alpha^2\ d\alpha} \tag{3.3}$$

et pour $\overline{\beta}_2\ (\theta)$ par l'expression correspondante ; $\overline{\beta}\ (\theta)$ est ensuite calculé par la demi-somme : $1/2\ \left(\overline{\beta}_1\ (\theta) + \overline{\beta}_2\ (\theta)\right)$. Le calcul de $\overline{Q}$ étant mené simultanément (3.1.2), l'inverse de $\overline{Q}$ fournit l'aire totale des sections des particules nécessaires pour produire cette diffusion totale unitaire. Par exemple, si $\overline{Q} = 0,5$, il est nécessaire que l'aire totale des sections des particules présentes dans un volume de 1 m^3 soit de 2 m^2, pour que le coefficient de diffusion b soit de 1 m^{-1}.

3.2 - PREVISIONS CONCERNANT LE RESULTAT DES CALCULS.

Elles sont possibles en se fondant sur les deux points suivants :
- l'évolution de la forme des indicatrices de Mie pour les tailles croissantes est retracée dans ses grands traits par les graphes des fonctions i (θ) α^{-4} ; les remarques antérieures concernant les pentes variées des courbes pour divers angles vont permettre de telles prévisions.
- la distribution des particules selon leur taille est supposée suivre une loi de Junge* exprimée par une fonction puissance : $F\ (\alpha)$ = Constante $. \alpha^{-m}$.

3.2.1 - Conditions de convergence : influence de la limite supérieure sur l'indicatrice.

Compte tenu de la forme donnée à $F\ (\alpha)$, les fonctions $i_T\ (\theta)\ \alpha^{-4}$ déjà représentées correspondent précisément à des distributions en α^{-4} et l'intégration de ces fonctions est celle figurant au numérateur des expressions 3.1 ou 3.3. Elle fournit (à la normalisation près) l'indicatrice pour la population considérée (caractérisée par la loi en α^{-4}). On voit immédiatement que les intégrales pour les divers angles vont toutes converger puisque les pentes moyennes des courbes sont voisines de $-2,3$ (cf. § 2.2.1, e), sauf l'intégrale relative à l'angle 0° pour lequel la pente moyenne de la courbe est nulle. Autrement dit, passé une certaine limite, la prise en compte des particules de plus en plus grosses ne modifiera pas la forme de l'indicatrice résultante sauf pour l'angle 0° et les angles immédiatement voisins où l'intensité continuera de croître. Ce fait illustré par la figure III.2 (3e partie) où la limite supérieure est portée de 50 à 200 sans que l'indicatrice change sensiblement sauf à 0°, et également vers 175-180°. Il n'en va plus de même si la limite supérieure devient inférieure à 50 (A. Morel, 1972 a, figure 2).

D'une façon générale, les fonctions i (θ) croissent avec la taille α selon des lois en α^{4+p} où p est la pente moyenne lue sur les graphes (bilogarithmiques) tels que II.3 ; pour rappeler (cf. § 2.2.1) ces pentes ont les valeurs :

$$
\begin{aligned}
&p = + 2 &&\text{pour } \theta \text{ quelconque, si } \alpha \text{ est petit} \\
& &&\text{pour } \theta = 0°, \text{ même lorsque } \alpha \text{ n'est pas petit, à condition que} \\
& &&\rho = 2\ \alpha(n-1) \text{ soit inférieur à } 4 \text{ ;} \\
&p = 0 &&\text{pour } \theta = 0° \text{ lorsque } \alpha \text{ est grand } (\rho > 4) \text{ ;} \\
&p \simeq -2,3 (\text{ou } -2) &&\text{pour } \theta \neq 0° \text{ lorsque } \alpha \text{ est grand ;} \\
&-2 < p < 1 &&\text{pour } \theta = 180°, \text{ lorsque } \alpha \text{ est grand ; la valeur de la pente dépend en} \\
& &&\text{fait de l'indice (il est difficile d'ailleurs d'apprécier} \\
& &&\text{une valeur moyenne).}
\end{aligned}
$$

Les fonctions $F\ (\alpha)\ i_T\ (\alpha,\theta)$ sont des fonctions en α^{4+p-m} si $-m$ est l'exposant de la distribution ; les intégrales à calculer sont donc des fonctions dont l'exposant est $5 + p - m$. Ces intégrales convergeront absolument si : $5 + p - m < 0$; (3.4) l'égalité à zéro n'entraine pas la convergence, l'intégrale étant un logarithme.

Si on porte en fonction de α les valeurs progressives des intégrales :

---

* On verra plus loin que des fonctions de ce type rendent bien compte des distributions effectivement observées pour les particules marines. En tout état de cause des distributions plus complexes peuvent toujours être décomposées et approximées dans chaque domaine par de telles formes, au moins théoriquement.

Diffusion de la lumiere par les eaux de mer

$$\int_{\varepsilon}^{\alpha} F(\alpha)\ i_T\ (\theta,\alpha)\ d\alpha \quad , \tag{3.5}$$

pour divers angles θ (figure II.12) on constate que :

a/ lorsque α est petit, quelle que soit la valeur de θ, toutes les courbes croissent selon une pente[*] dont la valeur est 7-m (soit 3,5 et 2 respectivement pour les deux cas présentés) ;

b/ pour des valeurs de α d'autant plus élevées que θ est plus petit (c'est à dire pour les valeurs de α correspondant au maximum des courbes i (θ) α^{-4}, cf. § 2.2.1 d et 2.2.3), les intégrales convergent asymptotiquement et les courbes présentent un palier asymptotique (lorsque la condition 3.4 est respectée) ;

c/ pour l'angle $0°$ (où p = 0) il n'y aura de palier que si m > 5, sinon l'intégrale continuera de croître avec la pente 5 - m (soit une pente égale à 1,5 pour la figure supérieure, et une pente tendant vers 0 pour la figure inférieure - branche de courbe logarithmique - qui constitue un cas limite) ;

d/ pour l'angle $180°$ l'évolution est plus compliquée du fait même que la pente moyenne p a tendance à s'annuler lorsque α croît (d'autant plus que l'indice est plus élevé, ce qui augmente la réflexion, cf. § 2.2.5) ; en général, après un palier, l'intégrale partielle recommence à croître.

En conclusion, puisque l'écartement entre les divers paliers caractérise l'indicatrice résultante, les prévisions suivantes peuvent être faites :
- si toutes les courbes présentent un palier (lorsque m > 5), l'écart entre les valeurs finales ne change plus et donc l'indicatrice ne se modifie plus si l'on continue de faire croître la limite supérieure d'intégration.
- si 2,7 < m < 5, la conclusion reste la même sauf en ce qui concerne l'angle $0°$, autrement dit le prise en compte de particules de plus en plus grosses ne change pas la forme de l'indicatrice sauf aux très petits angles (et également vers $180°$) où la croissance persiste.

Figure II.12 : Valeurs progressives des intégrales en fonction de la limite supérieure. L'exposant de la distribution a pour valeur $-3,5$ et -5 respectivement, l'indice de réfraction demeurant dans les deux cas 1,05.
La limite inférieure pour le calcul est 0,2 mais les courbes sont tracées après le premier pas d'intégration, c'est à dire à partir de $\alpha = 0,4$. La limite supérieure est 200.

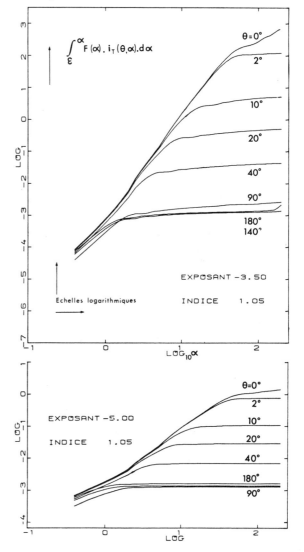

*Les légères sinuosités que l'on décèle sur les courbes ascendantes (au voisinage de $\alpha = 2$ par exemple) sont des artéfacts de calcul dus au changement de pas $(d\alpha)$ dans le calcul de l'intégrale. Ce pas est en effet déterminé par les calculs prélilinaires d'indicatrices (cf. § 2.2).

- si m < 2,7 aucune limite mathématique n'existe et le calcul n'a de sens que s'il existe une limite ayant une réalité physique.

3.2.2 - Rôle des petites particules : influence de la limite inférieure de taille sur l'indicatrice.

On vient de voir que dans certaines conditions qui ont été précisées, le calcul conserve sa signification, bien que la limite supérieure ait été posée arbitrairement. Un problème analogue se pose quant à la limite inférieure de taille. Celle-ci est physiquement inconnue et pour le calcul la limite inférieure de taille ne peut être, ici encore, qu'arbitraire. Il est nécessaire de prévoir quelle incidence a sur l'indicatrice la non prise en considération des très petites particules. Théoriquement le calcul peut être conduit depuis une limite inférieure hypothétique constituée par les particules de dimension "nulle". Pour étudier l'effet de la troncature * si l'intégration est conduite à partir d'une limite finie non nulle, deux cas sont à distinguer :

Cas où les particules négligées appartiennent au domaine de Rayleigh :
C'est le cas où l'intégration est faite à partir d'une limite inférieure $\alpha_m = \varepsilon$, au plus égale à 1. Une hypothèse défavorable consiste à admettre que les particules inconnues continuent d'être distribuées selon la même loi en α^{-m} ; cette hypothèse conduit à des nombres de particules toujours croissants (et à un nombre infini pour la taille "nulle"). La même loi de diffusion s'applique à toutes ces particules (cf. § 2.2.2), soit :

$$\begin{vmatrix} i_1(\theta) \\ i_2(\theta) \end{vmatrix} = \alpha^6 \Lambda^2 \begin{vmatrix} 1 \\ \cos^2\theta \end{vmatrix}$$

Pour chaque valeur de l'angle θ, l'intégration sur le domaine considéré - pour $i_T = 1/2(i_1 + i_2)$ - s'écrit :

$$1/2 \int_o^\varepsilon \Lambda^2 \alpha^{6-m} (1 + \cos^2\theta) \, d\alpha \quad . \tag{3.6}$$

Cet ensemble de particules présente donc une indicatrice de Rayleigh, exprimée par :

$$i_T(\theta) = 1/2 \, \Lambda^2 (1 + \cos^2\theta) \frac{1}{7-m} \varepsilon^{7-m} \quad , \tag{3.7}$$

c'est à dire une indicatrice de grandeur finie à condition que soit respectée l'inégalité** :

$$7 - m > o \quad , \tag{3.8}$$

sinon, on obtient $i_T = \infty$. Autrement dit, dans le premier cas, s'il y a troncature, les quantités $i_T(\theta)$ négligées sont finies et l'erreur résultante est calculable. L'erreur est infinie dans le second cas et le calcul n'a de signification que si une limite physique connue non nulle existe.

Dans le premier cas, les termes négligés varient avec l'angle selon la loi de Rayleigh (c'est à dire dans le rapport 1 à 2). Si on les compare à l'indicatrice très dissymétrique obtenue lorsque la limite supérieure est 200, on conçoit que l'erreur possible est maximale pour les angles égaux ou supérieurs à 90°. En valeur relative cette erreur est obtenue en formant le rapport des intégrales respectivement entre 0 et 0,2 et entre 0,2 et 200. Pour reprendre les deux exemples illustrés par la figure II.12, l'erreur résultant de la troncature (à 0,2) est inférieure à 0,1 % si la valeur de l'exposant est -3,5, et de l'ordre de 1 % si cette valeur est -5 (ceci pour les angles 90, 140 et 180° ; elle est évidemment inférieure aux angles plus petits). Cet ordre de grandeur demeure valable pour les autres cas. Dans l'annexe II sont présentées les valeurs calculées au moyen de l'expression (3.7) pour divers indices et pour des exposants variant de -3 à -5. Ces valeurs sont toujours négligeables comparées à celles que fournit l'intération dans l'intervalle de taille 0,2 - 200.

En conclusion on peut considérer que la limite inférieure de cet intervalle est fixée à une valeur suffisamment basse. L'indicatrice ainsi obtenue est significative car les particules de taille inférieure à cette valeur n'ont plus d'influence sensible (à condition que leur loi de distribution respecte la condition 3.8).

Cas où la troncature intervient pour des tailles plus élevées.
L'effet peut en être directement prévu sur les figures telles que II.12. En particulier, si la valeur de la limite inférieure est telle que la courbe relative à un angle donné θ a amorcé son palier, l'effet sera important ; pour cet angle les quantités négligées peuvent devenir supérieures aux quantités prises en compte dans l'intégration et la valeur finale (lorsque α = 200) est consécutivement diminuée. Les paliers intervenant pour des valeurs de α d'autant plus élevées que θ est plus petit (cf. § 3.2.1, b), cette diminution affecte d'abord la valeur pour l'angle 180°, puis pour 140°, et caetera.. La dissymétrie de l'indicatrice finale se renforce au fur et à mesure que la limite α_m inférieure s'élève :
si α_m = 1 par rapport aux valeurs obtenues précédemment avec α_m = 0,2 il apparait que les valeurs à 90°, 140° et 180° sont diminuées de 10 à 20 %.
si α_m = 2 les diminutions aux mêmes angles vont de 30 à 50 % environ tandis qu'à 40° un léger écart (- 10 %) apparait.
si α_m = 10 l'effet est encore accentué et d'autre part s'étend à un domaine angulaire plus large ; l'écart est notable dès l'angle 20°.

La figure II.13 met en évidence ce renforcement de la dissymétrie globale par diminution des valeurs aux grands angles, les valeurs aux petits angles (0° et 2°) n'étant pas affectées. Cette figure reprend la partie supérieure de la figure II.12 ; en outre sont représentées les courbes obtenues

---

* Ou l'effet de la méconnaissance des lois de distribution à utiliser pour les petites particules dont la présence, non observée, n'est que probable, par continuité.

** C'est évidemment la condition inverse de celle qu'exprime (3.4), avec p = 2, valeur de la pente dans le domaine de Rayleigh.

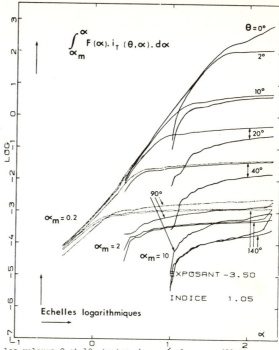

Figure II.13 : influence de la limite inférieure;
les courbes en pointillé sont reprises de la figure II.12 (partie supérieure, exposant -3,5).
Les deux autres groupes de courbes sont obtenues
lorsque la limite inférieure d'intégration est
successivement portée à 2 puis 10. Les trois
courbes non identifiées sont relatives à l'angle 180° ; elles sont caractérisées par une
croissance rapide pour les valeurs élevées de α.

lorsque la limite inférieure prend successivement les valeurs 2 et 10, toutes choses égales par ailleurs.
Cette évolution est plus complètement illustrée par le tracé complet des indicatrices présenté plus loin
(Figure III.12, 3e partie).

3.2.3 - Diffusion totale ; influence des limites sur le calcul.

Comme précédemment la question se pose de savoir si le calcul de la diffusion totale a un sens, alors
que seule est considérée une partie de la population des particules ; les limites inférieure et supérieure sont en effet posées pour les besoins du calcul mais ne traduisent pas une réalité physique. Le raisonnement, analogue à celui exposé ci dessus (3.2.1 et 3.2.2) doit cette fois être appliqué à l'intégrale :

$$\int_{\alpha_m}^{\alpha_M} F(\alpha) \; \alpha^2 \; Q(\alpha) \; d\alpha \qquad\qquad (3.9)$$

qui, à un facteur près, exprime la section efficace totale des particules distribuées selon la loi $F(\alpha)$
entre les tailles minimale et maximale α_m et α_M. Il suffit de rappeler que le facteur d'efficacité Q varie en α^4 dans le domaine de Rayleigh (équation 2.32), en α^2 dans le domaine de Rayleigh Gans (équation
2.35) et qu'enfin il est, en moyenne, indépendant de α dans le domaine de Mie et dans celui de la diffraction (§ 2.2.8).

Sans reprendre le détail on voit donc qu'en ce qui concerne l'influence de la *limite supérieure*
(dans le domaine de Mie), la diffusion totale tend vers une limite si l'intégrale de $F(\alpha) \; \alpha^2$ converge,
c'est à dire si la surface totale présentée par les particules tend elle même vers une limite. Ceci implique que soit respectée la condition :   m > 3   (3.10)
si la distribution est comme précédemment exprimée par $F(\alpha) = C \; \alpha^{-m}$. La surface totale et donc la diffusion croissent comme le logarithme de la limite supérieure α_M lorsque m = 3.

En ce qui concerne la *limite inférieure* (dans le domaine de Rayleigh), l'intégrale entre 0 et α_m
correspondant aux particules négligées, s'applique à une expression en α^{6-m}, tout comme dans le calcul
de l'indicatrice. En conséquence la même condition (3.8) :   m < 7   assure à cette intégrale une valeur
finie. Le calcul pour divers exposants et indices montre que la valeur de cette intégrale entre les limites 0 et 0,2, reste tout à fait négligeable devant la valeur de l'intégrale pour les particules prises en
considération (c'est à dire entre les limites de taille 0,2 et 200). Ces valeurs sont présentées dans
l'annexe II.

Ainsi qu'on l'a vu (§ 3.2.2) un nombre infini de particules peut néanmoins présenter une indicatrice
finie. De façon équivalente, la diffusion totale, ou plus précisément la section efficace totale, présentée par les particules de tailles comprises entre 0 et α_m peut également être finie, alors que l'aire totale des sections géométriques de ces mêmes particules ne l'est pas ; cette dernière en effet n'est finie
que si :   m < 3   ; ainsi, lorsque l'exposant demeure entre les limites 3 < m < 7, le cas envisagé est
bien réalisé.

Comme cela a été fait pour les valeurs angulaires, on peut également porter en fonction de la limite
supérieure, la valeur progressive de l'intégrale (3.9) ; la figure II.14 fournit un tel exemple pour diverses valeurs de l'indice et pour la valeur -3,10 de l'exposant. La pente initiale des courbes a pour
valeur 7 - m, elle prend ensuite la valeur 5 - m et enfin tend à s'annuler, les courbes présentent alors
un palier asymptotique puisque la condition de convergence est respectée. L'amorce des paliers a lieu

69

lorsque Q devient (en moyenne) indépendant de α, c'est à dire pour des valeurs de α variables selon la valeur de l'indice ; ces valeurs sont en fait liées puisqu'en ce point le paramètre ρ = 2 α (n-1) est égal à 4,1 (cf. § 2.2.8 et figure II.10).

Figure II.14 : Valeur progressive de l'intégrale donnant la section efficace (3.9), lorsque la limite supérieure croît, jusqu'à la valeur finale α = 200. Les diverses courbes correspondent aux indices de réfraction indiqués et à une seule valeur de l'exposant de la distribution des particules.

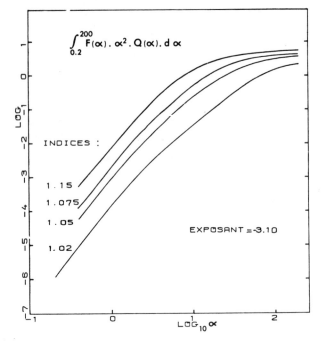

Figure II.15 : Figure analogue à la précédente, la variable ρ remplaçant la variable α. Les diverses courbes correspondent aux valeurs indiquées de l'exposant ; elles sont normalisées par leur valeur asymptotique (100 %). Les échelles sont logarithmiques.

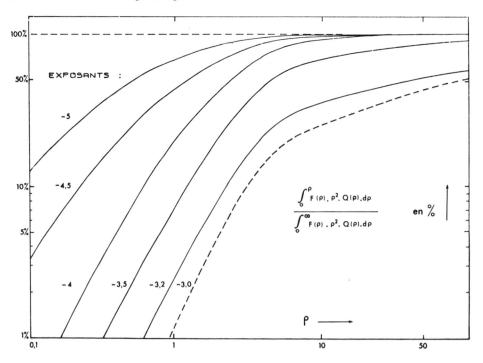

Dans la mesure où l'expression de Van de Hulst (2.29) est une approximation justifiée (c'est le cas avec les valeurs d'indice utilisées ici) un seul calcul suffit, la variable ρ remplaçant α. Les diverses courbes de la figure II.14 correspondent en réalité à une unique courbe mais tracée avec une échelle des abscisses plus ou moins dilatée. En conséquence de cette remarque, les intégrales calculées cette fois avec la variable ρ sont représentées par les courbes de la figure II.15 chacune correspondant à diverses valeurs de l'exposant. On peut calculer les valeurs asymptotiques de ces intégrales lorsque la limite supérieure croît à l'infini (lorsque la condition 3.10 est satisfaite) ces valeurs sont utilisées pour la normalisation du graphique ; toutefois la courbe relative à l'exposant -3 ne peut être placée qu'arbitrairement, puisqu'elle ne présente pas d'asymptote. Pour toutes les courbes l'infléchissement apparaît lorsque ρ dépasse la valeur 4 (Voir Annexe 2, pour le calcul des valeurs asymptotiques).

3.2.4 - Résumé concernant les domaines de validité du calcul.

On donnera à l'expression "domaine de validité" un sens précis : c'est le domaine pour lequel le calcul conserve sa signification indépendamment des valeurs attribuées aux limites inférieure et supérieure de taille. Ceci suppose donc que soient finies les quantités calculées lorsque l'on fait tendre les limites de taille vers zéro ou vers l'infini. Dans ces conditions la méconnaissance des limites physiques n'invalide pas le calcul. Si par continuité on suppose que la même loi de distribution (fonction puissance) régit la population quelle que soit la taille, les conditions assurant la validité concernent l'exposant de la loi. Ces conditions, ainsi qu'on l'a vu diffèrent selon les quantités à calculer. Elles sont rassemblées dans le tableau ci contre ; est indiquée, en regard de la valeur de l'exposant, la nature finie ou infinie des quantités calculées entre 0 et α_m (influence de la limite inférieure) ou entre α_M et ∞ (influence de la limite supérieure).

On peut remarquer qu'en ce qui concerne nombre, surface ou volume des particules, les conditions relatives respectivement aux limites inférieures et supérieures sont exclusives les unes des autres ; autrement dit, le domaine de validité comme défini ci-dessus n'existe pas ; le calcul n'est possible que si physiquement on peut admettre que des limites existent. Il n'en va pas de même pour les grandeurs relatives à la diffusion : les deux catégories de conditions peuvent être simultanément respectées dans certains intervalles d'exposant ; c'est ce que résume le schéma suivant où le domaine de validité est représenté par la zone ombrée.

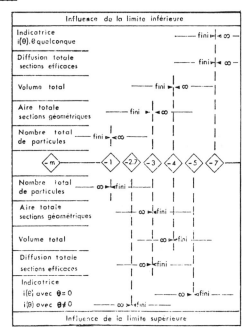

Il convient par contre d'examiner le cas de distributions continuement décroissantes, bien que différant de la loi puissance. Il s'agit donc de choisir des distributions plausibles, c'est à dire qui puissent, dans l'intervalle de tailles accessibles à l'expérience, rendre compte approximativement de la décroissance observée*. Mais au delà des limites de cet intervalle, l'hypothèse selon laquelle les lois puissance subsistent est abandonnée, les nouvelles distributions considérées se comportent différemment. Pour décrire des populations de particules (naturelles ou créées) on a souvent utilisé la fonction exponentielle :
$$F_1 (\alpha) = A_1 \exp(-B_1 \alpha) , \qquad (3.11)$$
ou plus fréquemment la fonction gausso-logarithmique (ou log-normale) :
$$F_2 (\alpha) = A_2 \exp \left(-B_2 (\log \alpha / \overline{\alpha})^2\right) , \qquad (3.12)$$

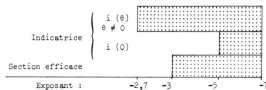

3.2.5 : Extension à des lois de distribution différentes de la loi de Junge.

Les conclusions précédentes concernant le domaine de validité sont valables seulement, tout au moins telles que présentées, pour des distributions exprimées par des lois puissance. Cependant elles peuvent être réexaminées et étendues à d'autres types de distribution. Le cas précédemment traité, mathématiquement commode, peut, précisément à ce titre, constituer un cas référence. Il est inutile de considérer le cas de distributions selon des fonctions du type creneaux (rectangulaire, triangulaire,..) pour lequel le problème des limites et de la signification du calcul ne se pose pas. Le cas de la distribution normale de Gauss s'y ramène, comme on le verra plus loin.

---

*Cette remarque concernant l'étude granulométrique des particules, ainsi que le choix fait ci-après de valeurs numériques pour les distributions, est une anticipation sur ce qui est exposé au début de la 3e partie, à laquelle on peut se référer.

où $\bar{\alpha}$ est la valeur correspondant au maximum de la distribution, et B_2 est l'écart type "géométrique". Pour les particules marines des lois exponentielles (ou de type exponentiel, comme la loi de Weibull) ont été proposées par Carder et al (1971). Egalement J.R. Zaneveld et H. Pak (1973) utilisent cette distribution, qui pour des raisons de convergence aux limites qu'on verra ci dessous, se révèle commode pout tout calcul théorique.

Il est pratique, pour la comparaison avec les lois puissance en $F(\alpha) = A\,\alpha^{-m}$, de considérer le graphe bilogarithmique des distributions précédentes. La première demeure exponentielle, deux points la déterminent complètement. La seconde devient simplement une parabole, deux points ne suffisent pas pour la déterminer, il en faut un troisième ou, aussi bien, il faut fixer, arbitrairement, la position du maximum $\bar{\alpha}$. On peut vérifier que la distribution normale de Gauss, au delà du maximum est représentée par une courbe exponentielle (mais de pente plus forte que celle de l'exponentielle 3.11 car l'argument est double). Sur la figure II.16, le graphe bilogarithmique de la fonction puissance en α^{-4}, choisie comme base de comparaison, est représenté par la droite de pente -4. On a imposé aux autres distributions de recouper la précédente à la fois pour $\alpha = 10$ et $\alpha = 100$ et on a choisi de centrer la distribution gausso logarithmique et également la distribution normale sur $\alpha = 1$.

Du côté des petites tailles la distribution exponentielle tend vers une valeur finie avec une pente nulle et la condition (3.8) de convergence est toujours respectée. La distribution gausso logarithmique tend vers zéro, la pente négative (entre $\alpha = 1$ et $\alpha = 10$) est nécessairement plus faible que celle de la sécante figurant la loi puissance ; la condition (3.8) est *a fortiori* respectée lorsque la loi puissance la respecte ($7 - m > 0$). Il faut des lois puissance à exposant négativement élevé pour que les petites particules, en nombre suffisamment élevé, n'aient pas un rôle négligeable ; pratiquement les distributions envisagées leur assignent un nombre faible et correspondant mathématiquement à une quasi-troncature pour la diffusion. La conclusion est identique en ce qui concerne la distribution de Gauss.

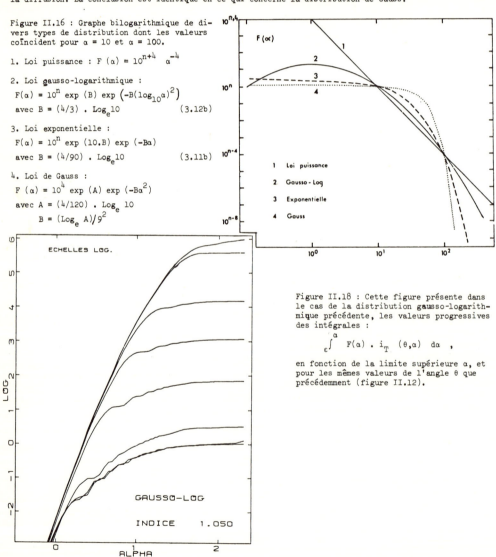

Figure II.16 : Graphe bilogarithmique de divers types de distribution dont les valeurs coïncident pour $\alpha = 10$ et $\alpha = 100$.

1. Loi puissance : $F(\alpha) = 10^{n+4}\;\alpha^{-4}$

2. Loi gausso-logarithmique :

$F(\alpha) = 10^n \exp(B) \exp\left(-B(\log_{10}\alpha)^2\right)$

avec $B = (4/3) \cdot \text{Log}_e 10$ (3.12b)

3. Loi exponentielle :

$F(\alpha) = 10^n \exp(10.B) \exp(-B\alpha)$

avec $B = (4/90) \cdot \text{Log}_e 10$ (3.11b)

4. Loi de Gauss :

$F(\alpha) = 10^4 \exp(A) \exp(-B\alpha^2)$

avec $A = (4/120) \cdot \text{Log}_e 10$

$B = (\text{Log}_e A)/9^2$

Figure II.18 : Cette figure présente dans le cas de la distribution gausso-logarithmique précédente, les valeurs progressives des intégrales :

$$\int_\varepsilon^\alpha F(\alpha) \cdot i_T\,(\theta,\alpha)\;d\alpha \;,$$

en fonction de la limite supérieure α, et pour les mêmes valeurs de l'angle θ que précédemment (figure II.12).

Diffusion de la lumiere par les eaux de mer

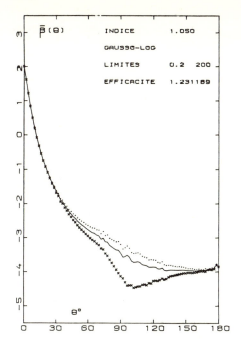
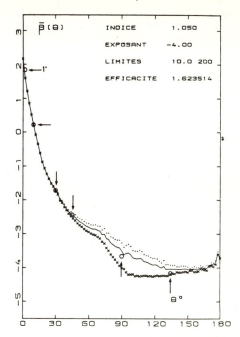

Figures II.17 : Indicatrices obtenues avec la distribution gausso logarithmique (équation 3.12 b, figure II.16) et avec la distribution de Junge d'exposant -4, tronquée inférieurement à α = 10 et supérieurement à α = 100 ; les indicatrices sont tracées en portant, en fonction de θ, le logarithme décimal du coefficient normalisé de diffusion $\overline{B}(\theta)$. Les valeurs de $\overline{B}_1(\theta)$ et $\overline{B}_2(\theta)$ relatives aux composantes polarisées verticale et horizontale respectivement sont figurées par des points et des croix.

Les points désignés par les flèches correspondent aux calculs effectués par O.B. Brown et H.R. Gordon (1971) pour le même exposant, pour les limites 8,6 et 86 (diamètres 1 à 10 μm, λ = 488 nm) et avec l'indice 1,05 - 0,01 i. Pour la comparaison les valeurs sont recalculées afin d'être présentées sous la forme de coefficient normalisé $\overline{B}(\theta)$. La limite supérieure est 86 et non 200 mais, compte tenu de l'exposant, la forme asymptotique est pratiquement atteinte (cf. § 3.2.1) et la comparaison est justifiée.

Du côté des grandes tailles, les pentes négatives toujours croissantes et supérieures à celle de la droite, font que se trouvent toutes satisfaites les conditions de convergence, quel que soit l'angle de diffusion θ (§ 3.2.1). Avec ces lois, la fréquence d'apparition des grosses particules est plus faible qu'avec les lois puissance correspondantes ; pratiquement cela équivaut comme précédemment, à une troncature. La figure II.17 illustre cet effet : elle permet de comparer les indicatrices obtenues avec la loi gausso logarithmique d'une part et avec la loi puissance d'autre part ; pour la première la distribution est étendue à l'ensemble des tailles de 0,2 à 200 (pour α) tandis que pour la seconde la distribution est tronquée inférieurement à $α_m$ = 10 et supérieurement à $α_M$ = 100. Par superposition, il apparait que ces deux indicatrices sont en fait confondues (ce qui n'est pas tout à fait le cas pour les composantes polarisées i_1 et i_2).

Il n'est pas utile de multiplier les calculs et exemples de cette nature, étant donné le caractère arbitraire des lois utilisées, en particulier en ce qui concerne la position du maximum de la distribution. Qu'il suffise de dire qu'avec les lois exponentielles et surtout log normales on obtient des résultats peu différents de ceux auxquels conduit la loi puissance équivalente (et dans le second cas, aussi voisin que l'on veut, selon les paramètres choisis). Avec la loi de Gauss, même tronquée au délà du maximum, il n'en est par contre jamais ainsi.

Enfin entre les tailles (10 et 100 pour l'exemple donné ici) où les distributions sont mises en coïncidence, les courbes présentent une convexité ; en conséquence, par rapport à la loi puissance, ces distributions favorisent en quelque sorte les particules de taille intermédiaire (20 à 50) qui joueront, comparativement, un rôle plus important dans la diffusion. Un exemple de cette conséquence sera donné plus loin (3e partie, § 3.1.2).

73

TROISIEME   PARTIE

Dans le calcul des propriétés diffusantes d'un système polydispersé, interviennent d'une part les paramètres liés à la loi et aux limites de la distribution et d'autre part, l'indice relatif qu'on appellera "moyen" puisqu'il est supposé être le même pour toutes les particules. A titre d'exemple, diverses applications numériques ont été présentées dans les développements antérieurs mais sans qu'ait été discuté le choix des valeurs attribuées aux paramètres mentionnés. Un certain nombre de cas théoriques ont été systématiquement traités en vue de la comparaison avec les résultats expérimentaux et un choix a dû être fait qu'il convient d'examiner ici.

Pour *l'indice de réfraction* des particules on ne peut guère faire que des hypothèses car aucune mesure directe ne semble avoir été faite. A partir du contenu minéral des particules en suspension (carbonate de calcium, silice, alumino-silicates, divers hydroxydes,...) supposé sous forme cristalline, on peut déduire une valeur moyenne d'indice. Les valeurs souvent avancées à partir de ces considérations sont de l'ordre de 1,15 ou 1,20*(valeurs relatives par rapport à l'eau, c'est à dire 1,53 - 1,60 en valeurs absolues) ; des calculs ont été faits pour les argiles en suspension avec la valeur 1,15 (et une partie imaginaire égale à 0,001 ; H. Pak, R.V. Zaneveld, G.F. Beardsley, 1971). D'une part il n'est probablement pas réaliste d'attribuer à toute la particule, même si elle est essentiellement minérale, l'indice valable pour la forme cristalline. Certains insolubles ou certains précipités peuvent être présents sous des formes colloïdales fortement hydratées ou bien la partie minérale d'une particule détritique peut n'être par exemple qu'une coque ; de telles particules minérales présenteraient un indice "moyen" plus proche de celui de l'eau que   celui   déduit de la stricte composition (et en excluant l'eau). D'un autre côté, la part des matières organiques dans le matériel en suspension est toujours importante et pour ces substances, l'indice est très proche de celui de l'eau. Récemment K.L. Carder et al. (1972) ont, à partir de mesures de diffusion sur des cultures d'algues unicellulaires (Isochrysis galbana), déduit l'indice relatif qui serait de l'ordre de 1,026 à 1,036 pour ce matériel organique. Par rapport à la masse totale des particules, la part organique**, prépondérante pour les eaux de surface au large, demeure élevée même dans les couches profondes ; entre la surface et des profondeurs allant jusqu'à 4000 m, l'abondance relative de la matière organique particulaire resterait supérieure à 25 % (D.C. Gordon, 1970), varierait de 40 à 60 % (L. A. Hobson, 1967), de 40 à 100 %***(P.J. Kinney et al., 1971), de 40 à 88 % (C. Copin, G. Copin, 1972), de 26 à 49 % (J.E. Harris, 1972). Ceci incite à choisir des valeurs proches de 1 pour l'indice relatif "moyen" des particules. Quoiqu'il en soit, les calculs préliminaires d'indicatrices individuelles ont été faits pour 5 valeurs de l'indice couvrant un domaine suffisamment large depuis des valeurs faibles (1,02 et 1,05) jugées plus vraisemblables, jusqu'à la valeur caractéristique de la fraction minérale (1,15), les valeurs intermédiaires choisies étant 1,075 et 1,10.

Quant à *la loi de distribution*, contrairement à l'indice, des résultats expérimentaux existent. En particulier l'utilisation d'un compteur granulométrique électrique (Coulter-counter) a permis d'obtenir des données nouvelles (L.A. Hobson, 1967 ; R.W. Sheldon et T.R. Parsons, 1967 ; K.L. Carder, 1970). Les mesures de H. Bader (1970), comme celles de J.C. Brun Cottan (1971), font apparaître que la loi de Junge (c'est à dire une loi puissance en d⁻ᵐ, d étant le diamètre), proposée précédemment pour les particules atmosphériques (C.E. Junge, 1963) semble bien décrire la distribution des particules marines. Ce compteur permet l'étude granulométrique des particules dont les diamètres équivalents vont de 1 μm à 15 ou 20 μm****; dans de nombreux cas la décroissance du nombre de particules, lorsque le diamètre croît, se ferait selon deux lois successives, l'une d'exposant compris entre -3,3 et -3,9 dans l'intervalle de 1 μm à 4 ou 5 μm, suivie, au delà de cette taille, d'une seconde loi d'exposant plus élevé (en valeur absolue) de -4 à -5 approximativement.

De nombreuses mesures effectuées dans des zones très différentes de l'Atlantique et du Pacifique (R.W. Sheldon et al., 1972) montrent que les distributions sont, d'une façon très générale, telles que pour des intervalles de tailles logarithmiques égaux (par exemple de 1 à 2 μm, puis de 2 à 4 μm, puis de 4 à 8 μm, ...), le volume des particules appartenant à ces classes reste *grosso modo* constant. On peut aisément vérifier que cette propriété d'équipartition logarithmique des volumes est celle de la distribution d'exposant -4. D'après ces auteurs, elle s'étendrait à une gamme de tailles très étendue. Pour les eaux superficielles et dans les régions productives, à la distribution précédente se superposent souvent des maxima (vers 16 - 20 μm par exemple) traduisant la présence de particules phytoplanctoniques. Pour les calculs, les exposants (de -3 à -5 avec un incrément de 0,1) ont été choisis dans un intervalle couvrant largement les valeurs expérimentales. Des résultats systématiques pour 8 valeurs d'exposant dans cet intervalle et pour les 5 valeurs d'indice sont présentés par ailleurs (A. Morel, 1973) ; également y figurent les résultats obtenus avec des distributions log - normales et exponentielles. Même s'il se révèle que les lois puissance sont un peu simplificatrices par rapport à la réalité, des distributions plus complexes peuvent être décomposées en plusieurs lois successives de ce type. En tout état de cause, elles sont commodes pour le calcul et peuvent également, comme on l'a vu, servir de base pour prévoir le résultat au cas où d'autres distributions doivent être envisagées (cf. § 3.2.5, 2e partie).

---

*Voir par exemple W.V. Burt (1956), Y.E. Otchakovsky (1965), N.G. Jerlov (1968).

**C'est le carbone organique particulaire qui est en fait dosé, le poids des substances organiques en est déduit en multipliant, en général, par un facteur 2 (D.C. Gordon utilise le facteur 1,80).

***De 19 à 55 % pour le carbone particulaire -cf. remarque ci-dessus-.

****D'autres intervalles peuvent être étudiés, mais correspondent à des tailles plus élevées ; 1 μm est approximativement le seuil inférieur de détection du système Coulter. Les diamètres déduits des mesures sont ceux des sphères équivalentes, c'est à dire présentant le même volume que les particules réelles, indépendamment de la forme de celles-ci.

Il n'y a pas semble-t-il de résultats concernant les lois régissant la distribution des particules de dimensions inférieures à 1 µm ; pour celles de dimensions supérieures à 20 µm des lois d'exposant -m continueraient de s'appliquer (m pourrait devenir un peu supérieur à 4, car les volumes relatifs aux classes d'ordre croissant ont tendance à décroître, R.W. Sheldon et al., 1972). Des limites, quelqu'arbitraires qu'elles soient du point de vue physique, doivent être posées pour les nécessités du calcul. Aussi leur influence, dans la mesure où elle est prévisible, a-t-elle précédemment été étudiée. Les calculs préliminaires d'indicatrices individuelles ayant été menés pour des valeurs de α allant de 0,2 à 200, les intégrales à calculer pour les systèmes polydispersés l'ont été entre ces limites*.

D'un point de vue pratique ceci amène à revenir sur la correspondance entre paramètres relatifs α et ρ et diamètres d. Pour le rappeler : $\alpha = \pi d\, n_e / \lambda_o$ et $\rho = 2\alpha (n_r - 1)$, où d est le diamètre de la sphère, λ_o la longueur d'onde dans le vide, n_e l'indice de réfraction de l'eau (1,33), n_r l'indice relatif de la particule par rapport à l'eau. Les calculs effectués entre α = 0,2 et α = 200 correspondent à des particules dont les diamètres vont de 0,02 µm à 20 µm, lorsque λ = 419 nm (on a alors α = 10 d), ou bien de 0,04 µm à 40 µm si λ = 838 nm. On aura de plus ρ = d si on prend pour n_r la valeur 1,05, ceci lorsque λ = 419 nm ; ce cas est souvent pris comme exemple. Ainsi afin de comparer les diverses distributions, celles-ci avaient été mises en coïncidence pour α = 10 et α = 100 (2e partie, § 3.2.5), soit pour d = 1 et 10 µm, si λ = 419 nm, c'est à dire pour les valeurs de diamètre à l'intérieur de la gamme couverte par les mesures. A cet égard, la loi gausso logarithmique utilisée aurait pu être approximée par deux lois puissance, l'une d'exposant -3,4 jusque α = 40 (d = 4 µm), l'autre d'exposant -4,5 entre α = 40 et 200 (d de 4 à 20 µm), ce qui constitue un exemple plausible (voir ci dessus).

Pour orienter les comparaisons entre indicatrices expérimentale et théoriques il est nécessaire au préalable d'examiner comment ces dernières varient avec l'indice et l'exposant de la distribution (Chapitre 1). Diverses comparaisons sont effectuées, et à partir des concordances observées, sont déduites des valeurs probables d'exposant et d'indice (Chapitre 2). Diverses applications de la théorie sont recherchées, en particulier pour expliquer la sélectivité spectrale de la diffusion et la polarisation. Des conséquences en sont également tirées en ce qui concerne les rôles différents des diverses particules, et les relations entre la concentration du matériel en suspension et la diffusion (Chapitre 3).

1 - VARIATIONS THEORIQUES DE L'INDICATRICE.

1.1 - INFLUENCE DE L'EXPOSANT DE LA LOI DE DISTRIBUTION SUR L'INDICATRICE.

La figure II.12 présentée dans la seconde partie a déjà montré dans quel sens s'exerce cette influence. L'écartement entre les divers paliers, qui caractérisent l'indicatrice finale, varie en effet avec l'exposant ; la façon dont il varie peut être précisée. L'amorce des différents paliers a lieu pour des valeurs fixes de α, qui dépendent de l'angle θ concerné **. En conséquence les écarts sont directement déterminés par la pente initiale qui est la même pour toutes les courbes (quel que soit l'angle) et dont la valeur est 7-m. Ceci permet de chiffrer*** ce qui qualitativement est prévisible : lorsque les particules de grande taille sont proportionnellement plus abondantes, c'est à dire lorsque m décroit en valeur absolue, la pente initiale des courbes est plus forte, les paliers sont donc plus distants et l'indicatrice est globalement plus dissymétrique. Cette évolution de l'indicatrice est montrée par la figure III.1. Celle-ci est tracée en portant en fonction de m (variant de 3 à 5) les valeurs des intégrales :

$$\int_{\alpha_m}^{\alpha_M} i_T(\theta,\alpha)\ \alpha^{-m}\, d\alpha \ ,$$ pour divers angles θ ; les limites inférieures et supérieures demeurent constantes (0,2 et 200 respectivement).

L'intensité à 0° sert de valeur de normalisation. On constate sur cette figure que la variation la plus importante est celle du rapport $i_T(2°) / i_T(0°)$ mais elle n'a pas de signification puisque, ainsi qu'on l'a vu, l'intégrale pour 0° ne converge pas. Ce sont les dispositions respectives des courbes pour les angles autres que 0° qui sont significatives. Ainsi l'indicatrice (de 2° à 180°) couvre 5 ordres de magnitude pour la valeur -3 de l'exposant et 3 ordres pour la valeur -5. Un autre point peut être constaté qui aura son importance : entre les valeurs -3 et -4 de l'exposant les diverses courbes demeurent approximativement parallèles. Ceci signifie que dans ce domaine l'indicatrice est peu sensible aux variations de la loi de distribution, au moins pour les angles "moyens" (car l'angle 180° fait quelque peu exception, et l'on ne peut rien dire de l'angle 0°).

Il convient également d'examiner le cas où la distribution suit, non une loi unique, mais une loi où l'exposant prend successivement deux valeurs m1 et m2. Comme cela a été observé, la seconde valeur est supérieure (en valeur absolue) à la première ; elle correspond à la loi qui s'applique aux particules de diamètre supérieur à 5 µm (soit α > 50). En fait, les intégrales pour les divers angles, sauf pour 0°, ont pratiquement atteint leur valeur asymptotique pour α = 50 (voir figure III.2). Par suite, que les particules soient distribuées selon une loi en -m1 ou -m2 au delà de la valeur 50 du paramètre α, n'a pas d'influence sur la forme finale de l'indicatrice, sauf pour l'angle à 0°. La figure III.2 le démontre *a fortiori* : pour deux valeurs d'indice, sont tracées les indicatrices qui correspondent à des populations s'étendant de α_m = 0,2 à α_M = 200 ou bien de α_m = 0,2 à α_M = 50 ; dans ce dernier cas tout revient à dire qu'à partir de α = 50, la valeur du second exposant m2 est infinie. Or les indicatrices sont pratiquement confondues sauf à 0°. En conclusion, avec les lois de distribution considérées, les grosses particules jouent un rôle faible et leur plus ou moins grande raréfaction (m2 plus ou moins élevé) affecte peu la forme de l'indicatrice.

---

* Egalement le calcul a été fait entre la limite théorique α = 0 (taille nulle) et α = 0,2 (cf. annexe 2 et § 3.2.2, 2e partie).

**Ce sont les valeurs de α pour lesquelles les courbes $i_T(\theta)\alpha^{-4}$ amorcent leur décroissance et qui sont, pour le rappeler, indépendantes de l'indice, sauf si l'angle θ est petit (de 2° par exemple). Cette propriété trouve sa justification dans la théorie de Rayleigh Gans (cf. 2e partie, § 2.2.3 et § 3.2.1 b).

***Par exemple, si α1 et α2 sont les valeurs de α pour lesquelles le premier anneau sombre atteint, d'après la théorie de Rayleigh Gans, les angles θ1 et θ2, le rapport des intensités diffusées à ces angles i(θ1) / i(θ2) variera avec l'exposant, proportionnellement à : $(\alpha_1 / \alpha_2)^{7-m}$. Un tableau (2e partie, § 2.2.3) donne les valeurs de α et de θ correspondantes.

1.2 – INFLUENCE DE L'INDICE DE REFRACTION SUR L'INDICATRICE.

La valeur de l'exposant agit sur la forme de l'indicatrice résultante, mais la forme des indicatrices individuelles figurant dans les intégrations est, elle, régie par la valeur de l'indice de réfraction. La figure III.3 fournit un exemple de cette influence : l'exposant étant fixé (m = 3,5), les rapports $i_T(\theta) / i_T(0°)$ sont portés cette fois en fonction de l'indice, les limites inférieure et supérieure des intégrales conservent les mêmes valeurs que précédemment (soit 0,2 et 200). Il avait été noté dans la seconde partie (§ 2.2.5 et § 2.2.6) que pour les indicatrices individuelles, l'influence de l'indice est pratiquement restreinte aux domaines extrêmes des petits angles (0°-10°) ou des grands angles (au voisinage de 180°). Les sommations effectuées ne modifient pas cette propriété qui permet de rendre compte des remarques suivantes :

- la dissymétrie globale de l'indicatrice décroît lorsque la valeur de l'indice croît. Ainsi le rapport des intensités diffusées à 2° et 140° correspond à 6 ordres de magnitude lorsque n = 1,02 et seulement à 4 ordres si n = 1,15.
- mais on peut constater que cette variation de la dissymétrie est presque uniquement à imputer à la variation du rapport $i_T(10°) / i_T(2°)$. Par contre les courbes relatives aux angles allant de 10° à 140° (et même la courbe relative à 160° non figurée) restent sensiblement parallèles, en conséquence la partie "moyenne" (10°-160°) de l'indicatrice dépend peu de l'indice de réfraction *.
- enfin, la courbe correspondant à l'angle 180° montre qu'en valeur relative, la rétrodiffusion s'accentue au fur et à mesure que l'indice croît (par suite de l'augmentation du facteur de réflexion).

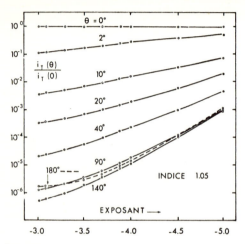

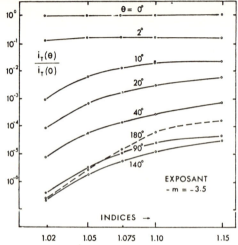

Figure III.1 : Influence de l'exposant de la loi de distribution des particules sur l'indicatrice ; l'indice de réfraction est 1,05. Les paramètres de taille correspondant aux limites de la distribution sont 0,2 et 200.

Figure III.3 : Influence de l'indice de réfraction sur l'indicatrice ; l'exposant de la loi de distribution est –3,5. Les limites de taille sont inchangées.

On peut noter dès maintenant qu'à partir des constatations précédentes des critères existent, qui permettraient, au moins théoriquement, de dégager des mesures, une valeur plausible de l'indice. Le rapport des intensités diffusées à 2° et 10° est un critère très sensible, en particulier si l'indice varie entre 1,02 et 1,075 ; pour les valeurs plus élevées, 1,075 et 1,15, la rétrodiffusion en constituerait un autre (malheureusement peu exploitable expérimentalement). Une particularité, que met en évidence le tracé complet de l'indicatrice, pourrait aussi servir de critère . On constate en effet l'existence d'un faible maximum relatif dont la position angulaire est directement liée à la valeur de l'indice (vers 60° pour n = 1,05 et vers 85° pour n = 1,10, figure III.2) **. Le maximum est d'autant plus accusé que le calcul prend en compte des particules plus grosses, elles seules en effet sont susceptibles de provoquer cet effet.

D'un point de vue pratique, les mesures d'indicatrice ont rarement été effectuées avec un intervalle angulaire suffisamment petit pour que ce phénomène soit à coup sûr mis en évidence. On peut relever cependant que l'indicatrice moyenne obtenue par Jerlov (1961) et celle présentée par Otchakovski (1965) présentent une convexité nette vers 60 – 70° (qui correspondrait à un indice de 1,05 environ). Un appareil, enregistrant de façon continue le coefficient de diffusion en fonction de l'angle, permettrait de savoir si cet effet existe de façon plus ou moins permanente et donc s'il est possible à partir de son observation d'en déduire un indice de réfraction moyen. A cet égard, les déterminations d'indicatrice faites par J.W. Reese et S.P. Tucker (1970) montrent que très fréquemment un faible maximum apparaît entre 60° et 75°.

Peut-on trouver vis à vis de l'exposant des critères tels que ceux qu'on vient d'indiquer pour l'indice ? En reprenant l'examen de la figure III.1, on constate que l'amplitude totale de l'indicatrice est *a priori* la grandeur vraiment variable avec l'exposant. Mais pour des raisons de non convergence déjà données, la valeur à 0° ne peut pas être prise en considération. De plus le domaine angulaire 2 – 10° est, comme on vient de le voir, particulièrement sensible à la valeur de l'indice. Utiliser le rapport des intensités à

---

* Ceci ne pourrait pas être généralisé ; cette propriété est liée aux valeurs considérées de l'indice, suffisamment proches de 1 pour rendre valide et applicable l'approximation de Rayleigh Gans (cf. 2e partie, § 2.2.3).

** Voir également les indicatrices présentées dans l'annexe II. La relation qui lie l'angle où apparaît cette concentration d'énergie et l'indice a été présentée graphiquement (2e partie, fig. II.8).

2° et 140° pour définir l'amplitude, conduirait à une grandeur pour laquelle les influences de l'exposant et de l'indice ne pourraient être séparées. Par contre, ainsi que le montre la figure III.3, le rapport i(10°) / i(140°) est pratiquement indépendant de l'indice. L'amplitude de l'indicatrice, non plus totale, mais restreinte au domaine 10° - 140° pourra bien constituer le critère recherché.

Figure III.2 : Les courbes en trait plein correspondent aux indicatrices calculées lorsque les limites de taille sont α_m=0,2 et α_M=200, les indicatrices en tireté lorsque la limite supérieure α_M est ramenée à 50. Le positionnement des courbes marquées 1 , relatives à l'indice 1,05 , et 2 , relatives à l'indice 1,10,est arbitraire. Par contre dans chacun de ces deux cas, les deux courbes, en trait plein et en tireté, sont correctement placées l'une par rapport à l'autre, de façon que le coefficient total de diffusion soit le même (c'est à dire qu'elles sont correctement placées si l'on considère les coefficients angulaires normalisés). Les croix et les triangles correspondent à la composante polarisée horizontale i_2, respectivement lorsque α_M=200 ou lorsque α_M=50 ; les points correspondent à la composante verticale i_1, sans que soient distinguées les valeurs obtenues dans les deux cas.

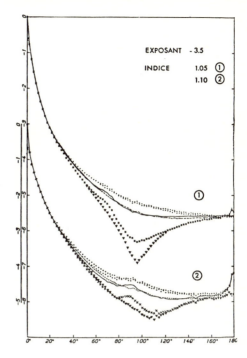

1.3 - CONSEQUENCES A TIRER DES VARIATIONS THEORIQUES DE L'INDICATRICE.

Une première conclusion peut être tirée de ce qui précède. En effet lorsque l'indice de réfraction varie dans d'assez larges limites, l'indicatrice de diffusion est peu modifiée, au moins en ce qui concerne les angles "moyens" (10° à 140°, pour fixer des valeurs) ; il en va de même lorsque l'exposant de la distribution varie, dans des limites un peu plus restreintes il est vrai (de –3 à –4 approximativement). Ceci peut constituer une première explication à ce qui est un fait établi expérimentalement, à savoir la faible variabilité de l'indicatrice des particules marines - on peut noter que ce point a été établi principalement à partir de mesures aux angles moyens -. Cette explication sera développée et nuancée au cours du chapitre suivant. Le cas des petits angles où la variabilité est plus grande sera également examiné.

2 - INTERPRETATION DES OBSERVATIONS ET APPLICATIONS.

Il convient d'abord de montrer que les indicatrices théoriques assez peu variables dont on vient de parler coïncident avec l'indicatrice expérimentale, et d'examiner ensuite si dans la gamme des variations possibles, certaines rendent plus précisément compte de l'expérience. Par ce procédé, des valeurs probables d'indice de réfraction et d'exposant de la distribution pourraient être dégagées.

Pour effectuer les comparaisons, on utilisera (§ 2.1) les propriétés - comme les rapports β(10°) / β(2°) et β(140°) / β(10°) - dont on a vu qu'elles étaient suffisamment variables, soit avec l'indice, soit avec l'exposant, pour pouvoir servir de critères. On utilisera (§ 2.2) également certains rapports de coefficients angulaires au coefficient total - c'est à dire des coefficients normalisés $\bar{\beta}(\theta)$ - dont les valeurs dépendent conjointement de l'indice et de l'exposant. Si la variabilité expérimentale de ces rapports se révèle plus faible que la variabilité théorique, cela signifiera qu'indices et exposants réels varient dans un domaine plus restreint que celui envisagé pour les calculs. Chaque propriété conduira à délimiter un tel domaine ; si l'ensemble de ceux ci sont compatibles ou présentent au moins une partie commune, on peut espérer ainsi déduire de l'expérience des valeurs plausibles d'indice et d'exposant. Des indicatrices calculées avec ces valeurs sont comparées à l'indicatrice expérimentale (§ 2.4).

Des méthodes d'évaluation du coefficient total de diffusion à partir de mesures angulaires ont été proposées par divers auteurs. Examiner la variabilité théorique de certains coefficients normalisés (à 90°, 45°, 6° et 4° - § 2.2 et 2.3) afin de préciser l'indice et l'exposant est en même temps une façon de reconsidérer le bien fondé de telles méthodes.

2.1 - APPLICATION DES CRITERES VIS A VIS DE L'INDICE ET DE L'EXPOSANT DE LA LOI DE DISTRIBUTION. RAPPORT DE COEFFICIENTS ANGULAIRES.

Le rapport des coefficients de diffusion à 10° et 2° est, comme on l'a vu (§ 1.2) beaucoup plus sensible aux variations d'indice que d'exposant ; à l'inverse le rapport des coefficients à 140° et 10° est dépendant de l'exposant presque exclusivement ; les variables indice et exposant sont en quelque sorte séparées. Les variations théoriques de ces rapports avec les deux variables considérées sont présentées sur la figure III.4. Des valeurs expérimentales des mêmes rapports sont groupées dans la table I. Les mesures effectuées conjointement à 2° et 10° ou à 10° et 140° sont assez peu nombreuses. De plus, pour être utilisées ici, les mesures à 140° doivent être exprimées en valeur absolue afin que la part de la diffusion moléculaire, souvent non négligeable, puisse être retranchée ; c'est à cette condition seulement que le rapport $\beta(140°)$ / $\beta(10°)$ est réellement significatif. En adoptant les valeurs 2.10^{-2} et $5,5.10^{-2}$ pour fixer les limites expérimentales des variations du rapport $\beta(10°)$ / $\beta(2°)$ et en les reportant sur le graphique (III.4) figurant les variations théoriques, se trouve délimité, par intersection, un domaine. Celui-ci correspond aux valeurs combinées à la fois d'indice et d'exposant, pour lesquelles le calcul rend compte de l'expérience. En raison même de la nature de ce critère, le domaine laisse imprécisées les valeurs de l'exposant, mais restreint les valeurs possibles de l'indice ; ainsi les valeurs extérieures à l'intervalle 1.02 - 1.06 semblent devoir être écartées. De la même façon, en choisissant comme limite les valeurs expérimentales $0,6.10^{-3}$ et $2,3.10^{-3}$ pour le second rapport, un autre domaine peut être déterminé qui renseigne peu sur l'indice (inférieur cependant à 1.10) mais précise les valeurs de l'exposant ; par exemple, si l'indice est supposé égal à 1.05 les valeurs de l'exposant compatibles avec les résultats expérimentaux sont le -3,8 à -4,3 approximativement.

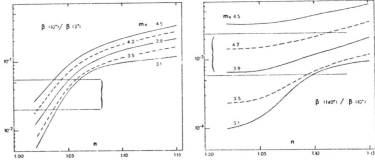

Figure III.4 : Variations théoriques des rapports $\beta(10°)$ / $\beta(2°)$ et $\beta(140°)$ / $\beta(10°)$ en fonction de l'indice de réfraction et pour les valeurs d'exposant indiquées. Les valeurs expérimentales sont situées dans les bandes désignées par les accolades.

Pour la clarté, les deux domaines ainsi définis sont reportés sur un diagramme m-m (c'est à dire dont les coordonnées sont l'exposant et l'indice, voir figure III.6). La partie commune correspond donc aux valeurs plausibles d'indice et d'exposant compte tenu des résultats expérimentaux disponibles.

Spilhaus (1968) avait envisagé d'utiliser le rapport $\beta(30°)$ / $\beta(45°)$ pour caractériser la forme de l'indicatrice. Les figures III.1 et III.3 montrent que les courbes tracées pour 20° et 40° restent sensiblement parallèles quand l'indice ou l'exposant varie. Il apparait donc que ce rapport, théoriquement peu variable, est tout compte fait assez mal choisi pour être l'index escompté. Au demeurant, les valeurs théoriques corroborent parfaitement les valeurs expérimentales (à condition que là encore, la diffusion moléculaire soit soustraite ; c'est d'ailleurs une difficulté dans l'interprétation de ce rapport que Spilhaus, semble-t-il, n'avait pas considérée). La table II montre à la fois les valeurs théoriques et un certain nombre de valeurs expérimentales de ce rapport. Même si statistiquement les valeurs sont significativement distinctes (A.F. Spilhaus, 1968), on voit qu'il est difficile de tirer des conclusions des faibles variations observées, sauf peut être en ce qui concerne l'exposant : ces valeurs expérimentales montreraient que celui-ci peut varier essentiellement entre -3,5 et -4,2, la valeur la plus fréquente étant -3,9, ce qui n'est pas du tout en contradiction avec les conclusions tirées de l'application des critères précédents.

2.2 - RELATIONS ENTRE $\beta(45°)$ ou $\beta(90°)$ et b

Jerlov avait dès 1953 (N.G. Jerlov, 1953) préconisé de déduire le coefficient total de diffusion b de mesures à 45°. Ultérieurement, de nombreux auteurs ont avancé que la preuve théorique de la constance du rapport entre $\beta(45°)$ et b avait été faite par D. Deirmendjian (1963). En réalité ce travail théorique traitait du cas de la diffusion par les brouillards et les nuages, le calcul étant fait avec l'indice de réfraction de l'eau et des lois de distribution particulières. La conclusion de Deirmendjian selon laquelle $\overline{\beta}(40°)$ est pratiquement constant (et d'ailleurs égal à 10^{-1}) n'a pas à être étendue à tous les cas, car il s'agit, non d'une preuve à proprement parler, mais de la constatation d'une propriété numérique qui tient aux distributions adoptées. Ainsi la figure III.5 montre comment varie $\overline{\beta}(44°)$ avec l'indice et ce

Table I : Valeurs expérimentales des rapports β(10°) / β(2°) et β(140°) / β(10°).

β(10°) / β(2°) X 10²

5	2,6 (2,2 à 3,0)	2,7	3,0	3,2	4,7	4,8 (4,5 à 5,4)	3,8
Médit.	Manche et Médit.	Argus Island	Long Island Sound	Méditerranée		Atlant.	Pacif.
(1)	(2)	(3)		(4a)	(4b)	(5)	

β(140°) / β(10°) X 10³

1,35 (1,00 à 1,70)	1,05	1,97
	0,60 à 1,90	1,66 à 2,28
Manche-Médit.	Atlant.	Pacif.
(2)	(5)	

(1) Rapport calculé à partir des valeurs lues sur la courbe présentée par Y.E. Otchakovsky (1965).
(2) Valeurs moyennes des rapports correspondant à l'indicatrice moyenne des particules. Celle ci est obtenue en associant les valeurs aux angles allant de 30° à 150° (A. Morel, 1965) aux valeurs relatives aux petits angles, de 1,5° à 14° (D. Bauer, A. Morel, 1967).
(3) R.E. Morrison (1970), rapports mesurés sur les courbes tracées selon ce que l'auteur appelle l'extrapolation "Duntley" (figure 4, in ref.).
(4) Rapport correspondant à la moyenne des mesures *in situ*, à trois longueurs d'onde ; (4b), à la moyenne des mesures *in vitro* à deux longueurs d'onde (laser) (F. Nyffeler, 1970). Pour les mesures *in situ*, il est nécessaire d'extrapoler de 3° vers 2° pour effectuer le rapport.
(5) Trois mesures en Atlantique (Bahamas) et deux en Pacifique (au large de San Diego) (T.J. Petzold, 1972). Pour 140°, la diffusion moléculaire a été retranchée des valeurs absolues données par l'auteur, ceci avant de calculer le rapport.

Table II : Valeurs théoriques et expérimentales du rapport β(30°) / β(45°).

Valeurs théoriques.

Exposant	Indice de réfraction			
	1,02	1,05	1,075	1,10
-3,5	3,748	3,875	4,005	4,006
-3,9	3,382	3,407	3,441	3,435
-4,2	3,053	3,057	3,066	3,055
-4,5	2,725	2,724	2,725	2,714

Valeurs expérimentales (et écarts-types).

3,70 ± 0,40 Manche-Médit.(1a)	3,33 ± 0,22 côtière	3,69 ± 0,67 Bermudes
3,85 ± 0,25	3,32 ± 0,14 N.W.Atlant.	3,53 ± 0,67 (Argus Island)
3,63 ± 0,40	3,53 ± 0,28 Gulf stream	3,28 ± 0,19
3,77 ± 0,75	3,61 ± 0,26 M. Sargasses	3,52 ± 0,33 Long Island
Océan Indien (1b)	2,91 ± 0,06 Bermudes	3,32 ± 0,24 sound
	(2)	(3)

3,43±0,39 Baltique	3,00	3,28	3,17	3,38
3,38±0,61 Médit.	2,83 (a)	3,48	3,18 (b)	3,64
	2,97		3,33	
(4)	Atlant.	Pac.	Atlant.	Pac. (5)

(1a) Valeur moyenne pour 27 mesures en Manche et 40 en Méditerranée (A. Morel, 1965).
(1b) Valeurs moyennes, respectivement aux longueurs d'onde 546, 436, 366 nm, pour 17 échantillons d'océan Indien (Madagascar, non publié, A. Morel, 1967).
(2) Valeurs moyennes pour 5 types d'eau entre Woods Hole et la mer des Sargasses (A.F. Spilhaus, 1968).
(3) R.E. Morrison, 1970.
(4) 12 mesures en mer Baltique (exclue station 1, 25 m) à 4 longueurs d'onde (655, 632, 525 et 450 nm), G. Kullenberg (1969).
16 mesures en Méditerranée (633 nm), G. Kullenberg et N.B. Olsen (1972). Ce sont les rapports des coefficients de diffusion pour les particules seules qui sont présentés ici.
(5a) Rapports obtenus à partir des valeurs observées par T.J. Petzold (1972).
(5b) Rapports obtenus à partir des valeurs précédentes mais diminuées de la part revenant à la diffusion moléculaire (1,67.10⁻⁴ m⁻¹ à 90° pour λ = 510 nm, A. Morel, 1968).

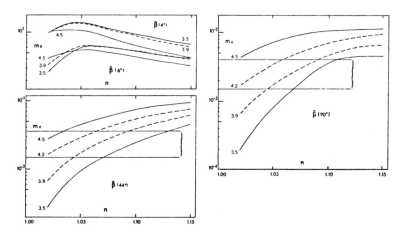

Figure III.5 : Variations théoriques avec l'indice et l'exposant des coefficients angulaires normalisés pour 4°, 6°, 44° et 90°. Les accolades correspondent comme dans la figure III.4 aux variations observées des mêmes coefficients. Ces valeurs théoriques, comme celles de la figure III.4 sont calculées pour des particules dont les tailles relatives α s'étendent de 0,2 à 200.

pour quatre valeurs de l'exposant de la loi de distribution [*]. Contrairement aux distributions envisagées par Deirmendjian les lois puissance entrainent, à valeur constante d'indice, une possibilité de variation importante pour le coefficient normalisé $\overline{\beta}(44°)$.

La table III présente un certain nombre de valeurs obtenues pour le rapport $\beta(45°)$ / b - c'est à dire $\overline{\beta}(45°)$ - valeurs assez assurées dans la mesure où le calcul de b par intégration s'appuie sur des valeurs expérimentales aux petits angles, et non sur des valeurs extrapolées, dans la mesure aussi où l'on peut, quand il y a lieu, retrancher de $\beta(45°)$ la part qui revient à la diffusion moléculaire. Ces remarques sont valables également pour les valeurs expérimentales de $\overline{\beta}(90°)$ figurant dans cette même table. De l'examen de ce tableau, il ressort que les valeurs expérimentales de $\overline{\beta}(45°)$ se situent dans un intervalle plus étroit que les valeurs théoriques, impliquant que les variations effectives d'indice et d'exposant sont plus faibles que celles considérées dans le calcul. Comme précédemment, les domaines correspondants sont obtenus en confrontant aux valeurs calculées les valeurs expérimentales choisies comme limites, soit $1,5.10^{-2}$ et $3,6.10^{-2}$ pour $\overline{\beta}(45°)$, $1,5.10^{-3}$ et 4.10^{-3} pour $\overline{\beta}(90°)$. Les domaines, d'ailleurs pratiquement confondus sont reportés sur le diagramme m-n (figure III.6).

En *définitive*, si la relation entre b et $\beta(45°)$ présente effectivement une certaine invariance, celle ci n'est pas, comme on a voulu le voir, théoriquement justifiée. A l'inverse, elle prouverait plutôt une relative stabilité des propriétés des particules marines, à la fois quant à leur distribution et à leur composition moyenne, dont dépend l'indice.

Table III : Valeurs expérimentales de $\overline{\beta}(45°)$ et $\overline{\beta}(90°)$ et écart-types.

$\overline{\beta}(45°)$ X 10^2

2,0 ± 0,5 Moyenne Manche et Méditerranée (1)	2,4 ± 0,9 Baltique (2a)	2,5 ± 0,7 Médit. (2b)	1,6 ± 0,4 Atlant. (3)
2,3 ± 0,7 Médit.	1,9 ± 0,3 Atlant. (4)	3,4 ± 0,2 M. Noire	2,58 1,82 3,05 1,78 2,80 Atlant.Pacif. (5)

$\overline{\beta}(90°)$ X 10^3

2,25 ± 0,4 (1)	3,0 ± 1,3 (2b)	2,9 ± 0,6 (3)	3,16 2,00 2,13 1,63 1,85 (5)

(1) Indicatrice moyenne des particules (A. Morel, 1965) rapportée à la valeur 444 de l'intégrale donnant b, calcul fait en posant $\beta(90°) = 1$ (A. Morel, 1968, et calcul corrigé en 1970) ; rapportée à la valeur 337 donnée par F. Nyffeler (1969), pour la même intégrale, les valeurs $2,0.10^{-2}$ et $2,25.10^{-3}$ deviennent respectivement $2,6.10^{-2}$ et $3,0.10^{-3}$. 27 mesures en Manche et 40 en Méditerranée.
(2a) 16 mesures en mer Baltique (G. Kullenberg, 1969).
(2b) 14 mesures en mer Méditerranée (G. Kullenberg, N. Olsen, 1972) ; la mesure à 150 m, station A2 a été écartée des moyennes.
(3) Inverses des rapports b/β (θ) données par R.E. Morrison (1970), b étant la valeur obtenue en incluant les mesures aux petits angles (appelée "Sp" par l'auteur).
(4) 23 mesures en Méditerranée, 90 en Atlantique, 104 en mer Noire ; mesures de V.I. Mankovski (1971).
(5) Mesures de T.J. Petzold (1972) ; avant de former les rapports, la diffusion moléculaire a été retranchée des valeurs présentées par l'auteur.

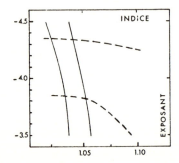 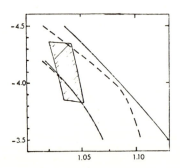

Figure III.6 : Diagrammes tracés dans le plan m-n (n, indice de réfraction selon les abscisses ; m exposant négatif de la loi de distribution, selon les ordonnées) pour mettre en évidence les domaines où valeurs théoriques et expérimentales sont compatibles. Figure de gauche : la bande comprise entre les deux courbes pleines correspond à l'application du critère constitué par le rapport $\beta(10°)$/ $\beta(2°)$; celle entre les deux courbes en tireté à l'application du critère $\beta(140°)$ / $\beta(10°)$. La partie commune à ces deux domaines est reportée sur la figure de droite. Sur celle-ci sont également montrés les domaines obtenus par confrontation entre valeurs théoriques et expérimentales des coefficients normalisés $\beta(44°)$ (domaine délimité par les courbes en trait plein) et $\beta(90°)$ (domaine délimité par les courbes en tireté).

[*]La courbe relative à l'exposant -3,10 n'est pas représentée. En effet pour cette valeur de l'exposant, la valeur asymptotique du coefficient total de diffusion est loin d'être atteinte avec la limite supérieure ρ_M utilisée (soit $\rho_M=200$, ce qui correspond à ρ variant de 8 à 60 selon que l'indice varie de 1,02 à 1,15) ; par suite $\overline{\beta}(44°)$ est surévalué dans ce cas. Cet effet, à quelques pour-cent près, n'existe plus avec les autres valeurs de l'exposant (cf. 2e partie, § 3.2.3 et figure II.15). Le problème ne se pose pas pour la figure III.4 car les coefficients angulaires, dont on fait le rapport, ont atteint leur valeur asymptotique, même pour l'exposant -3,10.

2.3 - PROPRIETES PARTICULIERES DES COEFFICIENTS β(4°) et β(6°)

La propriété numérique mise en évidence par Deirmendjian pour l'angle de diffusion 40° a un équivalent dans le cas des indices et des lois de distribution de Junge envisagées ici. Si l'on examine les indicatrices obtenues pour les divers indices et exposants (A. Morel, 1973) lorsqu'elles sont tracées en valeurs normalisées $\overline{\beta}(\theta)$, on constate qu'elles présentent approximativement les mêmes valeurs vers 4° et 6° (respectivement de l'ordre de 12 et de 6). Les valeurs de $\overline{\beta}(4°)$ et de $\overline{\beta}(6°)$ sont portées en fonction de l'indice et pour diverses valeurs de l'exposant sur la même figure III.4. On peut observer que pour une valeur d'indice donnée, les variations de ces coefficients avec l'exposant sont faibles comparées à celles du coefficient $\overline{\beta}(44°)$. Si même on prend également en considération les variations d'indice, entre 1,02 et 1,10 tout au moins, la conclusion demeure ; ainsi par exemple, $\overline{\beta}(4°)$ varie de 9,6 à 14 environ. Ceci rejoint une observation récente faite par V.I. Mankovski (1971), à partir du tracé des indicatrices en coefficients normalisés $\overline{\beta}(\theta)$ (Figure III. 6B). Toutes ces indicatrices se recoupent en moyenne vers θ = 4,5° et l'étude statistique montre que le coefficient de corrélation a sa valeur la plus élevée lorsque dans la régression entre b et β(θ), θ a la valeur 4,5°. Cette régression fournit β(4,5°) = 9,0 b. En supposant (d'après la pente de l'indicatrice) que β(θ) varie en $\theta^{-1,5}$ dans cette région angulaire, la valeur précédente se transforme pour 4° en $\overline{\beta}(4°)$ = 10,7. Cette valeur est identique à celle présentée ici : 4750 / 444= 10,7 (± 17 %) (D. Bauer, A. Morel, 1967, et table II, 2e partie).

Le fait que le coefficient normalisé $\overline{\beta}(4°)$ soit peu dépendant de l'indice et de l'exposant entraine évidemment que la confrontation des valeurs expérimentales aux valeurs théoriques renseigne peu sur les indices et exposants plausibles. Cependant si l'on adopte par exemple la valeur 10,5 ± 2, cette valeur exclut la possibilité d'indices supérieurs à 1,10 ; d'un autre côté, si l'on suppose que l'indice a pour valeur 1,05, tous les exposants (entre -3,2 et -5) sont possibles. On peut ajouter que $\overline{\beta}(10°)$, déjà théoriquement plus variable que $\overline{\beta}(6°)$ ou $\overline{\beta}(4°)$, reste néanmoins voisin de 1 ; il varie de 0,8 à 1,8 lorsque l'indice est compris entre 1,02 et 1,05 et l'exposant entre -3,5 et -4,2, ce que, sans entrer dans le détail, les mesures confirment * (il devient par contre supérieur à 2, quel que soit l'exposant si l'indice est égal ou supérieur à 1,075). Pour cet angle 10°, coefficient angulaire et coefficient total présentent des valeurs numériquement voisines.

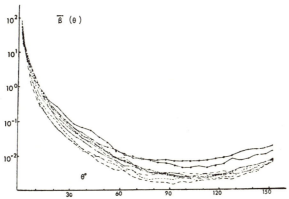

Figure III. 6B : Reproduit d'après V.I. Mankovski (1971). Indicatrices de diffusion tracées en valeurs normalisées $\overline{\beta}(\theta)$. Les mesures ont été effectuées en Atlantique nord.

2.4 - VALEURS PLAUSIBLES D'INDICE DE REFRACTION ET D'EXPOSANT DE LA LOI DE DISTRIBUTION

Les diverses confrontations qui viennent d'être faites conduisent chacune à délimiter un domaine d'indice et d'exposant rendant compatibles données expérimentales et valeurs théoriques. Ces domaines, comme le montre la figure III.6, présentent une partie commune pour laquelle donc tous les critères utilisés sont à la fois respectés. Ainsi se trouvent précisées les valeurs combinées d'indice et d'exposant qui peuvent être tenues comme plausibles. On peut remarquer que les valeurs d'exposant (de -3,8 à -4,2) sont en accord avec les valeurs fournies par numération directe des particules (cf. avant propos de la 2e partie) et également que les valeurs d'indice ainsi déduites, allant ** de 1,02 à 1,05 paraissent vraisemblables. Sur ce dernier point, récemment H.R. Gordon et O.B. Brown (1972) ont conclu que les indicatrices déterminées par G. Kullenberg (1968) en mer des Sargasses, peuvent être correctement interprétées si on donne à l'indice la valeur 1,05 - 0,01 i. Par une méthode très différente, fondée sur la sélectivité de la diffusion, J.R. Zaneveld et H. Pak (1973) calculent des valeurs de l'indice relatif approximativement comprises entre 1,02 et 1,04 pour les particules aussi bien des couches superficielles que profondes (3000 mètres). Il convient d'ajouter qu'il n'y a pas unicité de solution ; diverses combinaisons d'indice et d'exposant conduisent à des indicatrices voisines et acceptables compte tenu de la variabilité de l'indicatrice expérimentale. Quelques exemples sont montrés ci après. La méconnaissance des limites physiques de la population à introduire dans le calcul oblige à en examiner l'influence sur les conclusions précédentes.

---

* Voir page 12 $\overline{\beta}(10°)$ = 1,126 et également les valeurs de Mankovski : 0,7 et 1,3 approximativement, et de Petzold : 0,85, 1,13, 1,04 pour les trois mesures en Atlantique ; 0,88, 0,99 pour les deux mesures au large de San Diego.

**La valeur centrale de 1,04 est, pour le noter, celle qui permet les plus grandes variations d'exposant à l'intérieur du domaine de compatibilité. Les exemples présentés ci après, ne correspondent cependant qu'aux cas où l'indice est posé égal à 1,02 ou 1,05, ceci compte tenu des calculs préliminaires d'indicatrices individuelles (§ 2.2, 2e partie).

2.4.1 - Comparaison des indicatrices.

La comparaison peut être faite pour tous les angles en traçant les indicatrices calculées dans les conditions précédemment définies. Quelques exemples sont fournis ci-après (Figures III.7) ; est également présentée l'indicatrice expérimentale moyenne, tracée selon les mêmes échelles, avec les valeurs tabulées des coefficients normalisés (cf. 1ère partie). D'autres exemples de courbes théoriques,y compris pour les cas hors du domaine des valeurs vraisemblables, sont présentés de façon plus complète par ailleurs (A. Morel, 1973).

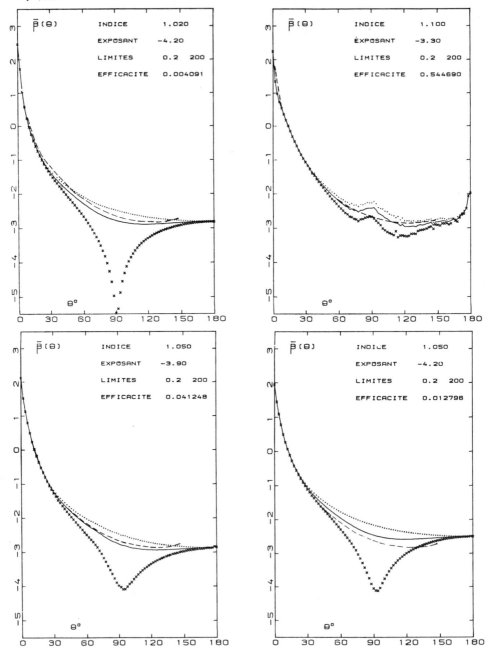

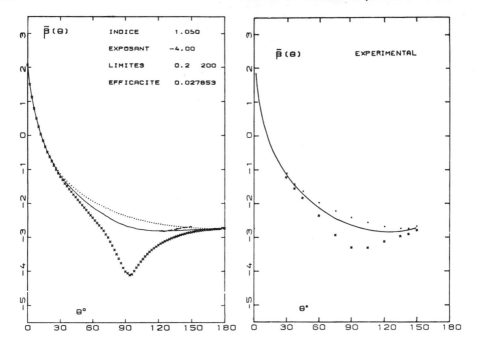

Figure III.7 : Les indicatrices théoriques, calculées pour les valeurs d'indice d'exposant et de limites indiquées, sont tracées en trait plein, en portant en fonction de θ le logarithme décimal des coefficients normalisés $\overline{\beta}(\theta)$. Les points et les croix correspondent respectivement aux composantes polarisées verticalement $\overline{\beta}_1(\theta)$, et horizontalement $\overline{\beta}_2(\theta)$. L'indicatrice expérimentale est présentée de façon identique. Elle est en outre reportée en tireté sur les figures précédentes. "Efficacité" est une notation abrégée désignant le facteur moyen d'efficacité $\overline{Q}$ dont la valeur calculée (équation 3.2, 2e partie) est indiquée pour chaque cas.

   L'accord le meilleur est obtenu lorsque l'exposant est pris égal à -4, l'indice étant 1,05 ; dans ces conditions courbe expérimentale et courbe théorique sont pratiquement indiscernables l'une de l'autre. Cet accord, qui reste bon pour les valeurs voisines d'indice et d'exposant, concerne l'indicatrice en lumière naturelle. A ce sujet deux remarques doivent être faites :
a) la concordance entre valeurs expérimentales et théoriques pour l'intensité totale β(θ), n'entraine pas *ipso facto* celle des valeurs pour les deux composantes polarisées ; par exemple, si n = 1,05 et m = -3,90 d'une part et si n = 1,02 et m = -4,20 d'autre part, les valeurs calculées de $\overline{\beta}(\theta)$ sont pratiquement identiques, mais la polarisation est beaucoup plus accentuée dans le second cas (elle est totale à 90° à 10^{-3} près). Cette question sera examinée en détail ultérieurement (§ 4).
b) en se reportant à la figure III.6, on constate que trois critères sur quatre peuvent être satisfaits lorsque, l'indice ayant des valeurs plus élevées, l'exposant a des valeurs plus faibles (en valeurs absolues). En prenant par exemple l'indice 1,10 et l'exposant -3,3, on obtient une courbe satisfaisante * eu égard à l'indicatrice expérimentale, sauf évidemment pour le domaine des petits angles. Le rapport β(10°) / β(2°) théorique (0,101) est 2 à 5 fois plus élevé que le rapport expérimental, et la pente dans cette région angulaire est trop faible.

   Cette seconde remarque permet d'insister à nouveau sur l'utilité du critère que constitue le rapport β(10°) / β(2°) pour déterminer indirectement l'indice moyen des particules.

2.4.2 - Première remarque sur les limites plausibles de la distribution.

   La théorie montre qu'il y a cohérence entre les résultats expérimentaux relatifs d'une part à la distribution des particules et d'autre part à la diffusion, ceci à condition que soient données à l'indice de réfraction certaines valeurs ; celles ci, par ailleurs, semblent raisonnables. Toutefois pour parvenir à cette cohérence, une autre catégorie d'hypothèses a dû être faite ; ce sont celles qui touchent aux valeurs des deux limites, à propos desquelles on peut se demander si elles ne remettent pas en question les conclusions. Les effets possibles sont successivement examinés :

Limite supérieure : la valeur qu'on lui attribue influant très peu sur la forme de l'indicatrice finale, sauf pour l'angle 0°, les résultats qui concernent le domaine de validité, obtenus sans faire intervenir la valeur à 0°, demeurent. Le cas particulier de l'angle 0° sera examiné plus loin (§ 3.1.3), de même que le problème de la polarisation (§ 3.3) pour laquelle la valeur donnée à cette limite est, comme on le verra, déterminante.

---

* En faisant abstraction du maximum autour de θ = 85° dont la position est liée à la valeur de l'indice (2e partie, § 2.2.5) et dont l'amplitude croît lorsque l'exposant décroît en valeur absolue.

Limite inférieure : lorsque la troncature[*] progresse vers des tailles de plus en plus élevées, la dissymétrie de l'indicatrice s'accentue (par diminution de la diffusion d'abord aux grands angles, puis aux angles de plus en plus petits - cf. 2e partie, § 3.2.2 -). Parmi les rapports choisis pour servir de critères le premier à être affecté est $\beta(140°) / \beta(10°)$, ensuite $\overline{\beta}(90°)$; $\overline{\beta}(44°)$ ne l'est que si la limite inférieure α_m dépasse 5 ; enfin $\beta(10°) / \beta(2°)$ ou $\overline{\beta}(4°)$ ne le sont pas tant que α_m reste inférieur à 10 (cf. figure II.12). Dans ces conditions, le rapport $\beta(140°) / \beta(10°)$ étant diminué, quels que soient par ailleurs l'indice et l'exposant, les courbes qui en montrent les variations (figure III.4) subissent toutes approximativement une translation, de plus d'un ordre de magnitude, si $\alpha_m = 10$; de la même façon, les courbes établies pour $\overline{\beta}(90°)$ sont elles aussi déplacées, de presque un ordre de magnitude. Comparées aux valeurs théoriques ainsi diminuées, les valeurs expérimentales ne s'expliqueraient alors qu'avec à la fois des indices élevés, supérieurs à 1,15, et des exposants élevés (en valeur absolue), supérieurs à 5 ou 6. En conséquence les divers domaines que montraient la figure III.6, vont être disjoints avec ces nouvelles hypothèses, et la partie commune n'existera plus. Il s'ensuit qu'aucune indicatrice ne peut être calculée susceptible de rendre compte des valeurs expérimentales, simultanément pour toutes les régions angulaires.

En conclusion, la cohérence évoquée précédemment disparait elle aussi. Pour la conserver il faut fixer la limite inférieure α_m du calcul à une valeur non supérieure à 10. Sans que cet argument puisse constituer une preuve irréfutable, il y a tout lieu de penser qu'il en va bien ainsi réellement ; il suffit de supposer que la loi de distribution établie expérimentalement jusqu'aux particules de diamètre 1 µm se prolongerait en fait jusqu'à des particules de diamètre 0,1 µm.

3 - AUTRES CONSEQUENCES ET APPLICATIONS DES CALCULS THEORIQUES

Le premier apport de la théorie a été de montrer qu'à partir des données granulométriques, et moyennant certaines hypothèses dont la vraisemblance a été démontrée, la forme générale de l'indicatrice reçoit une explication satisfaisante. L'analyse théorique fait également entrevoir d'autres conséquences en ce qui concerne les rôles différents que les diverses classes de particules joueraient dans la diffusion, certaines ayant un rôle prépondérant (§ 3.1). Cette idée de classes "efficaces" est essentielle pour aborder le problème de la sélectivité spectrale et en faciliter l'interprétation (§ 3.2). Les questions touchant à la polarisation et qui viennent d'apparaître lors de la comparaison entre indicatrices expérimentale et théoriques font l'objet d'un examen particulier (§ 3.3). Enfin les relations entre teneur en particules et diffusion telles qu'elles peuvent être déduites de la théorie, sont examinées d'un point de vue pratique (§ 3.4).

3.1 - ROLES DIFFERENTS DES DIVERSES CLASSES DE PARTICULES

3.1.1 - Cas de la diffusion totale.

On peut concevoir que dans le cas d'une population de particules distribuées selon une loi continuement décroissante avec la taille, c'est à dire dans le cas où l'on ne peut donc définir une taille moyenne, existe néanmoins une classe de particules jouant un rôle prépondérant dans le phénomène de diffusion. Ceci découle en fait de ce qui a été examiné dans la seconde partie (§ 3.2.3) et qui a conduit à préciser le domaine de validité du calcul. On a vu que les "petites" particules, bien que nombreuses (et même en nombre infini) peuvent présenter une diffusion totale finie et faible, parce que le facteur d'efficacité Q est faible ; à l'inverse, les "grosses" particules, pour lesquelles Q est constant et égal à 2, sont d'après les lois de distribution, en nombre trop faible pour contribuer notablement à la diffusion totale. Entre ces deux extrêmes, il existe donc un cas favorable, c'est à dire une catégorie de particules auxquelles la diffusion est due pour l'essentiel, parce qu'à la fois leur nombre et leur efficacité sont suffisamment élevés. Pour préciser les tailles de ces particules, il suffit de se reporter à la figure II. 15, où l'on peut lire - pour diverses valeurs de l'exposant - la valeur du paramètre ρ qui correspond à une fraction donnée de la diffusion totale (celle ci étant calculée pour des particules dont les tailles s'étendent de zéro à l'infini). Ces valeurs peuvent être présentées d'une autre façon qui met plus explicitement en évidence le rôle prépondérant de certaines classes de particules. Ainsi sur la figure III.8 sont portées, en fonction de l'exposant, les valeurs de ρ qui correspondent à des fractions données de la diffusion totale (1 %, 5 % , 10 %, 50 %, 90 %, 95 %, 99 %).

Toutes les courbes tendent vers l'infini pour la valeur -3 de l'exposant, puisqu'avec cette valeur et avec l'hypothèse d'une population non bornée, la diffusion est elle même infinie. Par contre pour les autres valeurs, on voit que la plus grande part de la diffusion (90 % par exemple, ce qui correspond à la bande délimitée par les courbes relatives à 5 % et 95 %) est provoquée par des particules qui appartiennent au domaine intermédiaire : pour donner un exemple, si l'exposant de la distribution est -4, les particules de diamètre compris entre 0,4 et 20 µm sont responsables de 90 % de la diffusion, celles de diamètre compris entre 0,6 et 10 µm, de 80 % (les diamètres sont calculés en supposant l'indice de réfraction égal à 1,05 et la longueur d'onde égale à 419 nm). La table III.4 donne ces valeurs de façon plus complète ; y figurent également les valeurs de ρ qui correspondent à 50 %. Ces dernières valeurs définissent en quelque sorte une classe médiane vis à vis de la diffusion : l'ensemble des particules de dimensions supérieures à celle de cette classe, et l'ensemble des particules de dimensions inférieures contribuent également à la diffusion. Comme cela est prévisible (2e partie, § 3.2.5) les lois de distribution log normale ou exponentielle conduisent comparativement à un intervalle plus restreint de taille pour une même fraction (80 %) de la diffusion.

---

[*] Ou ce qui équivaut à une troncature, si les distributions ne doivent pas être prolongées vers les petites tailles par des lois de Junge mais plutôt par des lois log normales ou exponentielles (cf. § 3.2.5, 2e partie).

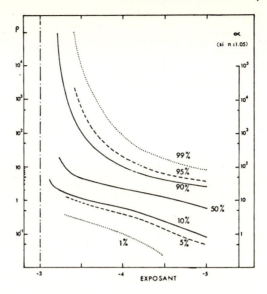

Figure III.8 : En fonction de l'exposant, valeurs du paramètre ρ qui correspondent à un pourcentage donné de la diffusion totale. Cette dernière est calculée pour une population non limitée et a donc sa valeur maximale possible. L'échelle pour le paramètre α correspond à la valeur 1,05 de l'indice de réfraction.

Table IV : Valeurs de ρ* correspondant à 10 %, 50 %, 90 % de la diffusion totale.

Exposants	−3,4	−3,6	−3,8	−4,0	−4,2	−4,5	GL**	EX.
10 %	1,42	1,04	0,78	0,60	0,42	0,22	1,20	1,79
50 %	5,60	3,80	3,01	2,46	2,10	1,22	2,70	3,20
90 %	317	48	19	10,2	6,6	4,3	.5,50	8,1

*si n = 1,05 et λ = 419 nm, le diamètre (en μm) est numériquement égal à ρ .
**les deux dernières colonnes correspondent respectivement aux valeurs de ρ dans le cas de distribution gausso-logarithmique et exponentielle dont les équations ont été données dans la seconde partie (légende de la figure II.16). Ces distributions sont à rapprocher de la distribution selon la loi puissance avec la valeur −4 de l'exposant.

   De la même figure III.8, on peut tirer d'autres conséquences. Ainsi par exemple, le fait que les petites particules ne soient pas accessibles à l'expérience et en particulier à la numération, ne rend pas impossible le calcul de la diffusion puisque une limite supérieure de l'erreur peut être donnée. Dans le même ordre d'idées, s'il s'avère que la loi puissance ne se prolonge pas vers les très petites tailles et que n'existent pas, ce qui est sûr, des particules toujours plus petites et plus nombreuses, l'abandon de cette hypothèse n'affecte que fort peu le résultat ; la position de la courbe relative à 1 % fixe l'ordre de grandeur de cet effet. Les résultats obtenus avec les autres lois de distribution confirment ce point.
   D'autre part, le coefficient total de diffusion qui intervient dans le phénomène de pénétration de la lumière dans la mer, c'est à dire dans un milieu infini, est probablement supérieur à celui qu'on peut mesurer avec un appareil où le volume diffusant est limité. Dans ce cas les grosses particules qui provoquent un signal intense et erratique ne sont pas intégrées dans la mesure. Cependant il est raisonnable de penser que le changement d'échelle correspond à un accroissement du coefficient de quelques pour cent seulement.

3.1.2 - Cas des coefficients angulaires de diffusion.

   On peut généraliser ce qui précède et envisager non plus le cas de la diffusion totale mais celui de la diffusion angulaire. La notion de classe de particules prépondérantes subsiste, mais les limites, ou la valeur médiane de cette classe, vont varier selon l'angle considéré. Les valeurs progressives des intégrales donnant les coefficients angulaires (2e partie, figure II.12) permettent de déterminer la valeur du paramètre α pour laquelle le coefficient de diffusion a atteint une fraction donnée de sa valeur finale (considérée comme asymptotique). On voit immédiatement qu'en ce qui concerne les grands angles (θ de 90° à 180°), la plus grande part de la diffusion est due aux petites particules puisque le palier asymptotique est très tôt atteint ; à l'opposé, la classe efficace sera déplacée vers des tailles plus élevées en ce qui concerne la diffusion aux petits angles. On peut donc dire que chaque partie de l'indicatrice est le reflet préférentiel d'une classe donnée de particules, les angles les plus petits correspondant aux particules les plus grosses. La même figure II.12 permet de prévoir que ces classes recouvrent un intervalle de taille d'autant plus large que la pente initiale des courbes est plus faible c'est à dire que l'exposant est plus élevé (en valeur absolue). Cet élargissement de la classe se fait par abaissement de la limite inférieure et déplacement vers les petites tailles.
   Il est commode pour situer cette classe dans l'échelle des tailles d'utiliser comme précédemment, la taille médiane (c'est la valeur α pour laquelle le coefficient angulaire considéré atteint 50 % de sa valeur finale). La figure III.9, valable pour l'indice de réfraction 1,05, montre les variations avec l'ex-

85

posant de cette taille médiane, ceci pour diverses valeurs de l'angle de diffusion θ. La taille médiane varie approximativement dans un rapport 30 lorsque l'angle de diffusion θ passe de 140° à 2° ; la valeur de ce rapport est assez peu affecté par la valeur de l'exposant, au moins s'il est inférieur à -3,70. La taille médiane correspond à un diamètre de 0,1 μm (si λ = 419 nm) pour la diffusion aux grands angles, 90° ou 140°, tandis que la classe de particules centrée sur 3 ou 4 μm constitue la classe prépondérante s'il s'agit de la diffusion à 2°.

En fixant maintenant la valeur de l'exposant, on peut, pour chaque angle préciser la classe "efficace" mieux que par son point médian. Pour cela on conviendra, par exemple, de limiter cette classe par les valeurs de α pour lesquelles le coefficient angulaire considéré atteint respectivement 10 % et 90 % de sa valeur finale. Ceci est illustré schématiquement par la figure III.10 (l'échelle des diamètres est déduite de l'échelle des α en posant λ = 419 nm).

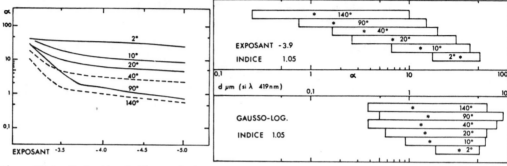

Figure III.9 : En fonction de l'exposant et pour les angles de diffusion indiqués, valeurs de α correspondant à la taille médiane (taille pour laquelle le coefficient angulaire considéré atteint 50 % de la valeur finale, lorsque α = 200). La valeur de l'indice de réfraction est 1,05.

Figure III.10 : Les extrémités des rectangles correspondent aux valeurs de α pour lesquelles le coefficient angulaire de diffusion considéré a atteint 10 ou 90 % de sa valeur finale (soit 100 %, lorsque α = 200). La taille médiane est figurée par l'astérisque. La partie supérieure correspond au calcul avec la loi puissance en -3,9, la partie inférieure au calcul avec la loi gausso-logarithmique dont l'équation a été donnée (2e partie, figure II.16).

En ce qui concerne la distribution selon la loi puissance (exposant -3,90 pour cet exemple), le déplacement de la classe efficace vers les petites tailles, ainsi que son élargissement, se font progressivement lorsque l'angle θ croît. On constate que ce sont les particules de diamètre inférieur à 1 μm qui provoquent la quasi totalité de la diffusion à 140° ou 90°, c'est à dire des particules qui, pour le noter, sont assez peu accessibles à l'expérience. Au contraire le rôle prépondérant est joué par les particules de 2 à 5 μm dans le cas de la diffusion à 2°. Si l'on considère une autre longueur d'onde ces valeurs doivent être modifiées proportionnellement, car les classes sont définies par rapport à α ; ainsi les valeurs précédentes du diamètre doivent être doublées si la longueur d'onde est elle même doublée (838 nm). Cette question directement liée au problème de la sélectivité de la diffusion sera reprise en détail plus loin (§ 3.2). Le résultat est assez différent avec la distribution gausso logarithmique envisagée. L'effet préférentiel est moins marqué que dans le cas de la distribution de Junge. Ceci se conçoit puisque par rapport à cette dernière, la première distribution tronque en quelque sorte la population à la fois pour les petites et les grandes tailles et, de ce fait, privilégie le rôle des particules intermédiaires dans la diffusion ; c'est une conséquence de ce qui avait été remarqué antérieurement (2e partie, § 3.2.5).

3.1.3 - Cas particulier de l'angle 0°.

Dans ce qui précède, il n'a pas été question de la diffusion à 0° et plus généralement de la diffusion aux très petits angles. En effet lorsque le calcul est effectué pour une population dont l'exposant, en valeur absolue, est inférieur à 5, l'intégrale donnant la valeur du coefficient de diffusion à 0° ne converge pas (cf. 2e partie, § 3.2.1). Cette absence de valeur asymptotique rend arbitraire la définition d'une classe efficace pour cet angle ; plus exactement et par continuité, cette classe correspondra toujours aux particules les plus grosses, c'est à dire aux dernières prises en compte dans l'intégrale. Cependant quelques enseignements peuvent quand même être tirés de la théorie. Sur les figures telles que II.12, où sont portées les valeurs progressives des intégrales : $\int i_T (\theta, \alpha) F(\alpha) d\alpha$,
la courbe relative à 0° continue de croître avec la pente 5 - m, tandis que la courbe relative à 2° amorce au contraire un palier pour une valeur α_2 du paramètre α, valeur variable selon l'indice *, et de l'ordre de 40 si par exemple l'indice est pris égal à 1,05. On peut donc immédiatement déduire ce que doit être la taille α_0 des plus grosses particules présentes, pour que le rapport i(0°) / i(2°) ait une valeur donnée k ; on a en effet : $(\alpha_0 / \alpha_2)^{5-m} \simeq i(0°) / i(2°)$
$$\simeq k$$

On peut tenter, au moins approximativement, d'appliquer cette relation aux mesures de T.J. Petzold (1971). Les résultats de cet auteur concernent non pas l'angle 0° mais 0,1°. Le rapport entre i(0,1°) et i(2°) est de l'ordre de 10², on aura donc, en supposant du même ordre le rapport k, (et en adoptant pour l'indice de réfraction la valeur 1,05) : $(\alpha_0 / 40)^{5-m} = 10^2$, ce qui conduit à : $\alpha_0 = 4000$, si m = 4 ou $\alpha_0 = 862$, si m = 3,5. Cette estimation, pour approximative qu'elle soit, montre néanmoins qu'il faudrait repousser la limite supérieure de l'intégrale bien au delà de α = 200, pour pouvoir rendre compte des résultats expérimentaux aux très petits angles.**

---

* Comme on l'a vu précédemment, cette valeur de α doit être telle que le paramètre ρ=2α(n-1) soit égal à 4,1 (2e partie, § 2.2.4 et 2.2.7).

** Sauf peut être pour les problèmes liés à la transmission d'images, cette possibilité de prévision rend assez inutile la prolongation du calcul pour α>200, très long, même avec un ordinateur rapide.

Ceci conduit aussi à souligner le caractère relatif ou mal défini de l'indicatrice dans ce domaine angulaire particulier, car la forme dépend probablement de l'échelle spatiale considérée. Ainsi l'indicatrice à faire intervenir dans les calculs de transmission à longue distance, ou bien l'indicatrice qu'on peut déduire de la fonction de transfert de modulation spatiale (W.H. Wells, 1970) pourraient présenter un lobe plus "pointu" que ne le fait l'indicatrice déterminée avec un dispositif expérimental utilisant un trajet optique plus court. De simples considérations géométriques montrent que cette conclusion n'est pas contradictoire avec le fait que le coefficient total de diffusion n'augmente guère lorsque le volume diffusant considéré s'accroît (cf. § 3.1.1). En effet, lorsque par suite du changement d'échelle, les grosses particules interviennent, ce lobe avant, de plus en plus pointu, devient aussi de plus en plus étroit. On peut concevoir par continuité que dans ce cas à la diffusion par les particules, s'ajoute celle due aux hétérogénéités d'indice, à l'échelle macroscopique cette fois, tout comme les fluctuations qui les provoquent.

3.2 - DEPENDANCE SPECTRALE DE LA DIFFUSION

Pour une population de particules distribuée selon une loi du type créneau, ou, ce qui revient pratiquement au même, distribuée selon des lois normale ou log normale (cf. 2e partie, § 3.2.5), le calcul de la sélectivité spectrale de la diffusion ne présente pas de difficulté théorique, puisqu'il n'y a pas de problème de limites. Pour une distribution établie par rapport à la taille vraie (diamètre), la modification de la longueur d'onde entraîne une modification inversement proportionnelle de la taille relative α - et également du paramètre $\rho = 2\alpha$ (n-1) -. Ce changement affecte à la fois la quantité à intégrer et les limites d'intégration dans les expressions donnant le coefficient d'efficacité moyen $\bar{Q}$ ou le coefficient angulaire de diffusion (respectivement équations 3.2 et 3.3, 2e partie). Que le calcul de $\bar{Q}$ soit effectué avec le paramètre α, ou avec le paramètre ρ pour faire intervenir l'expression approchée de Van de Hulst (2.29), il est nécessaire de toute façon d'attribuer une valeur à l'indice de réfraction.

Pour une population assez peu polydispersée, la sélectivité pour la diffusion (totale) est simple à prévoir. Elle dépend de la taille sur laquelle est centrée la distribution et, conjointement, de la valeur adoptée pour l'indice ; c'est à dire qu'il faut raisonner vis à vis du paramètre ρ et de la valeur que prend ce paramètre pour la taille moyenne. En se reportant aux figures II.10 et II.11 de la 2e partie, on voit que si la taille moyenne est telle que le changement de longueur d'onde λ conserve à ρ des valeurs toujours inférieures à 4, la diffusion sera sélective avec un exposant négatif, c'est à dire en λ^{-x}. Si dans ses variation avec λ, ρ demeure très petit, x sera égal ou voisin de 4 ; si ρ reste compris entre 0,1 et 2, x sera de l'ordre de 2, puis il deviendra inférieur à 2 et tendra vers 0 quand ρ s'approchera de 4. Si, enfin en raison du changement de longueur d'onde, ρ varie entre 4 et 7, la sélectivité s'inversera, c'est à dire s'exprimera par une loi en λ^{+x}. Tout ceci se déduit immédiatement des ondulations de la courbe Q (ρ). Pour finir, si ρ demeure toujours élevé la sélectivité sera pratiquement nulle. Si la population n'est plus faiblement polydispersée (par exemple si, étant distribuée selon une loi log normale, l'écart type géométrique est assez grand) les sélectivités précédentes peuvent subsister mais par effet de moyenne seront d'autant plus atténuées que la distribution sera plus polydispersée.

Des applications de ces principes ont été recherchées en optique océanographique, en particulier par W.V. Burt (1955,1956). L'idée directrice de l'interprétation cherchée consiste d'abord à supposer que la distribution est gausso logarithmique, et ensuite à centrer le maximum de cette distribution dans la zone où précisément ρ est inférieur à 4, ceci afin d'obtenir une sélectivité conforme à celle trouvée expérimentalement ; pour ce faire, l'indice a dû être fixé. L'inconvénient de cette approche est de fournir un résultat qui, d'une part, présuppose l'existence d'un maximum de la distribution et qui d'autre part, fait dépendre la position de ce maximum de la valeur attribuée à l'indice. Ainsi Burt conclut à des maxima du nombre de particules pour des diamètres allant de 0,6 à 1,2 μm, qui par exemple, seraient dix fois plus petits si l'indice avait été pris égal à 1,015 au lieu de 1,15. Si l'on s'abstient de faire ces hypothèses au sujet des limites de la distribution et de l'existence d'un maximum, le problème se pose de savoir si la théorie demeure faisable et si une quelconque sélectivité reste plausible. Egalement, avec ces nouvelles conditions, l'influence du choix de l'indice doit être examinée.

3.2.1 - Cas général : sélectivité dans le cas d'une distribution de Junge non limitée.

Cette distribution, en α^{-m}, est supposée s'étendre à toutes les particules depuis la taille nulle jusqu'à la taille infinie * ; l'indice de réfraction, sans être précisé, doit demeurer constant dans l'intervalle de longueur d'onde considéré, afin que pour une classe donnée de particules la variation des paramètres ρ ne soit due qu'à la variation de λ. On peut faire le raisonnement suivant : la population restant inchangée, si la longueur d'onde est modifiée, λ devenant λ', avec $\lambda' = k\lambda$ (k > 1 par exemple), ce sont les particules de dimensions k fois plus grandes que précédemment qui joueront le même rôle vis à vis de la diffusion. Plus précisément existait une classe de particules de diamètre d, dans un intervalle Δd pour lesquelles Q avait une certaine valeur ; lorsque la longueur d'onde est modifiée cette même valeur de Q est attribuable à la classe de particules de diamètre kd dans un intervalle kΔd. Selon la loi de distribution, le nombre de particules de dimension kd est plus faible que celui des particules de dimension d dans le rapport k^{-m}, mais la classe est elle même élargie dans le rapport k ; finalement le nombre de particules de la classe "kd" est k^{1-m} fois celui de la classe correspondante "d". Ces particules présentent une section géométrique k^2 fois supérieure, donc la surface totale présentée par la classe "kd" de largeur kΔd est donc k^{3-m} fois celle de la classe jouant initialement le même rôle. Or Q est inchangé ; donc la diffusion, produit de la surface par le coefficient d'efficacité, est elle même multipliée par k^{3-m}. Le raisonnement s'appliquant à toutes les classes, et puisque toutes les classes existent, la sélectivité de la diffusion pour l'ensemble de la population, suit en conséquence, une loi en λ^{3-m} **.

_____

*Le nombre de particules est évidemment infini ce qui n'empêche pas que la diffusion puisse être finie, sous certaines conditions comme on l'a vu précédemment (2e partie, § 3.2).

**On peut retrouver aisément ce résultat par le calcul en partant de l'expression donnant le coefficient d'efficacité moyen $\bar{Q}$ (2e partie, équation 3.2) et en prenant o et ∞ comme limites. Il suffit d'expliciter en fonction cette fois de d, n et λ, au lieu de ρ, puis de substituer kλ à λ . Le dénominateur (aire totale) reste inchangé, le facteur moyen $\bar{Q}$ est multiplié par k^{3-m} lorsque la distribution est exprimée par F(d) = cste . d^{-m} .

Ce résultat très simple peut paraître correspondre à un cas idéal et de ce fait ne pas être susceptible d'application. La notion de classe "prépondérante" vis à vis de la diffusion développée précédemment (§ 3.1.1) permet de préciser les possibilités d'application, en montrant l'influence des hypothèses relatives à l'indice et aux limites. On peut dès maintenant souligner le fait que la sélectivité n'est pas uniquement une propriété des suspensions faiblement polydispersées.

3.2.2 - Modification dans le cas de distribution non illimitée.

La loi en λ^{3-m} s'applique à l'intégrale $\int_0^\infty F(\rho) \, Q(\rho) \, \rho^2 \, d\rho$. Si les limites ne sont plus o et ∞ mais ρ_m et ρ_M, la même loi continue de s'appliquer mais un autre effet s'ajoute. En effet la valeur de l'intégrale est elle même modifiée puisque le changement de longueur d'onde entraine un changement des limites, liées aux diamètres d_m et d_M et non à ρ ; si λ devient $k\lambda$, ρ_m et ρ_M deviennent ρ_m/k et ρ_M/k. Il est possible de prévoir ce qu'il s'ensuit, en se reportant à la figure II.8, figure qui mettait en évidence les classes appelées "efficaces".

Pour prendre un premier exemple, supposons une population de particules s'étendant entre des tailles donnant à ρ les valeurs o et ρ_M, et supposons que pour une longueur d'onde λ donnée, la diffusion atteigne 95 % de sa valeur maximale possible (calculée quand ρ_M tend vers l'infini). Si la longueur d'onde est doublée, ρ_M étant divisé par deux, l'intégrale aura une valeur plus faible, correspondant à 90 % par exemple. En conséquence l'effet sélectif en λ^{3-m} sera légèrement modifié (si l'exposant 3-m est négatif, la modification est dans le sens du renforcement de la sélectivité).L'exemple opposé d'une population pour laquelle ρ varie de ρ_m à l'infini, montre que le même changement de longueur d'onde accroît la valeur de l'intégrale ; la sélectivité sera diminuée, de façon plus ou moins sensible selon la valeur initiale de ρ_m (diminuée, toujours dans le cas où 3-m est supposé négatif).

D'une façon générale, on conçoit que si la population de particules englobe largement les classes efficaces, la loi de sélectivité reste inchangée, sinon elle peut être plus ou moins affectée. Elle le sera par exemple, même si la gamme de taille est étendue, dans les deux cas suivants : ou bien si l'on suppose l'indice relatif très proche de 1 (ce qui conduit à de faibles valeurs pour ρ), ou bien si l'exposant de la distribution, ainsi que le montre la figure II.8, s'approche de la valeur -3 (car alors la notion des classes efficaces tend à disparaître). Pratiquement, et en suivant les conclusions relatives aux valeurs plausibles d'indice et d'exposant (§ 2.4), soit, par exemple, 1,05 et -4 respectivement, les classes prépondérantes, responsables de 98 % de la diffusion (de 1 % à 99 %) correspondent à des diamètres allant de 0,1 à 100 µm (pour λ = 419 nm). Dans ces conditions, probablement proches de la réalité, le changement de longueur d'onde, à quelques pour-cent près, conservera la valeur de l'intégrale et donc la loi de sélectivité qui sera approximativement exprimée par λ^{-1}. Des exemples numériques seront données ultérieurement (§ 3.4).

3.2.3 - Forme de l'indicatrice et longueur d'onde.

Cette discussion rejoint en pratique celle qui a été faite dans la 2e partie (§ 3.2.1 et 3.2.2) et qui concernait l'influence des limites sur le résultat du calcul d'indicatrice. Lorsque les conditions de convergence sont respectées et pourvu que la gamme de tailles soit suffisamment étendue la forme de l'indicatrice résultante, comme on l'a vu, n'est pas notablement influencée par les valeurs attribuées aux limites. Le changement de longueur d'onde qui entraine un changement de limites n'aura donc pas d'effet sensible, sauf encore pour la diffusion au voisinage immédiat de 0° (si m < 5). A cet égard la figure III.2 peut être considérée comme représentant la variation d'indicatrice pour une variation de longueur d'onde dans un rapport 4 (en toute rigueur, lorsque α_M passe de 200 à 50, α_m devrait devenir 0,05 et non conserver sa valeur 0,2 mais ceci n'aurait aucune influence sur le résultat - cf. 3.2.2, 2e partie -). Si effectivement la distribution de Junge n'a pas à être prolongée du côté des petites tailles et doit être tronquée, un effet prévisible serait l'accentuation de la dissymétrie de l'indicatrice lorsque la longueur d'onde diminue ; les valeurs numériques indiquées précédemment*permettraient de le chiffrer.

3.3 - POLARISATION **

Le problème de la polarisation a été jusqu'ici laissé de côté ; il a seulement été évoqué lorsque l'indicatrice expérimentale a été comparée aux indicatrices théoriques. En commentaire aux résultats auxquels conduit le calcul théorique on peut d'abord faire les remarques suivantes qui concernent à la fois les propriétés des indicatrices individuelles et celles des indicatrices pour les systèmes polydispersés :
a) les calculs faits pour les cas individuels on fait apparaître des oscillations pour les deux composantes polarisées $i_1(\theta)$ et $i_2(\theta)$ plus importantes encore que pour l'intensité totale $i_T(\theta)$ où, par effet de moyenne, elles sont quelque peu atténuées(cf. les exemples présentés dans l'annexe 2). La polarisation peut de plus être inversée, lorsque i_2 (composante "horizontale" ou "parallèle") est supérieure à i_1 (composante "verticale" ou "perpendiculaire") ; le taux de polarisation qui s'écrit (cf. 2e partie, § 1.2) :

$$p(\theta) = \frac{i_1(\theta) - i_2(\theta)}{i_1(\theta) + i_2(\theta)} \text{ , devient alors négatif.}$$

b) l'intégration qui permet de traiter le cas d'un système polydispersé de particules provoque un lissage. Les valeurs normalisées concernant la polarisation $\overline{p}_1(\theta)$ et $\overline{p}_2(\theta)$ (cf. équation 3.3, 2e partie) ne présentent guère plus d'irrégularité que la quantité $\overline{p}(\theta)$ elle même ; celles qui subsistent sont probablement dues au fait que dans l'addition pondérée qui pratiquement remplace l'intégration, le nombre de termes n'est pas encore suffisant.

---

*cf. 2e partie, § 3.2.2, "cas où la troncature intervient pour des tailles hors du domaine de Rayleigh".

** Comme cela a été annoncé dans la seconde partie, cette question n'est pas envisagée sous son aspect le plus général puisque dans les calculs préliminaires par la théorie de Mie, ont seulement été calculées les composantes i_1 et i_2 mais non i_3 et i_4. En conséquence peut être étudiée la polarisation de la lumière diffusée dans le cas où la lumière incidente est naturelle, mais non dans le cas où la lumière incidente est elle même polarisée (rectilignement, circulairement,...).

c) pour les indicatrices individuelles, la polarisation tend   d'une façon générale, à diminuer pour les
petits et pour les grands angles, et elle s'annule théoriquement à 0° et 180°. Elle est maximale à 90°
dans le cas des petites particules, c'est à dire pour celles qui appartiennent au domaine de Rayleigh
et de Rayleigh Gans. L'intégration étendue à l'ensemble de la population ne fait pas disparaître ces
traits généraux. On peut remarquer que la polarisation est, toutes choses égales par ailleurs (exposant, limites), d'autant plus forte que l'indice est plus proche de 1* ; c'est ce que dénote l'écart
entre $\overline{\beta}_1(\theta)$ et $\overline{\beta}_2(\theta)$, en particulier au voisinage de $\theta = 90°$. Comme le montrent les figures III.2 et
III.7, le minimum de $\overline{\beta}_2(\theta)$ très accentué est centré sur 90° pour l'indice 1,02, mais pour les indices
plus élevés, il est plus plat et situé à des angles un peu supérieurs à 90°. La figure III.11 met en
évidence cet effet.

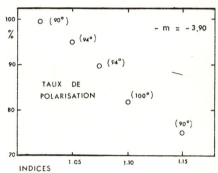

Figure III.11 : Valeurs théoriques du taux maximum
de polarisation en fonction de l'indice de réfraction, la distribution de la population ainsi que les
limites étant inchangées (exposant -3,9, limites
$\alpha_m = 0,2$ et $\alpha_M = 200$). Entre parenthèses figurent
les valeurs de l'angle θ pour lesquelles apparaît
le maximum de polarisation.

3.3.1 - Influence des limites et interprétation.

La figure III.2 illustre l'effet de la limite supérieure : lorsque la limite supérieure α_M est portée de 50 à 200, l'indicatrice pour la lumière naturelle $\overline{\beta}(\theta)$ est peu modifiée, il en va de même de la
composante verticale $\beta_1(\theta)$, par contre la polarisation maximale (et même la polarisation aux autres angles) décroît, car le minimum de la composante horizontale $\beta_2(\theta)$ est moins accentué.
La figure III.12 montre l'exemple opposé de l'influence de la limite inférieure α_m. Lorsque sa valeur passe de 1,2 à 10, $\overline{\beta}(\theta)$ est cette fois, et comme on l'a déjà vu, diminué sauf pour les petits angles, il en est de même de la composante $\overline{\beta}_1(\theta)$; par contre le minimum de l'indicatrice pour la composante $\beta_2(\theta)$ demeure inchangé.
En résumé, et pour prendre le cas typique de $\theta = 90°$, le taux de polarisation est dans le premier
cas diminué par augmentation de $\overline{\beta}_2(90°)$ tandis que dans le second cas c'est le résultat de la diminution
de $\overline{\beta}_1(90°)$.

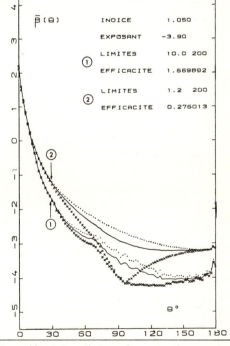

Figure III.12 : Indicatrices théoriques calculées
avec les valeurs indiquées, et tracées avec les
mêmes échelles et symboles que précédemment - points
pour $\overline{\beta}_1(\theta)$ et croix pour $\overline{\beta}_2(\theta)$ -. Les deux cas montrés
ici sont à rapprocher de celui présenté antérieure-
ment (figure III.7) où la limite inférieure était
$\alpha_m = 0,2$, indice et exposant étant les mêmes par
ailleurs.

*Ce qui s'explique puisqu'alors le domaine de Rayleigh Gans s'étend jusqu'à des tailles relatives α qui
peuvent être d'autant plus grandes sans que la condition $\alpha<1$ cesse d'être respectée (2e partie, §2.2.3).

L'interprétation de ces effets se fonde sur des raisonnements analogues à ceux faits pour l'intensité totale i_T (on se bornera à l'examen de la polarisation à 90°). Le graphe des fonctions $i_1(90°)\alpha^{-4}$ et $i_2(90°)\ \alpha^{-4}$ met en évidence le comportement différent de ces deux quantité (figure III.13). La décroissance de $i_1(90°)\ \alpha^{-4}$, sur laquelle se décalque *grosso modo* celle de $i_T(90°)\ \alpha^{-4}$ s'effectue donc selon une pente voisine de -2,3, pour cette valeur 1,05 de l'indice de réfraction. Pour la composante horizontale la quantité correspondante décroît plus lentement, avec une pente de l'ordre de -1 qui tend même à s'annuler pour les valeurs élevées de α, de l'ordre et supérieures à 100. Lorsqu' est effectué le calcul pour l'ensemble de la population, les deux intégrales ne convergeront pas simultanément (équation 3.3 et § 3.2.1, 2e partie). La convergence sera plus lente pour la composante horizontale (indice 2) que pour la composante verticale (indice 1) ; elle peut même ne pas avoir lieu puisque la condition : $5 + p - m < 0$ (condition 3.4, 2e partie), peut ne pas être respectée dans le premier cas (où $p \approx -1$ ou 0) tout en l'étant dans le second ($p \approx -2,3$). C'est approximativement le cas qu'illustre la figure III.14 (où $m = 3,9$); la courbe figurant la valeur progressive de l'intégrale continue de croître lorsqu'il s'agit de la composante horizontale; au contraire, la courbe relative à la composante verticale présente un palier asymptotique.

3.3.2 - Conséquences - seconde remarque sur la valeur des limites.

En application de ce qui précède, $\overline{\beta}_1(90°)$, comme $\overline{\beta}(90°)$, ne peut pratiquement être influencé que par le changement de la limite inférieure α_m , alors que $\overline{\beta}_2(90°)$ dont la valeur est déterminée par la limite supérieure α_M*n'est pas affecté par ce changement. C'est le cas que présentait la figure III.12. Au contraire (cas de la figure III.2) la modification de la limite supérieure α_M ne touche que la composante $\overline{\beta}_2(90°)$, à l'exclusion de $\overline{\beta}_1(90°)$ et conséquemment de $\overline{\beta}(90°)$; lorsque α_M croît, $\beta_2(90°)$ continue de croître et la polarisation diminue. En résumé, les influences de chacune des deux limites s'exercent sélectivement sur l'une ou l'autre des deux composantes polarisées.

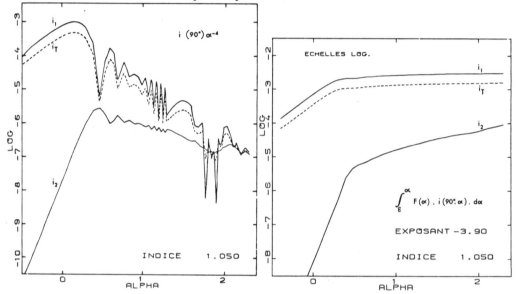

Figure III.13 : En fonction de la taille relative α, variations des fonctions : $i_1(90°)\alpha^{-4}$, $i_2(90°)\alpha^{-4}$ et $i_T(90°)\alpha^{-4}$ avec $i_T = (1/2)\ (i_1+i_2)$. Les échelles sont logarithmiques. Cette figure est à rapprocher de celles présentées dans la seconde partie (figures II.3 à II.5) ou de la figure analogue relative à l'indice 1,05 (A. Morel, 1972).

Figure III.14 : Figure analogue aux figures II.12 de la seconde partie, mais relative à l'angle $\theta = 90°$ seulement. En plus de la courbe montrant la valeur progressive de l'intégrale (en tireté), sont tracées les courbes relatives aux deux composantes polarisées.

On constate qu'il y a, et pour les mêmes raisons, analogie entre le cas de la diffusion à l'avant et le cas de polarisation à angle droit, plus précisément analogie dans le comportement de $\overline{\beta}(0°)$ et de $\overline{\beta}_2(90°)$ (cf. 3.1.3) ; ces grandeurs en effet continuent de croître avec la limite supérieure (sauf si $m > 5$).

Lorsque l'indicatrice expérimentale a été confrontée à diverses indicatrices théoriques il avait été remarqué que l'accord pouvait être à la fois satisfaisant pour les valeurs en lumière naturelle - $\overline{\beta}(\theta)$ - et ne pas l'être pour la polarisation. Ce désaccord ne saurait constituer un argument et n'a pas à être pris en considération dans les conclusions concernant les valeurs d'exposant et d'indice (§ 2.4) ; par exemple, le cas où les valeurs d'indice et d'exposant sont respectivement 1,02 et -4,2 conduit à un taux de polarisation théorique très élevé, lequel pourrait fort bien être réduit en repoussant la limite supérieure, sans pour autant que les valeurs de $\overline{\beta}(\theta)$ soient affectées. Ainsi les valeurs indiquées demeurent plausibles en dépit du désaccord quant à la polarisation. Ces considérations ont aussi un aspect négatif, car elles montrent les limites des possibilités dans l'interprétation des résultats expérimentaux. Suppo-

---

* Comme précédemment, il faut supposer que la limite supérieure α_M a quand même une valeur suffisante, c'est à dire que la population n'est pas faiblement polydispersée.

ser et même calculer une limite supérieure élevée, comme on l'a fait pour rendre compte de la diffusion observée aux petits angles, est un procédé sujet à caution en ce qui concerne la polarisation. En effet le prolongement de la courbe $i_2(90°)\alpha^{-4}$ n'est pas prévisible simplement comme l'est celui de la courbe relative à 0°, les conclusions au sujet de la limite supérieure demeureraient en tout état de cause hasardées. Il faut aussi rappeler que l'assimilation des particules marines à des sphères, si elle est justifiée pour l'intensité totale, l'est moins en ce qui concerne la polarisation (cf. avant propos, 2e partie et ref. Holland et Gagne, 1970).

3.4. RELATIONS ENTRE DIFFUSION ET TENEUR EN PARTICULES

Estimer la concentration des eaux de mer en matière en suspension est à juste titre, l'une des motivations souvent invoquées à l'exécution systématique de mesures de diffusion de la lumière. Cependant aussi bien expérimentalement que théoriquement, l'établissement de relations quantitatives comporte de nombreuses incertitudes. Elles proviennent des hypothèses qu'il faut faire sur le facteur d'efficacité moyen $\overline{Q}$, l'indice, les tailles extrêmes à considérer, éventuellement la masse volumique des particules etc.. Ces difficultés ont déjà été reconnues, par exemple, et pour ne citer que ces auteurs, par Y.E. Otchakovsky (1965 b) ou par G.F. Beardsley et al. (1970). Il est utile d'examiner l'influence des hypothèses, avant d'indiquer comment et avec quelles précautions les résultats des calculs théoriques peuvent être appliqués.

3.4.1 Remarque générale sur les variations possibles du facteur d'efficacité moyen $\overline{Q}$.

Le premier problème consiste à calculer le coefficient total de diffusion, lorsque le nombre de particules, compté entre deux limites de taille, est connu, ainsi que la loi qui en régit la distribution. Le problème inverse consiste à déduire des informations sur la quantité de particules (volume, surface, nombre) à partir d'observations du coefficient de diffusion. Dans le premier cas, la diffusion due aux particules dénombrées est parfaitement calculable dès lors qu'il l'indice est connu (ou est choisi) ; le problème reste néanmoins incomplètement résolu car la diffusion qui serait expérimentalement observée sur le même échantillon est *a priori* supérieure à celle qui est calculée, puisqu'elle fait intervenir des particules non dénombrées. Le fait qu'existent des classes "efficaces" (cf. § 3.1.1) montre que dans certains cas l'influence des limites est faible ; un ordre de grandeur de la différence entre diffusion calculable et diffusion réelle peut être estimé. La résolution du problème inverse implique des hypothèses plus nombreuses encore : à celles relatives à l'indice et aux limites s'ajoutent celles concernant la loi de distribution.

Les relations qu'on peut établir sont dépendantes de ces hypothèses. Mettre en évidence cette dépendance, revient à étudier les variations du facteur moyen d'efficacité $\overline{Q}$ avec l'indice n, l'exposant m, et les diamètres limites de la distribution d_m et d_M (ou les paramètres correspondants, α_m α_M ou ρ_m ρ_M). Pour les valeurs d'indice considérées, il est justifié d'utiliser l'expression limite de Van de Hulst pour calculer Q (2e partie, équation 2.29) et dans ce cas le facteur moyen, plus commodément défini par rapport à ρ, plutôt que par rapport à α, s'écrit :

$$\overline{Q} = \frac{\int_{\rho_m}^{\rho_M} F(\rho)\ Q(\rho)\ \rho^2\ d\rho}{\int_{\rho_m}^{\rho_M} F(\rho)\ \rho^2\ d\rho}$$

Ainsi d'un calcul unique, diverses applications peuvent être tirées, selon la valeur attribuée à l'indice. (Pour le rappeler - cf. 2e partie, § 2.2.10 - il faut évaluer Q avec les formules exactes lorsque ρ est petit, en pratique inférieure à 0,2). La figure III.15 (A. Morel, 1972 b) montre les variations de $\overline{Q}$ avec l'exposant de la distribution, les diverses courbes correspondant à différents couples de valeurs données aux limites ρ_m et ρ_M.

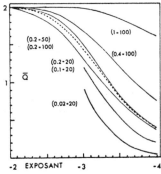

Figure III.15 : Variation du facteur moyen d'efficacité $\overline{Q}$ avec l'exposant de la distribution. Les valeurs entre parenthèses correspondent aux paramètres ρ_m et ρ_M calculées pour les limites de la distribution.

Lorsque l'exposant est égal à -2 toutes les classes de particules présentent des surfaces égales, la valeur moyenne $\overline{Q}$ est en fait la valeur moyenne de la fonction $Q(\rho)$ (cf. figure II.10, 2e partie), donc $\overline{Q}$ est voisin de 2 pourvu que la limite supérieure soit assez élevée (ρ_M = 10 ou 20 par exemple), et quelle que soit par contre la limite inférieure. $\overline{Q}$ peut être légèrement supérieur à 2 si l'intervalle ρ_m - ρ_M choisi est restreint et favorise un maximum de la courbe $Q(\rho)$. Lorsque l'exposant croît (en valeur absolue), l'importance relative croissante des petites particules entraîne la diminution de $\overline{Q}$. Lorsque m atteint et dépasse 4, les particules de taille élevée et donc la limite supérieure ρ_M interviennent peu, la valeur de $\overline{Q}$ n'est pratiquement dépendante que de la valeur de la limite inférieure.

Finalement le facteur $\overline{Q}$ peut varier beaucoup avec l'exposant, avec les limites et par suite avec l'indice et la longueur d'onde, puisque en effet ces paramètres réagissent sur les valeurs des limites. Ceci montre les difficultés qui surgissent dans les applications en raison des hypothèses qu'il faut faire, alors que pourtant le problème est parfaitement résolu sur le plan mathématique.

3.4.2 - Exemples d'application.

On peut préciser concrètement ces diverses influences en prenant l'exemple d'une population de particules distribuées selon une loi puissance dans l'intervalle de taille défini par les diamètres d_m = 1μm et d_M = 20 μm. Compte tenu des valeurs attribuées à l'indice et la longueur d'onde, à ces diamètres correspondent les valeurs α_m et α_M ou ρ_m et ρ_M du tableau suivant :

Table V : Correspondance entre d, α et ρ pour deux longueurs d'onde et trois valeurs d'indice.

d (m)	α		ρ		
	si λ = 419 nm	si n =	1,03	1,05	1,10
d_m = 1	α_m = 10	ρ_m =	0,6	1,0	2,0
d_M = 20	α_M = 200	ρ_M =	12	20	40
	si λ = 546 nm	si n =	1,03	1,05	1,10
d_m = 1	α_m = 7,7	ρ_m =	0,46	0,77	1,54
M = 20	α_M = 154	ρ_M =	9,24	15,4	30,8

Par l'expression rappelée ci dessus, on peut calculer le facteur moyen $\overline{Q}$ entre les limites constituées par les divers couples de valeurs ρ_m - ρ_M et ceci pour différents exposants de la loi de distribution. On obtient les résultats de la table VI. Les pourcentages indiqués * correspondent au rapport de la diffusion due aux particules de l'intervalle 1 - 20 μm à la diffusion hypothétique d'une population non bornée. Ceci montre que dans certains cas, en particulier si on doit attribuer une valeur élevée à l'indice, la part"expliquée"de la diffusion est faible. Dans un cas favorable (λ = 546 nm, n = 1,05, m = 3,9), aux particules de diamètre inférieur à 1 μm (ρ < 0,77) correspondrait théoriquement (et au maximum) 14 % de la diffusion, à celles de diamètre supérieur à 20 μm (ρ > 15,4) 9 %, la diffusion due aux particules entre 1 et 20 m formant 77 % du total théorique. Ceci revient à dire que dans ce cas les particules considérées formant l'essentiel des classes efficaces (cf. figure III.8).

Table VI : Pour divers indices et longueurs d'onde, facteur moyen d'efficacité $\overline{Q}$ valable pour une population de particules s'étendant de 1 à 20 μm et distribuée selon des lois puissance dont l'exposant est indiqué.

λ = 419 nm	Exposants							
	-3,6		-3,9		-4,2		-4,5	
n = 1,03	72,3	1,282	79,2	1,089	76,9	0,920	66,8	0,780
n = 1,05	73,0	1,756	74,7	1,626	67,4	1,496	54,5	1,374
n = 1,10	63,3	2,307	57,5	2,338	46,0	2,355	31,3	2,365
	%	$\overline{Q}$	%	$\overline{Q}$	%	$\overline{Q}$	%	$\overline{Q}$
λ= 546 nm								
n = 1,03	69,7	1,028	78,3	0,828	79,3	0,663	72,6	0,533
n = 1,05	72,7	1,539	77,1	1,375	71,9	1,222	60,4	1,086
n = 1,10	67,8	2,171	64,4	2,148	54,0	2,111	40,6	2,064

La figure III.16 présente pour les deux longueurs d'onde les variations du facteur moyen $\overline{Q}$ en fonction de l'indice et pour les divers exposants. La figure III.17 met en évidence le fait que la sélectivité de la diffusion dépend de l'exposant. La courbe tracée est celle qui correspond à la loi théorique applicable à une population non bornée, c'est à dire à la loi en λ^{3-m} (cf. § 3.2.1), les points figurent les rapports $\overline{Q}_{419}$ / $\overline{Q}_{546}$ caractéristiques de la sélectivité pour la population limitée * (1 - 20 μm) considérée.

Problème direct : calcul de la diffusion.

En application pratique de ce qui précède on peut prendre un exemple : supposons connu le nombre total de particules par unité de volume entre deux limites de taille correspondant aux diamètres d_m et d_M, c'est à dire le nombre :

$$N = \int_{d_m}^{d_M} F(d) \, d(d) \quad , \text{ et supposons la distribution exprimée par } F(d) = A d^{-m}.$$

La surface totale S des sections géométriques, ainsi que le volume total V de ces particules peuvent être calculés. Pour faciliter les applications, il est commode de rapporter ces grandeurs S et V au nombre N ;

---

* Les pourcentages indiqués au tableau III.6 permettent en théorie de retrouver les valeurs qui correspondraient à la population non limitée. Ainsi sont calculées les valeurs figurées par les croix qui normalement devraient se trouver sur la courbe de la figure III.17. Une légère imprécision, prévisible d'après le mode de calcul, fait surestimer les pourcentages et explique le léger désaccord.

On obtient alors :

$$\frac{S}{N} = \frac{\pi}{4}\frac{m-1}{m-3}\frac{d_M^{m-3} - d_m^{m-3}}{d_M^{m-1} - d_m^{m-1}}(d_m\, d_M)^{m-1}$$

et

$$\frac{V}{N} = \frac{\pi}{6}\frac{m-1}{m-4}\frac{d_M^{m-4} - d_m^{m-4}}{d_M^{m-1} - d_m^{m-1}}(d_m\, d_M)^{m-1}$$

(la première équation suppose m > 3, la seconde m > 4 ; si ces conditions ne sont pas satisfaites, les expressions modifiées se trouvent aisément). En remplaçant d_m et d_M par 1 et 20 µm et en adoptant m = 4,2, il vient : S = N x 2,04 µm² et V = N x 5,77 µm³. En choisissant une concentration qui est celle d'une eau limpide sous la couche superficielle N = 10^{10} particules m⁻³ (soit 10 000 particules par cm³, cf. J.C. Brun Cottan, 1971, H.R. Gordon, O.B. Brown, 1972), il vient : S = 2,04 10^{-2} m²/m³ et V = 5,77 10^{-8} m³/m³ (remarquons que si on attribue aux particules la densité 1 cette concentration volumique correspond à 57,7 µg/litre), supposant l'indice égal à 1,05, avec les valeurs de $\bar{q}$ du tableau précédent on calcule le coefficient total de diffusion b = S x $\bar{q}$, soit : b_{419} = 3,05 x 10^{-2} m⁻¹ b_{546} = 2,49 x 10^{-2} m⁻¹.

En se reportant à la figure III.8, on voit que ces coefficients calculés pour les particules de 1 à 20 µm pourraient être inférieurs de 15 à 20 % aux coefficients expérimentaux qui, en principe, correspondent à une population plus étendue. De cette manière*et en supposant l'indice égal à 1,05, ont été calculés les coefficients de diffusion à 546 nm relatifs aux échantillons de Méditerranée dont la granulométrie avait été établie par J.C. Brun Cottan (1971) - le calcul est un peu plus complexe du fait qu'il faut en général calculer pour chaque population 2 facteurs d'efficacité l'un entre 1 et 4 µm, l'autre entre 4 et 20 µm -. Sur les mêmes prélèvements, des mesures du coefficient angulaire à 30°, ß(30°), avaient été exécutées, desquelles on peut déduire b (cf.1ère partie,§6). Ces valeurs de b sont d'ailleurs en très bon accord avec celles obtenues *in situ* grâce à l'appareil intégrateur (D. Bauer, A. Ivanoff, 1971). Les coefficients calculés sont confrontés aux coefficients déduits de la mesure (Figure III.18). L'accord assez satisfaisant, pour les mesures de novembre 1969, l'est moins pour celles de juin 1969, en particulier pour les prélèvements effectués de 100 à 400 mètres (l'indice 1,03 conviendrait mieux pour rendre compte de ces mesures).

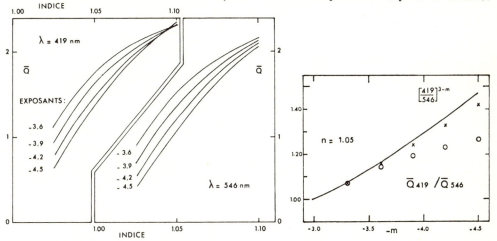

Figure III.16 : Pour les deux longueurs d'onde indiquées, variation en fonction de l'indice de réfraction du facteur d'efficacité $\bar{q}$. $\bar{q}$ est calculé pour une population de particules dont les limites extrêmes de taille sont 1 µm et 20 µm et dont la distribution est régie par des lois puissance dont l'exposant est indiqué.

Figure III.17 : Rapport de la diffusion à 419 nm à la diffusion à 546 nm mettant en évidence la sélectivité qui varie avec l'exposant de la loi de distribution (voir texte et remarque de la page précédente).

Problème inverse : utilisation des mesures de diffusion.

Trois hypothèses sont nécessaires à sa résolution. Ainsi à titre d'exemple, en choisissant l'indice égal à 1,05, en supposant la distribution régie par une loi d'exposant -4 et des limites de taille correspondant aux diamètres 0,5 µm et 50 µm (cette limite inférieure équivaut à une troncature comme celle que provoque approximativement une filtration sur millipore HA), on peut établir les relations suivantes proposées évidemment avec toutes les réserves qu'imposent les choix précédents. Le facteur moyen d'efficacité pour λ = 546 nm est : $\bar{q}$ (0,5 + 50, 546) = 0,70 (pour λ = 419 nm, sa valeur serait 0,92). L'aire totale des sections géométriques a pour valeur : S (m²/m³) = b (m⁻¹) x (1 / 0,70) de l'expression donnant S/N, on tire : N (m⁻³) = 10^{12} x 1,715 S , de l'expression donnant V/S - qui devient V/S = 1/3 Log 100, puisque dans ce cas particulier d'exposant -4, l'intégrale donnant V est un logarithme - on obtient V (m³/m³) = 10^{-6} x 1,535 S.

Si par exemple on veut comparer les résultats d'une mesure de diffusion avec ceux d'un comptage de particules effectué entre 1 et 20 µm et également ceux d'une filtration, supposée efficace entre les tailles 0,5 et 50 µm, il est intéressant de connaitre comment les grandeurs nombres, surfaces et volumes se répartissent dans les trois classes, respectivement de 0,5 à 1 µm, de 1 à 20 µm et de 20 à 50 µm. Supposons que la mesure de b à 546 nm fournisse par exemple la valeur 0,07 m⁻¹, l'aire totale des sections géométrique est donc 0,10 m²/m³ et l'on aura pour N, S et V, la répartition suivante exprimée en pourcentage du total :

---

* mais sans faire cette correction de 15 ou 20 % qui reste incertaine.

	classe 0,5 à 1μm	classe 1 à 20 μm	classe 20 à 50. μm	total 0,5 à 50μm
N (m^{-3})	87,5%	12,5%	≈ 0	$17,15.10^{10}$
S (m^2/m^3)	51%	48%	1%	0,10
V (m^3/m^3)	15,0%	65,0%	20%	154.10^{-9}

La relation entre b et V (b en m^{-1} = 0,455.10^{+6} V en m^3/m^3 est représentée sur la figure III.19 où sont également portés, en fonction des concentrations en poids sec de particules (10^{-9} g/g, c'est à dire en μg/litre) les coefficients de diffusion déterminés sur les même échantillons (voir également figure 6, A. Morel, 1970). Il apparait que le rapport entre poids sec et volume théorique est très inférieur à 1, ce qui en tout état de cause, s'expliquerait par une forte hydratation des particules. Vouloir préciser davantage serait illusoire, d'une part parce que la relation est établie avec des paramètres moyens (indice et exposant) et d'autre part parce que la filtration n'effectue pas une troncature franche (à d = 0,5 μm) mais au contraire progressive et peut être aussi variable. Ainsi les filtres Whatman G.F. présenteraient une efficacité de rétention diminuant de 100 % à 40 % entre les diamètres 4 μm et 0,7 μm respectivement (R.W. Sheldon, W.H. Sutcliffe Jr., 1969). Ce point ne doit pas être oublié lorsqu'il s'agit de comparer des mesures de poids sec, soit entre elles, soit avec des mesures de diffusion, surtout si l'on considère qu'une fraction importante des particules, en poids et en volume, se situe dans le domaine des petites tailles. Selon Lisitzin et Bodganov (1968), 40 à 60 % de la masse particulaire serait due aux particules de dimension inférieure à 1 μm, ce qui avec la loi d'exposant –4, en suppose la prolongation jusqu'à des diamètres de 0,02 μm ; avec cet exposant il y a équipartition logarithmique des volumes, ainsi par exemple les deux classes 0,02 - 1 μm et 1 - 50 μm, présentent le même volume total.

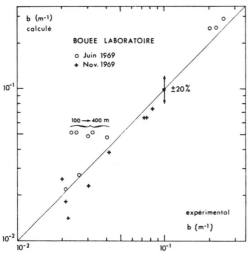

Figure III.18 : Coefficients de diffusion calculés à partir des données granulométriques, comparés aux coefficients mesurés (voir texte). Bouée laboratoire φ = 42°14 N   G = 05°35 E .

Figure III/19 : Poids sec des particules retenues sur filtres Whatman GF (mesures effectuées par G. Copin) comparées aux coefficients de diffusion mesurés –par l'intermédiaire de β(30) – sur les mêmes prélèvements. Les mesures en Atlantique ont été effectuées durant la campagne Harmattan (1971).

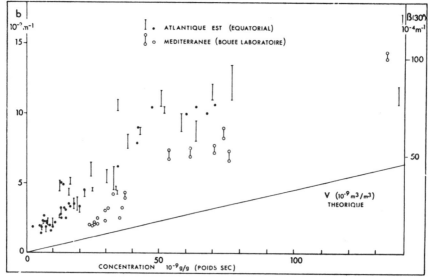

REFERENCES

Bauer, D., and A. Morel. 1967. Etude aux petits angles de l'indicatrice de diffusion des eaux de mer. *Ann. Geophys.* **23**(1):109–123.

Bauer, D., and A. Ivanoff. 1971. Description d'un diffusiomètre "intégrateur". *Cah. Océanogr.* **23**(9):827–839.

Beardsley, G. F., H. Pak, K. Carder, and B. Lundgren. 1970. Light scattering and suspended particles in the Eastern Equatorial Pacific Ocean. *J. Geophys. Res.* **75**(15):2837–2845.

Brun-Cottan, J. C. 1971. Etude de la granulométrie des particules. Mesures effectuées avec un Coulter Counter. *Cah. Océanogr.* **23**(2):193–205.

Burt, W. V. 1955. Interpretation of spectro-photometer readings on Chesapeake bay waters. *J. Mar. Res.* **14**(1):33–46.

✳Burt, W. V. 1956. A light scattering diagram. *J. Mar. Res.* **15**(1):5.

Carder, K. L., G. F. Beardsley, Jr., and H. Pak. 1971. Particle size distribution in the eastern equatorial Pacific, *J. Geophys. Res.* **76**:5070–5077.

Carder, K. L., R. D. Domlinson, and G. F. Beardsley. 1972. A technique for the estimation of indice of refraction of marine phytoplankters. *Limnol. Oceanogr.* **17**(6):833–839.

Copin, C., and G. Copin. 1972. Chemical analysis of suspended particulate matter collected in the North east Atlantic. *Deep Sea Res.* **19**:445–452.

Deirmendjian, D. 1963. Scattering and polarization properties of polydispersed suspensions with partial absorption. In *Electromagnetic Scattering*, Milton Kerker, ed., pp. 171–189. Pergamon Press.

Gordon, D. C. 1970. Some studies on the distribution and composition of particulate organic carbon in the Atlantic Ocean. *Deep Sea Res.* **17**:233–243.

Gordon, H. R., and O. B. Brown. 1972. A theoretical model of light scattering by Sargasso sea particulates. *Limnol. Oceanogr.* **17**(6):826–832.

Harris, J. E. 1972. Characterization of suspended matter in the Gulf of Mexico. 1, Spatial distribution. *Deep Sea Res.* **19**(10):712–726.

Hobson, L. A. 1967. The seasonal and vertical distribution of suspended particulate matter in an area of the Northeast Pacific Ocean. *Limnol. Oceanogr.* **12**(4):642–649.

Holland, A. C., and G. Gagne. 1970. The scattering of polarized light by polydisperse systems of irregular particles. *Appl. Opt.* **9**(5):1113–1121.

Jerlov, N. G. 1953. Particle distribution in the ocean. *Rep. Swed. Deep Sea Exped.* **3**:73–97.

Jerlov, N. G. 1961. Optical measurements in the Eastern North Atlantic. *Medd. Oceanogr. Inst. Göteborg,* **38**(11):4–40.

Jerlov, N. G. 1968. *Optical oceanography.* Amsterdam: Elsevier.

Junge, C. E. 1963. Air chemistry and radioactivity. New York: Academic Press.

Kinney, P. J., T. C. Loder, and J. Groves. 1971. Particulate and dissolved organic matter in the Amerasian basin of the arctic Ocean. *Limnol. Oceanogr.* **16**(1):132–137.

Kullenberg, G. 1968. Scattering of light by Sargasso sea water. *Deep Sea Res.* **15**(4):423–432.

Kullenberg, G. 1969. *Light scattering in the central Baltic.* Copenhague Univ. Inst. Phys. Oceanogr. Report no. 5.

Kullenberg, G., and N. B. Olsen. 1972. *A comparison between observed and computed light scattering functions.* Copenhague Univ. Inst. Phys. Oceanogr. Report no. 19.

Lisitsin, A., and Y. Bogdanov. 1968. Granulométrie et composition de la matière en suspension dans l'océan Pacifique (en russe). *Oceanogr. Res. Coll.* **18**:53. Science Publ. Moscou.

Mankovski, V. I. 1971. Relations entre coefficient total et coefficients angulaires de diffusion. *Akad. Nauk. Ukr. SSR. Mar. Hydrophys. Inst.* **6**(65):145–154.

Morel, A. 1965. Interprétation des variations de la forme de l'indicatrice de diffusion de la lumière par les eaux de mer. *Ann. Geophys.* **21**(2):281–284.

Morel, A. 1968a. Note au sujet des constantes de diffusion de la lumière pour l'eau et l'eau de mer optiquement pures. *Cah. Océanogr.* **20**(2):157–162.

Morel, A. 1968b. Relations entre coefficients angulaires et coefficient total de diffusion de la lumière pour les eaux de mer. *Cah. Océanogr.* **20**(4):291–303.

Morel, A. 1970. Examen des résultats expérimentaux concernant la diffusion de la lumière par les eaux de mer. *Electromagnetics of the Sea,* A.G.A.R.D. Conf. Proc. No. 77, Paris.

Morel, A. 1972a. Application de la théorie de Mie au calcul de l'indicatrice de diffusion de la lumière pour les eaux de mer. *C. R. Acad. Sci.* **274**:1387–1390.

Morel, A. 1972b. Au sujet de l'emploi du coefficient total de diffusion pour évaluer la teneur des eaux de mer en particules en suspension. *C. R. Acad. Sci.* **247**:1447–1450.

Morel, A. 1973. *Indicatrices de diffusion théoriques calculées par la théorie de Mie appliquée à des systèmes polydispersés de particules.* Centre Rech. Océanogr. Villefranche-sur-Mer, Report No. 10.

Morrison, R. E. 1970a. Experimental studies on the optical properties of sea water. *J. Geophys. Res.* **75**(3):612–628.

Nyffeler, F. 1970. Etude de la diffusion de la lumière aux petits angles par l'eau de mer. *Electromagnetics of the Sea,* A.G.A.R.D. Conf. Proceed. no. 77, Paris.

Otchakovsky, Y. E. 1965a. Comparison des indicatrices mesurées et calculées pour l'eau de mer. *Tr. Inst. Okeanol. Akad. Nauk. SSSR.* **77**:125–130.

Otchakovsky, Y. E. 1965b. Relations entre coefficient d'extinction et matière en suspension dans la mer. *Tr. Inst. Okeanol. Akad. Nauk. SSSR.* **77**:35–40.

Pak, H., J. R. Zaneveld, and G. F. Beardsley, Jr. 1971. Mie scattering by suspended clay particles. *J. Geophys. Res.* **76**(21):5065–5069.

Petzold, T. J. 1972. *Volume scattering functions for selected ocean waters.* Visibility Lab., San Diego, Cal., technical report S.I.O. ref. 72-78.

Reese, J. W., and S. P. Tucker. 1970. *Light measurements off the San Diego Coast.* N.U.C. technical publication no. 203:1–37.

Sheldon, R. W., and T. R. Parsons. 1967. A continuous size spectrum for particulate matter in the sea. *J. Fish. Res. Bd. Canada* **24**:909–915.

Sheldon, R. W., and W. H. Sutcliffe, Jr. 1969. Retention of marine particles by screens and filters. *Limnol. Oceanogr.* **14**(3):441–444.

Sheldon, R. W., A. Prakash, and W. H. Sutcliffe, Jr. 1972. The size distribution of particles in the ocean. *Limnol. Oceanogr.* **17**(3):327–340.

Spilhaus, A. F. 1968. Observations of light scattering in sea water. *Limnol. Oceanogr.* **13**:418–422.

Wells, H. G., and M. N. Todd. 1970. Loss of optical resolution in sea water by multiple small angle scattering. *Electromagnetics of the sea,* A.G.A.R.D. Conf. Proc. No. 77, Paris.

Zaneveld, J. R. V., and H. Pak. 1973. Method for the determination of the index of refraction of particles suspended in the ocean. *J. Opt. Soc. Am.* **63**(3):321–324.

Part III

TOTAL ATTENUATION COEFFICIENT:
INSTRUMENTS AND DATA

Editor's Comments
on Papers 5 Through 11

Following the publication of R. W. Preisendorfer's phenomenological theory of light in the sea, there has been a concerted effort to develop instrumentation for the measurement of the optical properties that his theory defines. The basic optical property, T, (the beam transmittance) is generally measured by means of a collimated beam of artificial light. Measurements of this property can thus be made either *in situ* or on collected samples.

Preisendorfer gives the basic operational definition for the total attenuation coefficient α (Paper 3, equations 17 and 18). S. Q. Duntley (Paper 2, equation 1) rewrites this equation in terms of "the residual power (P_r°) reaching a distant position r along the path without having been deviated by any type of scattering process'' and (P_o), the total flux content of the beam as it leaves the projector.

The strict concept that any radiant energy that is deviated by scattering should be counted as a part of the total attenuation is a relatively recent concept and one that has special significance to the phenomena of contrast reduction along paths of sight underwater. Many of the early experiments to determine the beam transmittance of ocean water (and also of distilled water) were concerned only with the transmittance of total flux within the cross-sectional area of the beam of radiant energy that was employed. Such measurements included forward-scattered flux, and the values of transmittance obtained consequently led to coefficients that were inconsistent with the strict concepts of attenuation.

The collection of spurious scattered flux by a beam transmissometer cannot be altogether avoided. The error introduced will, of course, be maximal in a spectral region for which scattering is large and absorption is small. In spectral regions of high absorption, the error due to forward scattering will be minimal. These and other factors that enter into the design of an oceanic transmissometer are treated in detail by R. W. Austin and T. J. Petzold in Paper 5. Austin's independent analysis of the precision of measurement of the volume attenuation coefficient is given in Paper 6. These considerations of accuracy and precision have recently been employed by R. C. Smith and J. E. Tyler (1976) in a critical analysis of the published data for the total attenuation coefficient for the wavelength range from 200 nm to 800 nm. On the basis of this analysis, the data given in Figure 1 provides a quick appraisal of our current knowledge of the spectral attenuation coefficient for ocean water.

In Papers 8–11 the authors have all performed essentially the same experiment to determine beam transmittance for monochromatic light at various wavelengths. However, E. O. Hulburt (Paper 8) uses his data to calculate an attenuation coefficient that is the sum of the attenuation due to both scattering and absorption, whereas S. A. Sullivan (Paper 9) has used his data to compute an absorption coefficient.

In the spectral region 610 to 700 nm, these two experiments led to values for distilled water that are within ±10% of the average value at 610 nm, decreasing to ±6% of the average value at 700 nm. In the spectral region from 400 to 450 nm, however, the two experiments gave results for distilled water that varied by as much as ±23% from the average value.

R. W. Austin's analysis (Paper 6) demonstrates that in order to achieve a relative precision of 1% in the measurement of the total attenuation coefficient (α) of distilled water (assuming $\alpha = 0.05 \ m^{-1}$), the optimum path length through the water should be 20 meters. Since the longest water path so far used for the numerical determination of the

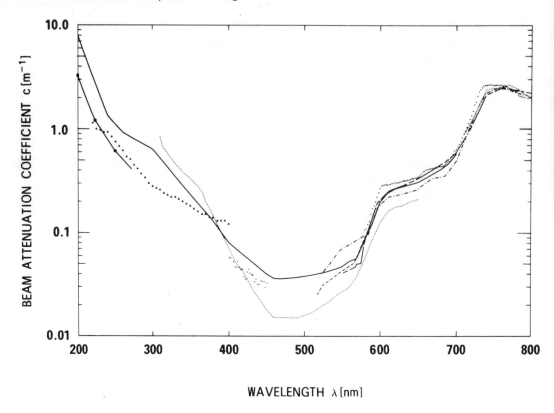

WAVELENGTH λ [nm]

··	Sawyer (1931)	310 — 650 nm
————————————————	Dawson and Hulburt (1934) and Hulburt (1945)	200 — 400 nm 400 — 700 nm
— — — — — — — — —	James and Birge (1938)	365 — 800 nm
··—·—·—·—·—·—·—·—	Clark and James (1939)	365 — 800 nm
··	Curcio and Petty (1951)	710 — 800 nm
▲ ▲ ▲ ▲ ▲	Lenoble and Saint-Guilly (1955)	220 — 400 nm
■ ■ ■ ■ ■ ■	Armstrong and Boalch (1961a,b)	200 — 400 nm
●●●●●●●●	Sullivan (1963)	400 — 450 nm 580 — 790 nm

Fig. 1. Selected data for the total attenuation coefficient, $c[m^{-1}]$, of pure water as a function of wavelength, $\lambda[nm]$. Only data judged to have relatively higher precision and accuracy are included in this figure. *Reprinted from R. C. Smith and J. E. Tyler, Transmission of Solar Radiation into Natural Waters, p. 134 in* Photochem. Photobiol. Rev. **1:**117–155 (1976). Copyright © 1976 by Plenum Press.

total attenuation coefficient has been 4 meters (J. LeNoble and B. Saint-Guilly, Paper 11), the values of the total attenuation coefficients for distilled water in the wavelength region of maximum transmittance are not accurately known.

L. F. Drummeter and G. L. Knestrick (1967) have used water paths of 9.7 meters and 19.5 meters in an effort to locate fine structure in the relative spectral attenuation of ocean water and distilled water. They found no discernible real structure.

During the SCOR Discoverer Expedition (May 1970), Austin obtained extensive monochromatic measurements of both beam transmittance data and simultaneous temperature data as a function of depth in various types of ocean water. A selection of his data recordings from the Data Report of the SCOR Discoverer Expedition is included in Paper 7. In the red and infrared regions of the spectrum, J. A. Curcio and C. C. Petty Paper 10, have successfully measured beam transmittance for the spectral region from 700 to 2500 nm.

REFERENCES

Armstrong, F. A. J., and G. T. Boalch. 1961a. The ultraviolet absorption of sea water. *Mar. Biol. Assoc. U.K. J.* **41**:591.

———. 1961b. Ultraviolet absorption of sea water and its volatile components. In N. G. Jerlov, ed., *Int. Union Geod. Geophys. Monogr.* **10**:63.

Clark, G. L., and H. R. James. 1939. Laboratory analysis of the selective absorption of light by sea water. *Opt. Soc. Am. J.* **39**:43.

Dawson, L. H., and E. O. Hulburt. 1934. The absorption of ultraviolet and visible light by water. *Opt. Soc. Am. J.* **24**:175.

Drummeter, L. F., and G. L. Knestrick. 1961. A high resolution investigation of the relative spectral attenuation coefficients of water. Part I, Preliminary. *U.S. Naval Research Laboratory Report 5642*, 7 pp.

James, H. R., and E. A. Birge. 1938. A laboratory study of the absorption of light by lake waters. *Wisconsin Acad. Sci. Trans.* **31**:1.

Sawyer, W. R. 1931. The spectral absorption of light by pure water and Bay of Fundy water. *Contrib. Can. Biol. Fish.* New Series, 7:74.

Smith, R. C., and J. E. Tyler. 1976. Transmission of solar radiation into natural waters. *Photochem. Photobiol. Rev.* **1**:117–155.

Tyler, J. E. (ed.) 1973. Data Report SCOR Discoverer Expedition—May 1970. University of Calif., Scripps Inst. of Oceanography, S.I.O Ref. 73-16, June 1973, 1009 pp.

5

This article was written expressly for this Benchmark volume.

CONSIDERATIONS IN THE DESIGN AND EVALUATION OF OCEANOGRAPHIC TRANSMISSOMETERS

R. W. Austin and T. J. Petzold

Visibility Laboratory of the Scripps Institution of Oceanography
University of California, San Diego

INTRODUCTION

The quantitative evaluation and intercomparison of in-water imaging systems requires a knowledge of those optical properties of the water that affect the transmission of a beam of collimated light. Although no single measurement can, by itself, provide the full description of the required properties, a knowledge of the volume attenuation coefficient α is generally found to be the most useful single property, and it is the property most readily obtainable if properly designed instrumentation is available. Although α may be obtained in a variety of ways, one of the simplest instrumental methods is from measurements of the transmittance of the water with an instrument known generically as the beam transmissometer. In the following we will develop the concepts for the determination of α from the measurement of the transmittance of radiance in a beam transmissometer. Factors involved in the design of these instruments, methods of calibration and sources of error will be discussed in the later sections.

DEVELOPMENT OF VOLUME ATTENUATION COEFFICIENT CONCEPTS

The equation of radiance transfer for a beam of light passing through a medium which absorbs but does not scatter may be written as

$$dN = -a\,N\,dr \tag{1}$$

where the radiance N (power $\cdot$ length^{-2} $\cdot$ solid angle^{-1}) changes by an amount dN on passing through a lamina dr of the medium. Equation (1) states that:

(a) the change of radiance dN is proportional to the magnitude N of the incident radiance and the length dr of the path through the lamina, and

(b) the radiance decreases in the direction of propagation, i.e., the negative sign indicates a loss of radiance.

The proportionality coefficient, a, is by definition the absorption coefficient and has the dimension of length^{-1}. Similarly, if only scattering and no absorption exists, then

$$dN = s\,N\,dr \tag{2}$$

where s is defined as the scattering coefficient. If both scattering and absorption are acting together to attenuate the radiance, which is the general case, then since the medium is linear the total change may be obtained by summing the two loss terms. Thus,

$$dN = -(a + s) N dr \qquad (3)$$

and since $a \equiv a + s$, we see that

$$\frac{dN}{dr} = -aN \quad . \qquad (3a)$$

The above assumes that there is no ambient light field in the medium. In the general case where ambient light interacts with the scatterers in the medium, a path radiance term $N_* dr$ must be added to account for the addition of radiance scattered into the direction $\vec{r}$. Thus, the more general relationship is

$$\frac{dN}{dr} = -aN + N_* \qquad (3b)$$

where N_* is the path radiance in the direction $\vec{r}$ and has the units of radiance per unit length or power · volume^{-1} · solid angle^{-1}. It will be assumed in what follows that the measurement path is suitably shielded from the effects of the surrounding light field so that $N_* \ll (a + s) N$, and we may neglect the contribution from N_*. This is an important consideration in the design of transmissometers, but it is not a difficult condition to meet.

Equation (3a) may be solved for the radiance N_r remaining after propagating an initial radiance N_o along a path of length r. Thus,

$$N_r = N_o e^{-ar} \quad . \qquad (4)$$

Observing that the transmittance T_a, of the radiance beam over the path of length r is

$$T_a = \frac{N_r}{N_o} = e^{-ar} \, , \qquad (5)$$

we find that the value of a may be obtained from a transmittance measurement by taking the natural logarithm of both sides of Eq. (5). Thus,

$$a = -\frac{1}{r} \ell n \, T_a \quad . \qquad (6)$$

The loss in radiance per unit input radiance which occurs in traversing a path of length r is defined as the attenuance of the path which we will denote by the symbol A. Thus,

$$A \equiv \frac{N_o - N_r}{N_o} = \frac{\Delta N_\alpha}{N_o} \qquad (7)$$

and from Eq. (5) and Eq. (7)

$$A = 1 - e^{-ar} = 1 - T_\alpha \ . \qquad (8)$$

By recasting Eqs. (7) and (3a) we find a similarity in form, i.e.,

$$\frac{\Delta N_\alpha}{N} = A \qquad (7a)$$

and

$$\frac{\Delta N_\alpha}{N} = -a\Delta r \ . \qquad (3c)$$

Equation (3c) is only valid when $a\Delta r \ll 1$ and the definitions of ΔN_α are such that in Eq. (7a), ΔN_α is the flux lost in traveling a distance Δr and in Eq. (3c), ΔN_α is defined as the flux gained in Δr. Thus, taking this difference in signs into account, we can equate Eqs. (7a) and (3c) and find

$$A = a\Delta r \qquad (9)$$

and as $a = a + s$,

$$A = a\Delta r + s\Delta r \ . \qquad (10)$$

The validity of Eqs. (9) and (10) is again restricted to the situation where $a\Delta r \ll 1$.

There are two analogous terms to attenuance which can be established and these are the absorptance $\mathcal{Q}$ and the scatterance S. The basic relationships are

$$\frac{\Delta N_a}{N} = \mathcal{Q} \qquad (11)$$

and

$$\frac{\Delta N_s}{N} = S \ . \qquad (12)$$

By developing Eqs. (1), (2), (11), and (12) in a manner similar to Eqs. (3) and (7), we find

$$\mathcal{C} = a\Delta r \tag{13}$$

and

$$S = s\Delta r \tag{14}$$

which when combined with Eq. (10) yields

$$A = \mathcal{C} + S . \tag{15}$$

This relationship is precisely true only in the limit, as $\Delta r \to 0$, since it was derived from the differential equation of radiance transfer. The general application of Eq. (15) over long paths or when A, $\mathcal{C}$, and S do not meet the requirement that they be very much less than one, leads to incorrect results as will be shown in the following.

It is intuitively useful to look at the processes of absorptance and scatterance from the point of view of photon survival. We may consider that $\mathcal{C}$ and S are the probabilities that a photon will be absorbed or scattered, respectively, in traveling a path r, or

$$P_a \equiv \mathcal{C} = 1 - T_a = 1 - e^{-ar} \tag{16a}$$

and

$$P_s \equiv S = 1 - T_s = 1 - e^{-sr} \tag{16b}$$

where T_a is the transmittance of a path of length r in a medium containing only absorption and T_s the transmittance in a medium containing only scattering. From these two equations we will develop a hierarchy of relationships helpful in understanding the measurement of beam transmittance.

The corresponding probabilities that a photon will not be lost, i.e., the probabilities of photon survival, are

$$\bar{P}_a = 1 - P_a = T_a = e^{-ar} \tag{17a}$$

and

$$\bar{P}_s = 1 - P_s = T_s = e^{-sr} . \tag{17b}$$

In a medium having both absorption and scattering, the probability that a photon will not be lost in traversing a path of length r is the joint probability that it will be neither absorbed nor scattered. If we denote this probability as $\bar{P}_\alpha$, then

$$\bar{P}_\alpha \equiv \bar{P}_a \cdot \bar{P}_s = T_a \cdot T_s = e^{-ar} \cdot e^{-sr} = e^{-(a+s)r} \tag{18}$$

and as $(a + s) = \alpha$,

$$\bar{P}_\alpha = e^{-\alpha r} = T_\alpha . \qquad (19)$$

The combined probability that photons will be lost by either scattering or absorption in traversing the medium will be

$$1 - \bar{P}_\alpha = 1 - \bar{P}_a \cdot \bar{P}_s = 1 - (1 - P_a)(1 - P_s) = P_a + P_s - P_a \cdot P_s \qquad (20)$$

and replacing the probabilities with their equivalents, we see that $1 - T_\alpha = \alpha + S - \alpha \cdot S$ or as $A = 1 - T_\alpha$,

$$A = \alpha + S - \alpha \cdot S . \qquad (21)$$

Comparing Eq. (21) with Eq. (15) we see that the former contains the additional term, $-\alpha \cdot S$, which Eq. (20) shows us to be the joint probability that a photon will be absorbed and scattered. As the photon can be lost only once, invoking both absorptance and scatterance independently without considering the probability of their jointly occurring results in too large a value for attenuance. We also see that as α and S become very small, their product approaches zero and Eq. (15) becomes an approximation to the more complete statement of Eq. (21).

MEASUREMENT AND DESIGN CONSIDERATIONS

There are many ways to measure beam transmittance and determine the volume attenuation coefficient. Some of these are described by Duntley who points out that the various methods, when well conceived and carefully executed, will give identical results within the precision of the measurement, over a wide latitude of water properties. The concern in this writing will be limited to instruments assumed to be portable and intended to be used to make *in situ* measurements of beam transmittance or the volume attenuation coefficient in natural waters.

All *in situ* transmissometers use a source of radiant energy and a photometric detector to determine the transmission of light energy of some spectral bandpass through a known path length of water. If we adhere to the defining equation $N_r = N_o e^{-\alpha r}$, the transmissometer must determine the ratio, N_r/N_o which is the transmittance over the path length r. A primary concern here is that N_r is the light energy which has traveled the path length r and which has not been scattered. In an ideal instrument no light which has been scattered, that is, its direction of travel changed, however slightly, will be allowed to reach the detector. Such an ideal instrument does not exist. All transmissometers will accept light scattered in the forward direction to some degree; consequently, some error results.

In an actual transmissometer system the received flux available to the detector will be proportional to the solid angle of acceptance Ω_R (up to a maximum angle determined by the divergence of the projector beam). Thus, in order that there be a finite flux for the measurement, the receiver must have a finite angular acceptance. To the extent that Ω_R is not zero, the receiver will also be accepting that scattered flux which falls within its field of view. This means that the apparent scatterance S' will be less than the total or true scatterance S for the path by some amount δ which implies an error in the measurement of T_a. Thus, from Eqs. (8) and (21), $T'_a = 1 - α - S' + αS' = 1 - α - (S-δ) + α(S-δ) = 1 - α - S + αS + δ(1 - α)$

or

$$T'_a = T_a + δ(1 - α) ,\tag{23}$$

and

$$ΔT_a = T_a - T'_a = -δT_a\tag{24}$$

where T'_a is the apparent or measured beam transmittance, T_a is the "true" transmittance which would apply to a measurement which did not include flux that had been scattered, and the factor $(1 - α)$ or T_a is the transmittance of the path due to absorption alone.

We can now determine how the error due to the acceptance of flux that has been scattered affects the determination of the volume attenuation coefficient. From Eq. (23) we find

$$\frac{T'_a}{T_a} = 1 + δ \frac{T_a}{T_a} \quad \text{or} \quad \frac{e^{-α'r}}{e^{-αr}} = 1 + δ \frac{e^{-ar}}{e^{-αr}}$$

hence

$$e^{-(α' - α)r} = 1 + δ\, e^{sr} .$$

By using Eq. (16b) and substituting $S - S'$ for $δ$ we obtain

$$e^{-(α' - α)r} = 1 + \frac{S - S'}{1 - S} = \frac{1 - S'}{1 - S} = e^{-(s' - s)r} .\tag{25}$$

where $1 - S' = T'_a = e^{-s'r}$ serves to define the apparent scattering coefficient, s', which is associated with the apparent scatterance S'. Taking the natural logarithm of both sides of Eq. (25) we find that

$$α - α' = s - s' \quad \text{or} \quad Δα = Δs\tag{26}$$

By the above definition, the error in alpha, i.e. $Δα$, is always positive, i.e. $α > α'$, and equal to the difference between the 'true and apparent scattering coefficients, while the error in beam transmittance $ΔT_a$ as expressed in Eq. (24) is less than the error $δ$ in scatterance, by the "absorption transmittance" factor and of opposite sign, i.e. $T_a < T'_a$. The term $Δs$ is the portion of the scattering coefficient, s, which is included in the solid angle of acceptance of the receiver, $Ω_R$. (See Eq. (27), p. 11)

TYPES OF OPTICAL SYSTEMS USED

Limiting the light which reaches the receiver to essentially mono-path light by restricting the acceptance of scattered light by the receiver is the goal of all well-designed transmissometers. Early investigators were not always conscious of this necessity and early transmissometers, as well as some instruments still in use, used a simple design composed of a lamp for a light source and a flat plate photo-voltaic detector.

More modern systems incorporate the use of imaging optics to limit the divergence of the light beam projected from a source and to restrict the field of view (or angular acceptance) of the receiver. Most instruments use a "collimated" optical system. Such a system is illustrated in Fig. 1.

COLLIMATED OPTICAL SYSTEM

In the system shown in Fig. 1 the light source, usually a tungsten filament, is imaged by the condensing lens on the projector field stop which is located at the focal plane of the projector objective lens. The objective lens forms an image of the field stop at infinity and a beam of light is formed which diverges as it leaves the projector. The angular size of this divergence is determined by the size of the field stop and the focal length of the lens. Light will be emitted from every point on the objective into a cone whose half apex angle (See Fig. 2) is:

$$\theta_{\nu} = \frac{\phi_{fs}}{2f_o} \text{ radians,}$$

where ϕ_{fs} is the diameter of the field stop, f_o is the focal length of the objective lens. Assuming that the field stop is completely filled with the image of the source, the maximum power in the beam is

$$P = \frac{\pi^2}{4} \cdot \phi_o^2 \cdot \sin^2 \theta_{\nu} \sum_{\lambda_1}^{\lambda_2} N_\lambda T_\lambda \text{ ,}$$

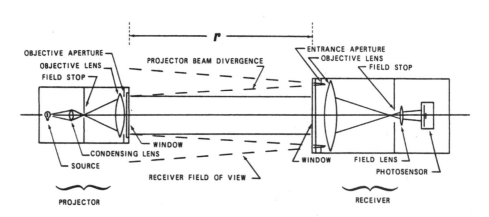

TRANSMISSOMETER COLLIMATED OPTICAL SYSTEM

Figure 1

where ϕ_o is the diameter of the exit aperture of the objective lens, N_λ the inherent radiance of the source, and T_λ the optical transmittance of the system.

The beam leaves the objective lens, passes through a window, and after traveling the measurement path r passes through the window of the receiver and enters the objective lens. The entrance aperture must be large enough to accept all of the light in the beam, allowing for the divergence of the beam plus some clearance. The receiver objective lens forms an image of the projector field stop in its focal plane. A receiver field stop larger in diameter than this image is located in the focal plane. The purpose of this field stop is to restrict the field of view, or acceptance angle, of the receiver. Similarly, the angular acceptance of the receiver is a function of its field stop diameter ϕ_{fs} and the focal length, f_o of the objective lens. Light entering any point on the receiver objective lens will pass through the receiver field stop only if it enters within the angular limits defined by $\theta_{1/2} = 1/2 \ \phi_{fs}/f_o$ for the receiver. To arrive at the detector of the receiver a ray of light must meet two criteria, (1) it must enter the aperture of the receiver objective lens and (2) it must enter at an angle less than $\theta_{1/2}$ of the receiver. The angular acceptance of the receiver should be larger than the divergence of the projector to insure that all of the beam energy will arrive at the receiver detector.

The divergence angle of the projected beam and the acceptance angle of the receiver, i.e., field of view, determined as described, will be the limits of the system in air. These angular limits will be reduced when the instrument is submerged in water. This change is caused by the difference in the index of refraction between air n_a and water n_w. The amount of this change is $\theta w/\theta a = n_a/n_w$ and if n_a is assumed to be 1.000 and n_w to be 1.333, then $\theta w/\theta a = 0.750$.

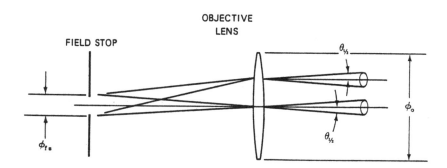

OBJECTIVE
LENS

FIELD STOP

Figure 2

CYLINDRICALLY LIMITED SYSTEM

This type of optical system is somewhat different from a collimated system in that it produces a cylindrically limited beam rather than a diverging collimated beam. In the projection system (Fig. 3), a condensing lens forms an image of the light source in the field stop. The projector's objective lens then images the field stop, in air, at the receiver entrance aperture stop. The projector field stop is of such a diameter that its image is the same size as the projector's objective aperture stop. All rays in the illuminated sample path fall within the cylinder defined by the projector aperture and the image of the field stop.

The receiver optics are similar (Fig. 4). Behind a window is an aperture, a lens, a field stop, and the detector. Again the lens forms an image of the receiver field stop at the aperture of the projector, but this time in water. This is important to insure that all of the projected flux will get through the system both in air and in water. If, for example, the receiver were designed so that the projector aperture was imaged in the plane of the receiver field stop in air, then when the instrument was submerged this image will fall beyond the field stop and be larger. Some of the rays would then be clipped by the receiver field stop and an erroneous transmittance reading will result. The entrance aperture and field stop of the receiver are sized to give the receiver a field of view in water which is a cylinder just slightly larger and encompassing the cylindrical volume illuminated by the projector (Fig. 5).

TRANSMISSOMETER PROJECTOR OPTICS

Figure 3

TRANSMISSOMETER RECEIVER OPTICS

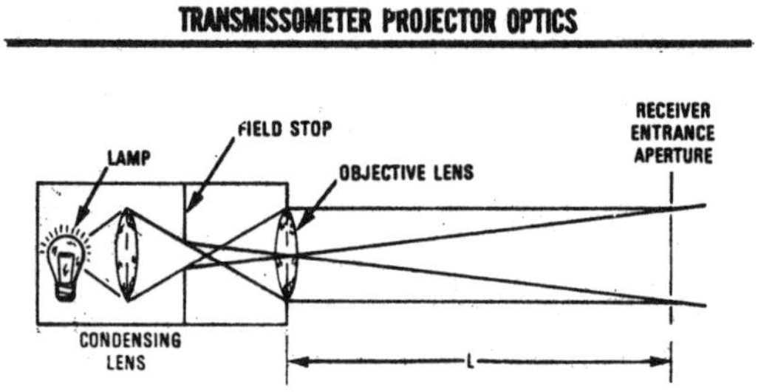

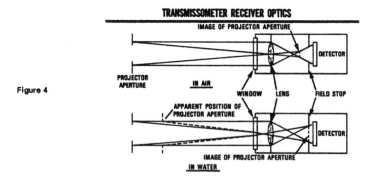

Figure 4

112

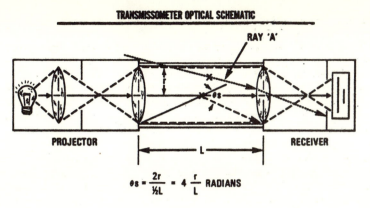

Figure 5

Only those rays which appear to be coming from the projector aperture and enter the receiver aperture will arrive at the detector. Ray "A" which may be a ray of scattered light from the projector beam or ambient light, will enter the receiver lens as if it were coming from outside the projector aperture and will be stopped by the receiver field stop. All rays from the projector are not parallel to the optical axis. The sample volume is traversed by all rays which can be drawn between the area of the projector aperture and receiver aperture. The greatest angle a ray may be deviated and still be accepted by this type of system is indicated by θ_s in Fig. 5. This angle is specified by the ratio of the beam diameter to one half the beam length. This upper limit is a special limiting case. The preponderance of scattered light accepted will have a deviation angle much less than this.

Measured alpha approaches true alpha as the ratio of radius to length (r/L) approaches zero. The radiant flux in the beam is also a function of r/L and is given by the expression:

$$P = \pi^2 \times T \cdot N \frac{r^4}{L^2}$$

where:

P = Radiant flux arriving at the detector

T = Transmittance of optical system including filters

N = Inherent radiance of the source

r = Radius of beam

L = Length of beam

The usable lower limit of the r/L ratio depends upon the inherent brightness of the source and the sensitivity of the detector. Using a very sensitive detector, such as a photomultiplier tube, allows the r/L ratio to be small, and the instrument will give a better α_t measurement in turbid waters.

SOURCES OF ERROR

SCATTERING

The acceptance by the receiver of light which has been scattered from the beam into the forward direction causes the transmission measurement to be high and the computed beam attenuation coefficient to be low as discussed previously. A well-conceived transmissometer is designed to minimize this error. The magnitude of this error depends upon the scattering properties of the water and upon the optical geometry of the transmissometer, i.e., the divergence of the light rays in the projected beam and the angular acceptance allowed by the receiver optics.

Measurements of the scattering properties of ocean waters allow an estimate of this error, $\Delta\alpha$. Thus, as $\Delta\alpha$ is caused by the acceptance of low angle forward scattered light it may be computed from

$$\Delta\alpha = \Delta s = 2\pi \int_0^\theta \sigma(\theta)\,\sin\theta\,d\theta\ (m^{-1})\ , \tag{27}$$

where θ is the angle above which scattered light is not accepted by the system and $\sigma(\theta)$ is the volume scattering function in absolute units, sr^{-1}, m^{-1}. We may now compute the fractional or relative error in transmittance and α by differentiating Eq. (5). Thus,

$$T = e^{-\alpha r},\quad \frac{dT}{d\alpha} = -re^{-\alpha r} = -rT \quad \text{and} \quad \frac{dT}{T} = -r\,d\alpha \quad \text{or} \quad \frac{\Delta T}{T} = -r\Delta\alpha\ .$$

The relative error in the volume attenuation coefficient is $\Delta\alpha/\alpha$. However, we have seen that the error in alpha is due to the error in s, thus

$$\frac{\Delta\alpha}{\alpha} = \frac{\Delta s}{\alpha}\ .$$

Figures 6 and 7 show the magnitude of errors which can occur due to the acceptance of scattered light by a transmissometer system. Figure 6 shows the percent error, due to the acceptance of scattered light, in the transmittance measurement as a function of acceptance angle θ for three greatly different types of water. Figure 7 shows the resultant error in the volume attenuation coefficient α for the same particular waters. The curves indicate clearly the necessity to keep the acceptance of low angle forward scattered light small in making a transmittance measurement.

CALIBRATION METHODS

Three methods for obtaining the basic calibration of transmissometers of known photometric linearity are commonly used.

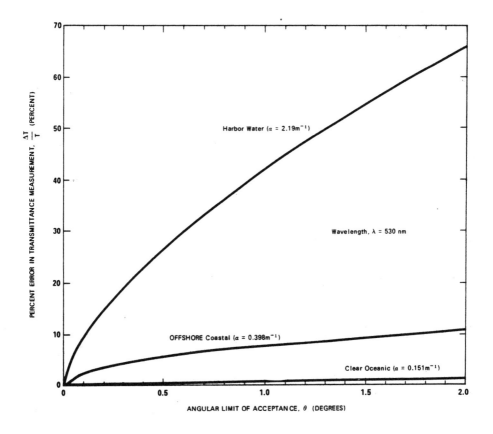

Figure 6

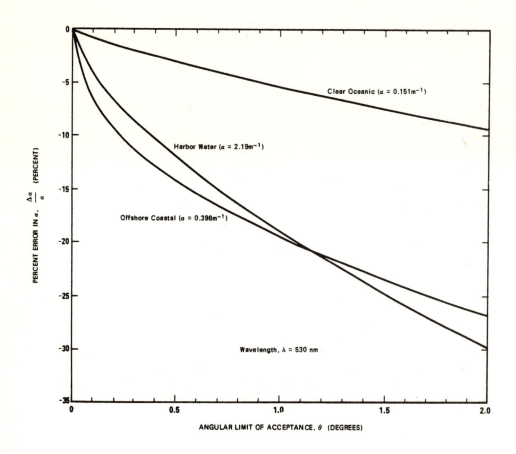

Figure 7

PURE WATER METHOD

This method, although in common use in the past, has drawbacks which make it generally unsatisfactory for the calibration of general field transmissometers. Obtaining or producing optically pure water is extremely difficult. Even if such water were available, placing an instrument into it will usually introduce enough contamination to change its transmittance. Its use is not recommended at the present time for absolute transmittance calibration.

TWO PATH LENGTH METHOD

The transmittance of water can be obtained by measuring the relative radiance N_r transmitted over two path lengths r_2 and r_1. The transmittance over distance $(r_2 - r_1)$ is the ratio

$$T_{r_{1,2}} = \frac{N_{r_2}}{N_{r_1}} \ .$$

The attenuation coefficient is

$$\alpha = - \frac{1}{(r_2 - r_1)} \cdot \ell n \frac{N_{r_2}}{N_{r_1}} \ ,$$

and the transmittance for a unit path length is

$$T_1 = \left(\frac{N_{r_2}}{N_{r_1}} \right)^{1/(r_2 - r_1)} \ .$$

This is a direct measurement of the transmittance. The transmission losses in the optics used are constant and these losses and the gain of the photometer cancel when the ratio,

$$\frac{N_{r_2}}{N_{r_1}} \ ,$$

is taken. The only variable is the difference in transmittance between the two distances. This is a fundamental method of determining the transmittance of water. Water so measured can then be used for the calibration of transmissometers.

AIR CALIBRATION METHOD

Air (relative to water) can be considered to have no transmission losses over the distances normally used to make water measurements.

If air is to be used to calibrate a transmissometer two things must be considered:

(1) All of the rays of light leaving the projector must be accepted by the receiver both in air and water. The divergence of the projector, the field of view of the receiver, and the alignment will all change when the instrument is submerged in water.

(2) The transmission losses through the window surfaces which become the interface with the water will be less in water than in air and this change must be accounted for. The Fresnel reflection loss at a glass-to-air interface is higher than at a glass-to-water interface. If the index of refraction of the glass is known, this loss can be accurately computed.

The reflection as light passes through an interface where a change of index occurs is given by the Fresnel equation which for the case where the light is travelling normal to the interface reduces to:

$$R = \left(\frac{n_2 - n_1}{n_2 + n_1}\right)^2$$

where R is the relative portion of the light reflected at the boundary and n_1 and n_2 are the indices of refraction of the media on the two sides of the interface. A sample calculation will demonstrate the magnitude of this effect. Let

R_a = reflectance in air for one glass-to-air surface

R_w = reflectance in water for one glass-to-water surface

T_a = $1 - R_a$ = transmittance in air for one surface

T_w = $1 - R_w$ = transmittance in water for one surface

n_g = index of refraction of the glass

n_a = index of refraction of air

n_w = index of refraction of water

N = the number of surfaces involved.

Then, for one surface

$$R_a = \left(\frac{n_g - n_a}{n_g + n_a}\right)^2, \; R_w = \left(\frac{n_g - n_w}{n_g + n_w}\right)^2, \text{ and } \frac{T_a}{T_w} = \frac{1 - R_a}{1 - R_w} = \frac{1 - \left(\frac{n_g - n_a}{n_g + n_a}\right)^2}{1 - \left(\frac{n_g - n_w}{n_g + n_w}\right)^2} \, .$$

For N surfaces

$$\left(\frac{T_a}{T_w}\right)^N = \left| \frac{1-\left(\dfrac{n_g - n_a}{n_g + n_a}\right)^2}{1-\left(\dfrac{n_g - n_w}{n_g + n_w}\right)^2} \right|^N .$$

If $n_a = 1.000$ $n_g = 1.516$, and $n_w = 1.340$, then $R_a = 0.04206$ and $R_w = 0.003798$. Consequently, $T_a = 0.9579$ and $T_w = 0.9962$, making the air-to-water transmittance ratio for one surface $T_a/T_w = 0.9616$. Most transmissometers have two or four water interface windows. Thus for two surfaces

$$\left(\frac{T_a}{T_w}\right)^2 = 0.9247 ,$$

and for four surfaces

$$\left(\frac{T_a}{T_w}\right)^4 = 0.8550 .$$

This calculation can be verified by direct measurement of a piece of glass of known index using a transmissometer. To do this one can measure the change in transmittance when the glass is inserted into the beam. This measurement is performed first in air and again with the transmissometer in water, making certain that the sample window is very clean. There will be two surfaces involved (both sides of the window). Thus

$$T_a^2 = \frac{T \text{ (air and glass)}}{T \text{ (air without glass)}}$$

$$T_w^2 = \frac{T \text{ (water and glass)}}{T \text{ (water without glass)}}$$

$$\left(\frac{T_a}{T_w}\right)^2 = \frac{T \text{ (air and glass)}}{T \text{ (air without glass)}} \cdot \frac{T \text{ (water without glass)}}{T \text{ (water and glass)}}$$

When this is done carefully, the calculated and measured values should agree within ±0.1 percent.

If a transmissometer optical system is designed to operate properly both in air and in water the instrument can be calibrated in air by using the computed or measured value of the air-to-water transmittance ratio for the appropriate number of glass surfaces, i.e.,

$$\left(\frac{T_a}{T_w}\right)^N .$$

If the unit has an adjustable gain it may be calibrated by cleaning the windows carefully and adjusting its sensitivity until the indicated transmittance is the required air-to-water transmittance ratio. The transmissometer will then indicate the absolute transmittance of the water in which it is immersed.   This calibration procedure can easily be performed each time the instrument is operated.

6

This article was written expressly for this Benchmark volume.

PRECISION CONSIDERATIONS IN THE MEASUREMENT
OF VOLUME ATTENUATION COEFFICIENT

R. W. Austin

Visibility Laboratory of the
Scripps Institution of Oceanography
University of California, San Diego 92152

The precision in the determination of the volume attenuation coefficient, α, derived from measurements made with a transmissometer having a fixed path length and fixed photometric precision, is a function of the magnitude of the α of the water being measured. Conversely there is an optimum measurement path length which is dependent upon the size of α. It will be shown that this optimum path length, i.e., the one which will minimize the error in α, is in fact equal to the attenuation length, L, where $L = 1/\alpha$. At this optimum point the relative error in α will be $E_{min} = \Delta\alpha/\alpha = -e\Delta T$ where $\Delta\alpha$ is the absolute error in α caused by the photometric error, ΔT, in the measurement of beam transmittance.

The equation

$$T_r = e^{-\alpha r} \tag{1}$$

provides the relationship between the three variables of concern where r is the path length and T_r is the transmittance measured over that path length. By differentiating we obtain

$$d\alpha = -\frac{1}{r}\frac{dT}{T_r} \tag{2a}$$

or in incremental form

$$\Delta\alpha = -\frac{1}{r}\frac{\Delta T}{T_r} . \tag{2b}$$

Thus, $\Delta\alpha$, the absolute precision in the determination of α, will depend directly upon the absolute photometric precision ΔT in the measurement of T_r and will vary inversely with the product of r and T_r. Note also that as r is increased, $T_r = e^{-\alpha r}$ will decrease, and the product for a given α will be a maximum, hence $\Delta\alpha$ will be a minimum, when $r = 1/\alpha$.

121

Further, the fractional or relative precision form of 2(b) is

$$E_r = \frac{\Delta a}{a} = -\frac{1}{ar}\frac{\Delta T}{T_r} \, , \tag{3}$$

and substituting Eq. 1 in Eq. 3 we obtain

$$E = -\frac{e^{ar}}{ar}\Delta T \, . \tag{4}$$

Differentiating E with respect to a and solving for the minimum we find that the optimum path length is the attenuation length, i.e., when $ar = 1$ the fractional error is a minimum and has a magnitude,

$$E_{min} = -e\Delta T \, . \tag{5}$$

In a given transmissometer the path length r and the precision with which T_r can be measured, i.e., ΔT, will be fixed and Δa will vary inversely with T_r. We can take advantage of this fact and plot Δa vs T_r using $\Delta T/r$ as a parameter. We should expect ΔT for an instrument of reasonable design to be around .005 to .01. It is unlikely that ΔT, would be less than .002 under the best of circumstances and we would expect not more than .02 for the worst. The range of path lengths is, of course, arbitrary, but the values of the parameter $\Delta T/r$ which were chosen for the accompanying figure allow for paths of 0.2 to 20 units (meters). A dotted vertical line is shown at $T = .37$, $(ar = 1)$, to recognize the finding of the previous paragraph.

A curve for T_r vs ar is also shown in the same figure as a help in relating T_r to a for various instrument path lengths. Two examples might be helpful in explaining the application of the chart.

Example No. 1. Assuming an instrument with a 1-meter path length and a photometric precision, ΔT, of .005 ($\Delta T/r = 0.005$) what are Δa and $\Delta a/a$. for various a values? Taking values for a of .05, .10, .17, .20, 1.0 and 2.3 we obtain the following:

a	ar	T_1	Δa	$\Delta a/a$
.05	.05	.951	.00526	.105
.10	.10	.905	.00553	.0553
.17	.17	.844	.00593	.0349
.20	.20	.819	.00611	.0305
1.0	1.0	.368	.0136	.0136
2.3	2.3	.10	.499	.0217

Example No. 2. For a 20-meter path length instrument with a ΔT of .004 ($\Delta T/r = 0.0002$) the values for Δa and $\Delta a/a$ when $a = .04, .05, .10, .17,$ and $.20$ are:

a	ar	T_{20}	Δa	$\Delta a/a$
.04	0.8	.449	.00045	.0111
.05	1.0	.368	.00054	.0109
.10	2.0	.135	.00148	.0148
.17	3.4	.0333	.00599	.0353
.20	4.0	.0183	.0109	.0546

(Note that for practical purposes when $ar \leq 0.10$, $T_r = 1.00 - ar$ within 0.005 and it is not necessary to use the $T = e^{-ar}$ curve. Also the range of the figure may be extended by noting the repetitive nature of the Δa vs T_r curves.)

The two examples show clearly that the selection of the optimum path length for a transmissometer to be used for volume attenuation coefficient measurements depends upon the magnitude of the value of a to be measured. In the examples, the 20-meter instrument is clearly better for measurements in water where the a is less than 0.17m^{-1}, i.e., attenuation lengths greater than 6 meters, but in cases where the a exceeds 0.17m^{-1} the balance is in favor of the one meter instrument even though we have assumed a slightly better photometric precision for the 20-meter instrument.

The analytical expression which establishes the criteria for choosing between instruments of path lengths r_1 and r_2 having the same photometric precision is

$$a_c = \frac{1}{r_1 - r_2} \ln \frac{r_1}{r_2} \tag{6}$$

where a_c is the attenuation coefficient at which the measurement precision is equal for the two path lengths.

Listed below are the "crossover attenuation coefficient" values, a_c, and the "crossover transmissivities", T_c, (Transmittance per meter) for several cases of interest. Also listed are the absolute precisions Δa at these a_c values computed for a photometric precision $\Delta T - .005$.

r_1	r_2	a_c	T_c	Δa
1m	.25m	1.85m^{-1}	$.157\text{m}^{-1}$	$.032\text{m}^{-1}$
1	.5	1.39	.250	.020
2	1	.69	.500	.010
3	1	.55	.577	.0087
10	1	.256	.774	.0064
20	1	.158	.854	.0058
10	3	.172	.842	.0028
20	3	.112	.894	.00233
20	10	.0693	.933	.0010

The above concepts are useful in evaluating the errors which may be expected in data obtained with transmissometers of various path lengths where the photometric accuracy is known or may be estimated. The concepts are similarly useful in establishing the design requirements for transmissometers for field application where the range of attenuation coefficients may be anticipated.

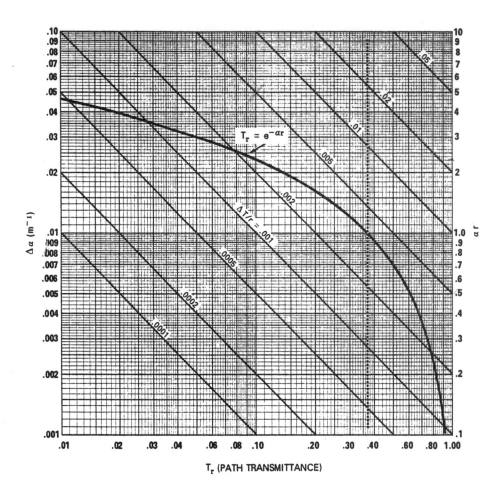

Fig. 1.   Attenuation coefficient error, Δa, shown as a function of path transmittance, T_r, with $\Delta T/r$ as a parameter. Bold line showing $T_r = e^{-ar}$ uses ar scale at right.

7

TRANSMITTANCE, SCATTERING, AND OCEAN COLOR

R. W. Austin

This article was rewritten by the author from Section J of the
Data Report of the SCOR Discoverer Expedition, 1970, Vol. 2,
SIO Reference Report 73-16, J. E. Tyler, ed., Scripps Institution
of Oceanography, 1000 pp.

TRANSMITTANCE MEASUREMENTS

Measurements of the optical beam transmittance as a function of
depth in the water column were obtained at S. C. O. R. Stations 02
through 23 inclusive. These continuous vertical profiles provided an
immediate visual indication of the degree of optical homogeneity of the
water column and more significantly, showed the precise depth where
changes in the transmittance occurred. As changes in the beam transmittance in the open ocean usually correlate with changes in the concentration of biogenous material, these profiles are useful for locating
the depth of potentially phytoplankton-rich water. The transmissometer
used for these measurements was equipped with a temperature sensor and
simultaneous records of temperature and transmittance versus depth were
obtained. From one to seven profiles were obtained on each station with
a total of 105 profiles for the cruise. Measurements were made in one
of five spectral bands.

The vertical profiles were obtained with a transmissometer of the
type described by Petzold and Austin (1968). Briefly, the instrument
has a one-meter optical path length through the water. The light flux is
confined to a right cylinder 19 millimeters in diameter. All the flux
leaving the projector in either air or water is contained within this
cylinder and, if not attenuated, is collected and measured by the receiver. As a result, the instrument may be calibrated by making the proper adjustment to its sensitivity in air. The one meter of air is
assumed to have unit transmittance; however, the flux loss caused by the
reflectance of the glass surfaces in contact with water will increase
when the instrument is in air. Consequently, the air adjustment of instrument sensitivity must take this change into account. As the optical
beam is folded by a porro prism at mid-path, there are four of these

glass-water interfaces. The resultant air calibration setting is 0.855 for the particular glass used.

The transmissometer is equipped with a filter wheel containing 5 Wratten spectral filters. Figure 1 shows the relative spectral response of the optical system obtained by individually combining these filters with the spectral emittance of the 2500° K tungsten-iodide lamp, the spectral sensitivity of the silicon photocell, and the transmittance of the 2 millimeter Schott BG-18 infrared-blocking glass filter used in the instrument. The centroid wavelengths, $\bar{\lambda}$, shown in Fig. 1 are slightly different than those shown on the individual curve sheets due to a recomputation using new filter measurements. The differences are not significant for most purposes considering the half-power bandwidths, $\Delta\lambda$, of the filters.

The transmittance per meter (transmissivity) and temperature outputs were simultaneously recorded against depth on a two pen, flat bed x, y_1, y_2 recorder. Figure 2 shows a typical example of such a recording. Each of the 105 original curve sheets was carefully retraced with fluctuations in transmittance reproduced in their essential detail. These fluctuations are not the result of instrumental noise but indicate the passage of discrete particles through the optical path causing a momentary reduction in transmittance. On various occasions visual inspection of the water surrounding the instrument either by in situ observation or by obtaining a sample from the depth in question, has shown a large number of zooplankton to be present when such fluctuations are seen.

On the left margin of the original curve sheets (0-meters depth) a mark was placed at 85.5%/meter showing the air calibration setting prior to immersion (See Fig. 2). At 79%/meter the internal reference ("REF") reading for that filter is shown. A mark at 0% per meter shows that the instrument zero was in proper adjustment. At the right hand margin (100-meters depth) marks at 79% and 0% per meter show that no zero shift or change in sensitivity occurred during lowering. The accuracy of the transmissivity profiles was better than ±1% for the 491, 535, and 581 nanometer measurements. Due to a reduction in the sensitivity of the instrument at the shorter wavelengths, there was a greater amount of gain and zero drift present in the instrument receiver at 473 and especially at 450 nanometers. As a result, the accuracy of the uncorrected trans- missivity profiles for these wavelengths was somewhat less. On most of the profiles, however, corrections were made to compensate for these drifts when tracing the original data sheets.

The temperature trace is useful for showing the location of thermo- clines, inversions, etc. The absolute temperature accuracy was about ±0.5°C, and the time constant of the thermistor sensor was 3.5 seconds. In water having a pronounced thermocline the rate of instrument descent was reduced in order to lessen the dynamic error in indicated tempera- ture which this time constant introduced.

The temperature trace is identified on each sheet. In all cases it is the smoother of the two curves. Each curve is identified with the local apparent time (L. A. T.) of the lowering, the latitude and longi- tude of the station, the water depth in fathoms as provided by the ship, the centroid wavelength for the measurement and the surface water tem- perature as determined by a bucket thermometer.

Three of the retraced curves have been selected for reproduction here as representative examples of the transmissivity and temperature profiles obtained on the SCOR DISCOVERER expedition. The first curve (Fig. 3) shows the high transmittance, uniformly mixed water typical of the low productivities found in areas such as the Sargasso Sea. The second curve (Fig. 4) shows evidence of large amounts of particulate material dispersed in the vertical water column. It is postulated that zooplankton and/or fish larvae were responsible for the observed per- turbations of the transmittance record. The third curve (Fig. 5) was obtained in the Gulf of Panama in highly productive water with a sur- face mixed layer and a marked thermocline starting at about 15 meters depth. As is frequently observed in such situations, the transmittance is essentially constant in the surface mixed layer with a sharp drop in transmittance in the thermocline indicating the presence of a layer of turbid water. Below this the transmittance slowly increases until re- latively clear water is reached at a depth of 70 meters. It is interest- ing to note that the profile of chlorophyll "a" concentration for this and most other stations reflects the same features shown in the trans- mittance profile. In fact, Kiefer and Austin (1974) have shown that the volume attenuation coefficient as derived from the beam transmittance (after subtracting the attenuation coefficient for pure water) is highly correlated with the chlorophyll "a" concentration.

REFERENCES

Petzold, T. J. and R. W. Austin, 1968, "An Underwater Transmissometer for Ocean Survey Work." S.I.O. Ref. 68-9, Scripps Institution of Oceanography.

Kiefer, D. A. and R. W. Austin, 1974, "The Effect of Varying Phytoplank- ton Concentration on Submarine Light Transmission in the Gulf of California." Limnol. Oceanogr., 19, pp. 55-64.

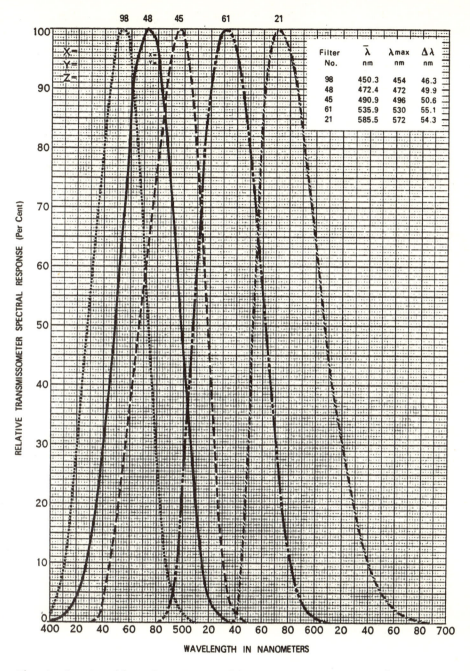

Fig. 1 Reprinted from *Data Report of the SCOR Discoverer Expedition*, p. J-4.

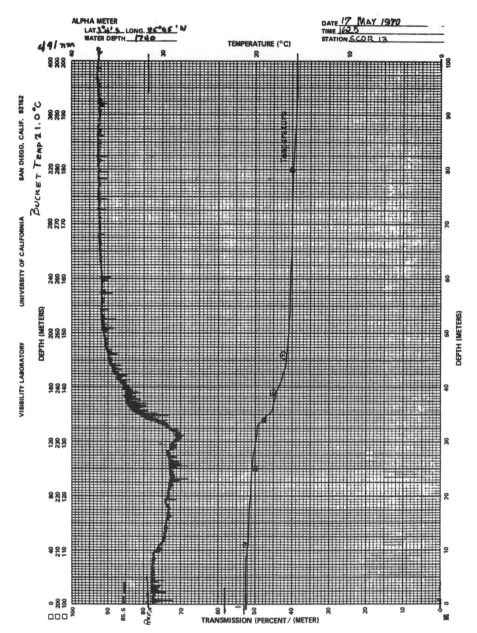

Fig. 2   Reprinted from *Data Report of the SCOR Discoverer Expedition*, p. J-6.

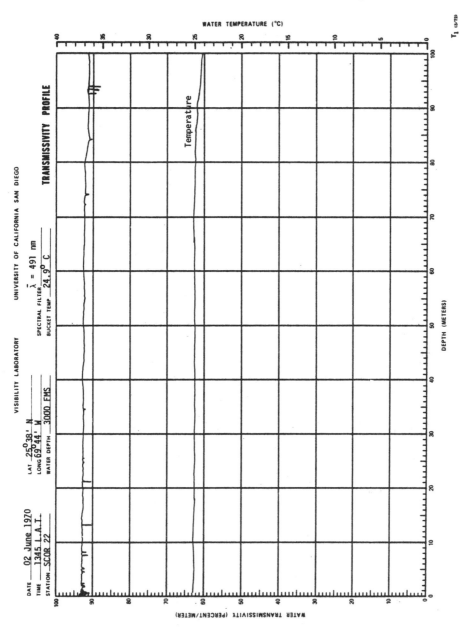

Fig. 3 Reprinted from *Data Report of the SCOR Discoverer Expedition*, p. J-106.

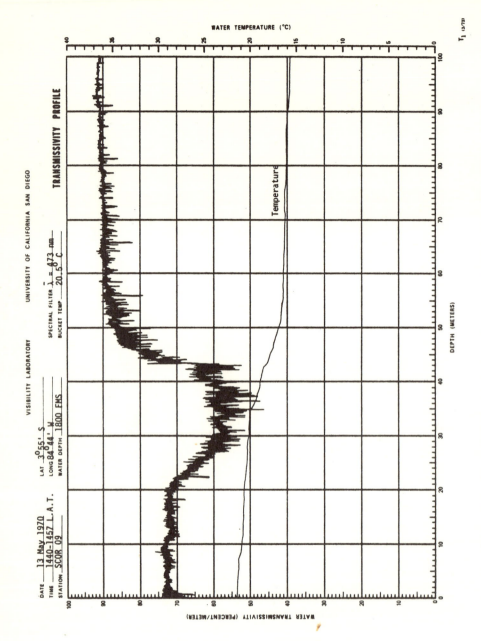

Fig. 4 Reprinted from *Data Report of the SCOR Discoverer Expedition*, p. J-43.

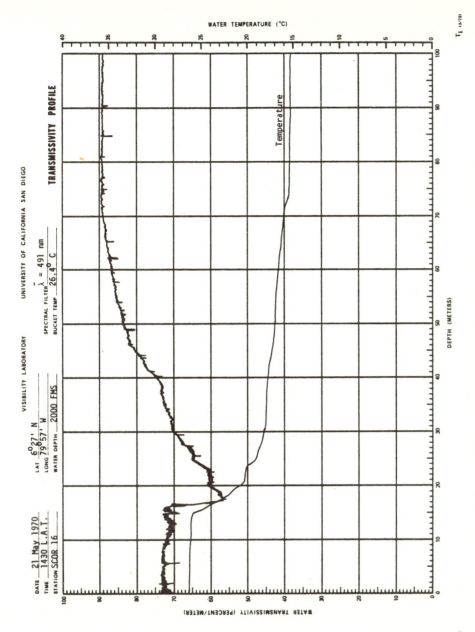

Fig. 5 Reprinted from *Data Report of the SCOR Discoverer Expedition*, p. J-75.

8

Reprinted from pp. 698–699, 700 of *Opt. Soc. Am. J.* **35**(11):698–705 (1945)

Optics of Distilled and Natural Water

E. O. Hulburt

Naval Research Laboratory, Washington, D. C.

(Received September 6, 1945)

INTRODUCTION

THE sea viewed in the daytime by an observer looking downward from a ship or an airplane has a certain color and brightness which, apart from the surface reflection, is caused by the portion of the daylight that has penetrated into the sea and has returned upward. The character and intensity of the up-welling light depends on the optical properties of the sea water. It is the purpose here to describe laboratory measurements of the absorption and scattering of distilled water and of the water of the Chesapeake Bay and the Atlantic Ocean, and to relate these through theory with observations of the Bay and the Ocean.

MEASUREMENTS IN THE LABORATORY

Apparatus was arranged for measuring the attenuation and scattering of samples of water for light of the visible spectrum. For the attenuation a glass tube was used 364 cm (12 feet) in length and 4.2 cm in internal diameter, closed with glass end-plates that were flat and perpendicular to the axis of the tube. A beam of light from a tungsten lamp, rendered parallel by a lens, passed through the tube being limited by diaphragms to a circular cross section of 2.5 cm so that it did not touch the walls of the tube. The beam passed through a Nutting photometer and came to a focus on the slit of a glass prism spectrograph. The long glass tube could be moved out of the beam and replaced by a short tube 20 cm in length, both tubes being filled with water from the same sample. In this way the attenuation κ of the water was determined throughout the visible spectrum from 420 to $700\mu\mu$.

κ is defined by

$$i = i_0 e^{-\kappa x}, \tag{1}$$

where i and i_0 are the respective intensities of a parallel beam of light of wave-length λ that enters and emerges from a thickness x cm of the water. Now

$$\kappa = \beta + \sigma, \tag{2}$$

where β and σ are the attenuations due to true absorption and to scattering, respectively.

In order to measure the scattering σ the water was placed in a spherical flask of 8 cm radius. A parallel beam of tungsten light was passed through the center of the flask and the brightness of the beam at various angles ϕ to the forward direction of the beam was measured with a Macbeth illuminometer. The flask was removed and replaced by a diffusing white plaque normal to the beam and in the position previously occupied by the center of the flask. The brightness of the white plaque was measured. From these measurements the values of σ_ϕ were determined, where σ_ϕ is the fraction of the light of the beam that was scattered per unit volume of water per unit solid angle in the direction ϕ.

The attenuation σ of parallel, or collimated, light due to scattering is given by

$$\sigma = 2\pi \int_0^\pi \sigma_\phi \sin \phi \, d\phi. \tag{3}$$

Of the light that is scattered from the parallel beam portions σ_F and σ_B are scattered in the forward and backward directions, respectively, where

$$\sigma = \sigma_F + \sigma_B, \tag{4}$$

$$\sigma_F = 2\pi \int_0^{\pi/2} \sigma_\phi \sin \phi \, d\phi, \tag{5}$$

$$\sigma_B = 2\pi \int_{\pi/2}^\pi \sigma_\phi \sin \phi \, d\phi. \tag{6}$$

Quantities of interest here are η and $1-\eta$, the fractions of the light scattered in the forward and backward directions, respectively. Then

$$\eta = \sigma_F/\sigma, \tag{7}$$
$$1-\eta = \sigma_B/\sigma$$

If the values of β, σ, and η are known throughout the visible spectrum for a sample of water the upcoming daylight from a uniform sea of such water may be calculated.

Distilled water was investigated first, because earlier measurements of its optical properties were not in complete agreement, and a knowledge of pure water was important to an understanding of natural waters. The preparation of approximately chemically pure water is relatively simple, but to prepare it free from optical impurities, that is, "dust free," is well known to be difficult. A good way is by slow distillation in quartz. There was no opportunity to do this in the case of the considerable quantity of water required in the present experiments. The procedure used was to allow distilled water to stand for several weeks in five-gallon bottles. The upper two-thirds of the water in a bottle was siphoned directly into the absorption tube and scattering flask for rinsing and finally for filling. Motes were visible in the settled-out distilled water in the tube and flask, especially when viewed by forward scattering, but after standing for several days many, but not all, settled out. After ten days $\sigma_{\pi/2}$, the scattering coefficient at right angles for tungsten light, diminished to

$$\sigma_{\pi/2} = 1.44 \times 10^{-6},$$

and did not decrease any more with further standing in the flask.

The theoretical value for water for tungsten light viewed with the light adapted eye is

$$\sigma_{\pi/2} = 1.21 \times 10^{-6},$$

as calculated from the density fluctuation expression with polarization defect.[1]

[Editor's Note: Material has been omitted at this point.]

The values of $\kappa = \beta + \sigma$ for distilled water were measured throughout the visible spectrum. They are listed in Table I, Column 2, and are plotted in Fig. 1 in which for clarity the scale of ordinate for $\lambda < 580\mu\mu$ is 10 times the scale for $\lambda > 580\mu\mu$. By reflecting the beam of light back through the 12-foot absorption tube, so that the beam traversed 24 feet of distilled water, the shallow minimum of absorption in the blue green could be seen visually to lie in the region 460 to 470$\mu\mu$.

[1] L. H. Dawson and E. O. Hulburt, J. Opt. Soc. Am. 31, 554–557 (1941).

TABLE I. Absorption and scattering of distilled water and Chesapeake Bay.

	Distilled water			Chesapeake Bay		plankton
λ	κ	σ	β	κ	β	β
400$\mu\mu$	8.0×10^{-4}	3.57×10^{-4}	4.4×10^{-4}	—	—	—
410	7.0	3.23	3.8	—	—	—
420	6.1	2.95	3.1	80.5×10^{-4}	62.5×10^{-4}	60×10^{-4}
430	5.3	2.69	2.6	70.5	52.5	49
440	4.6	2.45	2.1	62.8	44.8	43
450	4.0	2.25	1.7	55.6	37.6	36
460	3.6	2.04	1.6	51.2	33.2	31
470	3.6	1.89	1.7	47.5	29.5	28
480	3.65	1.72	1.8	44.7	26.7	25
490	3.7	1.59	2.1	41.9	23.9	22
500	3.8	1.47	2.3	38.8	20.8	18
510	3.9	1.35	2.6	36.8	18.8	16
520	4.0	1.25	2.8	35.1	17.1	14
530	4.2	1.17	3.0	33.7	15.7	13
540	4.4	1.09	3.3	33.1	15.1	12
550	4.7	1.00	3.7	32.3	14.3	11
560	5.3	.932	4.4	32.3	14.3	10
570	6.6	.868	5.7	32.3	14.3	9
580	8.4	.810	8.6	33.1	15.1	8
590	12.0	.756	11.2	36.5	18.5	7
600	19.7	.708	19.0	42.9	24.9	6
610	24.3	.662	23.6	46.5	28.5	5
620	26.5	.618	25.9	47.6	29.6	4
630	28.0	.581	27.4	48.8	30.8	4
640	29.2	.535	28.7	50.0	32.0	4
650	30.8	.507	30.3	51.8	33.8	4
660	33.5	.483	33.0	54.2	36.2	4
670	37.5	.455	37.0	56.3	38.3	2
680	40.6	.429	40.2	58.9	40.9	1
690	46.7	.404	46.3	63.	45.	0
700	57.6	.380	57.2	74.	56.	0

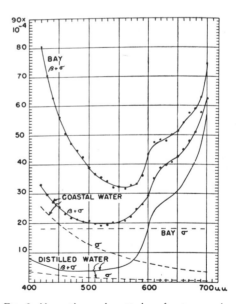

FIG. 2. Absorption and scattering of water samples.

[Editor's Note: Material has been omitted at this point.]

9

Reprinted from Opt. Soc. Am. J. **53**(8):962–968 (1963)

Experimental Study of the Absorption in Distilled Water, Artificial Sea Water, and Heavy Water in the Visible Region of the Spectrum*

Seraphin A. Sullivan†

The Catholic University of America, Washington, D. C.

(Received 19 February 1963)

INTRODUCTION

THE transmission of light through water has been the subject of many investigations, as is borne out in the literature.[1,2] Several studies made of the absorption of light in liquid water in the visible region of the spectrum show some variation in the results reported.[3] In the near infrared region, Curcio and Petty[4] reported five prominent absorption bands between 0.7 and 2.5 μ.

The object of the present experiment is to investigate the absorption spectrum of distilled water in the visible region from 790–580 mμ, with the possibility of uncovering similar bands which represent the higher harmonics of the fundamental vibrational frequency. At the same time, the experiment gives an opportunity for looking at the spectrum with greater resolution than previously reported. Finally, the experiment affords a means of comparing the absorption of distilled water, artificial sea water, and heavy water in parts of the visible spectrum.

EXPERIMENTAL DETAILS

Apparatus

The dispersing element used was a single-beam single-pass prism spectrometer. It was essentially the same high-resolution spectrometer designed by Nielsen et al.[5] for measurements in the infrared. It was adapted for use in the visible region by substituting a dense flint glass 60° prism with a 12.5-cm base in the Wadsworth-Littrow mounting. Suitable masks and baffles were used to screen off unused optical surfaces which might reflect stray light. Further adaptations were made in the scanning mechanism of the spectrometer so that in the region observed it was possible to scan at speeds not faster than 3 mμ per min.

Lines from mercury, neon, and argon spectra were used to calibrate the wavelength scale of the spectrometer. The wavelength scale was then related to the readings on a Veeder counter on the wavelength drive by means of a dispersion curve. The settings of the wavelength drive were observed to be reproducible within ± 0.2 mμ in the red end of the spectrum, the reproducibility improving with decreasing wavelength. Spectrometer bandwidths varied from 0.5 mμ in the far red to 0.1 mμ in the violet.

A tungsten strip-filament lamp was used as a source. This 18-A 6-V dc lamp was powered by three heavy-duty storage batteries to eliminate line fluctuations. A blower was kept on the lamp housing during the experiment to achieve temperature stability. To insure a constant intensity during the experiment, a current-monitoring device was used. By means of a suitable shunt and a galvanometer circuit, it was possible to detect changes of 0.1% in the lamp current. By using a variable resistor in series with the lamp, the current could be maintained within this limit. It was found that with this control, the recorder signal of the lamp intensity was stable to within 1%.

The light from the source was focused on a 1-mm pinhole in an opaque screen. The emitted light was then rendered parallel by a system of achromatic lenses. The circular beam was stopped down to a diameter of 22 mm to ensure that the beam would not touch the walls of the absorption cell. The light was chopped at 450 cps before entering the slit of the spectrometer. A reference signal from the chopper, generated by a magnetic pickup, was fed into a phase-comparing amplifier.

After passing through the spectrometer, the radiation was detected by an RCA 1-P21 photomultiplier tube placed at the exit slit of the spectrometer. The signal was then fed into a Baird-Atomic 450-cycle phase-comparator amplifier, rectified, and sent through a Leeds and Northrup strip-chart recorder. Prior to tak-

* A dissertation submitted to the Faculty of the Graduate School of Arts and Sciences of the Catholic University of America in partial fulfillment of the requirements for the degree of Doctor of Philosophy. This project was supported in part by the Geophysics Research Directorate, U. S. Air Force Cambridge Research Center.

† Present address: Saint Bonaventure University, Saint Bonaventure, New York.

[1] N. E. Dorsey, *Properties of Ordinary Water Substance* (Reinhold Publishing Corporation, New York, 1940), 341 ff.

[2] E. F. Dupre and L. H. Dawson, *Transmission of Light in Water: An Annotated Bibliography* (U. S. Naval Research Laboratory, Washington, D. C., 1961).

[3] E. O. Hulburt, J. Opt. Soc. Am. **35**, 698 (1945).

[4] J. A. Curcio and C. C. Petty, J. Opt. Soc. Am. **41**, 302 (1951).

[5] J. R. Nielsen, F. W. Crawford, and D. C. Smith, J. Opt. Soc. Am. **37**, 246 (1947).

ing the data, it had been ascertained under what conditions the response of the system was linear.

To hold the liquid samples, two cylindrical cells were constructed of nonflexible clear cast Acrylic plastic, one 60.0 cm and the other 132.0 cm in length. The ends of these tubes were milled and caps were made to hold 2-mm-thick circular Pyrex windows. Neoprene O rings were used as seals. The caps were fitted with set screws placed 120° apart on the circumference of the windows to allow adjustments to be made for the parallel settings of the windows.

Preparation of the Samples

The distilled water used in the experiment was triply distilled over quartz. To maintain the best possible optical purity and to remove as many dust particles as possible, it was decided to filter the water just before taking the data. Pyrex fritted-glass filter cylinders of ultrafine porosity were used; the average pore size of these filters was 0.9–$1.4\,\mu$. As an indication of the optical purity, the transmission of the sample in the 60.0-cm tube at the wavelength 470 mμ was noted. After filtration, it was found that a transmission of $88.5\%\pm1.0\%$ could be maintained; this figure includes the reflection losses at the air–glass and glass–water surfaces. After standing for 24 h, the same sample was found to have a transmission of 80% or less at the same wavelength.

The artificial sea water was made by combining chemicals with triply distilled water. The composition was such as to approximate sea water and the sample contained about 0.60 equivalents of positive and negative ions per liter.[6] The following chemicals were used: NaCl, $MgSO_4$, $MgCl_2$, $NaHCO_3$, $CaCl_2$, KCl, and KBr. The water was filtered immediately after being made up. Its specific gravity was measured and found to be 1.025 ± 0.002.

The heavy water was composed of 99.7% deuterium oxide. Its density at 20° was 1.1057 ± 0.0001 g/cc. This density showed no measurable change after filtration.

Procedure

In taking the data for the various samples, a uniform procedure was followed. Before placing the cell in the optical path, a signal trace of the source spectrum was recorded. Care was taken to mount the cell the same way each time in the optical path. In order to ensure uniformity in this and to avoid any displacement of the image due to the insertion of the cell, a system consisting of an optical lever and a projecting lens was used to check the image position. With the cell out of the mount, an image of the pinhole magnified approximately 25 times was projected on a screen and its position marked; the cell was then mounted and position of the image again checked. At no time was the image displaced by

more than 4% of its diameter; in all cases it was adjusted, if necessary, by means of the set screws on the cell windows. This ensured reproducibility in mounting the cell.

Signal traces of the transmission of both empty cells were recorded to measure the air–glass reflection losses of the windows.

For all three types of water, at least three sets of data were taken. Prior to the observations, each sample was filtered in such a way that the filtrate flowed down the side of the flask, thus avoiding the formation of bubbles. The sample was then carefully siphoned into the absorption cell. All samples were tested within 3 h after filtration. For any longer delay, the water was not considered safe from some deterioration.

The source spectrum was recorded before the cell was mounted; this was repeated after the data was taken with the cell mounted. It was found that the source spectrum was reproducible to within 1%. Throughout the tests, the current-monitoring galvanometer was checked at frequent intervals, and the variable resistor adjusted accordingly. The room temperature was checked and did not vary by more than 3°C over all the tests. The temperature of the sample was read before and after each observation; the mean temperature over all the observations was $23.0°\pm1.5°$C. In taking the data, the spectrum was broken up into several regions. The 60.0-cm cell was used for all three types of water in the region 790–695 mμ. The 132.0-cm cell was used for the distilled water and the artificial sea water from 720–580 mμ, and for the heavy water from 750–390 mμ. To minimize the effect of stray light, optical filters were inserted in the optical path. For the distilled water and the artificial sea water, Corning red filters 2030, 2404, and 2424 were used in the regions 790–695, 720–635, and 635–580 mμ, respectively; a Corning blue filter, 5543, was used in the region 450–400 mμ. For the heavy water, red filters 2030 and 2404 were used in the regions 790–695 and 720–640 mμ; no filter was used from 640–478 mμ; a blue filter, 5543, was used in the region 480–390 mμ.

CALCULATIONS AND RESULTS

The absorption coefficient, α, is defined by the equation

$$I = I_0 e^{-\alpha x}, \tag{1}$$

where I_0 and I are the intensities of the radiation, at a given wavelength, before and after passing through a path of length x through the sample. Since in this experiment no reference cell was used, corrections had to be made for the reflection losses at the air–glass and glass–water surfaces.

For this purpose we may write two similar forms of Eq. (1),

$$I = I_0 e^{-\alpha_t z} \tag{2}$$

[6] *Smithsonian Physical Tables* (Smithsonian Institution, Washington, D. C., 1954), 9th revised ed., p. 776.

TABLE I. Absorption coefficient α for distilled water, artificial sea water, and heavy water
in the region 790–580 mμ ($\times 10^4$ cm^{-1}).

Wavelength (mμ)	Distilled water	Artificial sea water	Heavy water	Wavelength (mμ)	Distilled water	Artificial sea water	Heavy water
790	205.2	209.7	7.4	715	111.4	110.7	4.9
8	207.0	210.6	7.2	4	106.2	105.2	4.9
6	210.2	213.7	7.0	3	102.6	102.2	5.1
4	209.7	216.4	7.5	2	97.9	98.1	5.3
2	217.7	224.3	7.2	1	93.8	94.9	5.2
780	234.6	237.6	6.7	710	90.4	90.8	5.1
8	237.5	241.4	6.4	9	87.1	86.4	5.3
6	240.3	243.8	6.9	8	83.6	83.9	5.1
4	246.8	246.8	6.4	7	80.4	80.2	4.9
2	248.5	250.7	6.6	6	77.9	77.0	4.9
770	251.9	254.0	6.5	705	75.0	74.5	4.7
8	252.5	256.7	6.3	4	72.7	73.3	4.7
6	261.5	261.0	6.8	3	70.4	70.7	4.7
4	263.3	270.1	6.8	2	68.3	68.8	4.6
2	265.3	271.7	6.8	1	66.5	66.4	4.5
760	268.5	272.7	6.9	700	64.8	64.7	4.5
9	268.8	273.7	6.4	9	63.2	63.3	4.3
8	269.2	273.3	6.8	8	61.5	61.8	4.3
7	268.5	273.1	6.7	7	59.6	60.4	4.5
6	268.8	272.4	6.9	6	58.1	58.5	4.4
755	269.6	271.4	6.3	695	56.9	57.0	4.3
4	270.9	270.5	6.4	4	55.5	55.2	4.1
3	267.9	271.8	6.6	3	54.3	54.2	3.7
2	268.3	270.5	6.7	2	52.9	53.5	3.8
1	268.7	271.0	6.9	1	51.8	51.2	3.4
750	268.3	269.8	6.9	690	51.1	50.1	3.3
9	270.3	271.2	6.7	9	50.2	49.4	3.5
8	270.5	273.3	6.9	8	49.1	48.6	3.5
7	270.0	274.0	6.9	7	48.0	47.9	3.5
6	272.3	276.3	6.6	6	47.7	47.9	3.5
745	271.2	275.3	6.4	685	46.7	47.4	3.5
4	271.3	275.7	6.5	4	45.8	47.3	3.3
3	271.1	275.1	6.2	3	45.5	46.8	3.3
2	270.4	274.2	6.7	2	45.2	46.6	3.3
1	271.8	271.8	6.3	1	45.0	46.0	3.3
740	269.8	272.9	6.7	680	44.7	45.0	3.2
9	269.7	269.2	6.7	9	44.5	44.9	3.4
8	268.6	269.3	6.4	8	44.4	44.3	3.5
7	265.7	266.4	5.9	7	44.0	44.0	3.4
6	264.3	264.7	5.9	6	43.9	43.7	3.3
735	260.8	260.0	5.9	675	43.8	43.2	3.3
4	255.3	255.5	5.5	4	42.8	43.1	3.3
3	249.2	248.3	5.6	3	43.1	43.1	3.4
2	244.7	245.3	6.1	2	42.9	43.0	3.3
1	239.5	239.8	5.8	1	42.5	43.1	3.6
730	230.9	230.8	5.9	670	42.5	42.2	3.5
9	221.1	219.1	6.0	9	42.5	42.2	3.4
8	210.1	210.4	5.7	8	42.2	42.3	3.5
7	197.3	194.6	6.0	7	42.0	42.1	3.4
6	186.3	186.5	6.1	6	41.9	42.4	3.7
725	175.6	175.9	6.0	665	41.9	41.9	3.4
4	167.6	166.4	5.5	4	41.7	41.6	3.1
3	159.0	159.7	5.4	3	41.5	41.4	3.3
2	151.2	150.2	5.4	2	41.1	41.1	3.4
1	144.7	144.6	5.4	1	41.0	40.6	3.3
720	137.5	137.2	5.4	660	40.7	40.5	2.9
9	132.5	133.4	5.4	9	40.3	40.3	3.0
8	127.4	127.0	5.3	8	39.5	40.6	3.0
7	122.1	121.7	5.5	7	39.2	40.7	2.7
6	116.8	117.3	5.4	6	39.3	39.9	3.2

TABLE I (continued)

Wavelength (mμ)	Distilled water	Artificial sea water	Heavy water	Wavelength (mμ)	Distilled water	Artificial sea water	Heavy water
655	38.7	38.9	3.1	615	30.0	30.1	
4	37.6	38.4	3.0	4	29.8	29.9	3.6
3	36.5	37.3	3.0	3	29.4	30.2	
2	35.9	37.0	2.7	2	29.6	30.0	4.3
1	36.0	35.7	3.1	1	29.8	30.1	
650	35.1	35.7	3.4				
9	34.7	35.2	3.2	610	29.3	29.9	4.0
8	34.3	34.2	3.2	9	29.4	29.7	
7	34.4	34.1	3.2	8	29.6	29.7	3.7
6	34.3	33.5	3.2	7	29.4	29.7	
				6	29.5	29.6	4.0
645	33.9	33.6	3.2				
4	33.9	32.9	3.3	605	29.3	28.9	
3	33.7	32.8	3.0	4	29.0	29.0	4.1
2	33.7	32.9	3.3	3	28.9	29.0	
1	33.6	32.6	3.1	2	28.4	28.7	4.3
640	33.4	32.6	3.2	1	28.0	28.1	
9	33.3	32.5					
8	32.6	32.3	3.3	600	27.2	26.5	4.2
7	33.1	32.3		9	26.1	24.6	
6	32.6	32.9	3.3	8	24.6	24.0	4.1
635	32.5	32.0		7	23.5	23.2	
4	32.3	32.1	3.4	6	22.2	21.7	4.1
3	32.6	32.0					
2	32.1	31.9	3.4	595	21.4	20.8	
1	32.1	31.8		4	20.1	19.8	3.8
630	32.0	31.2	3.3	3	19.0	18.8	
9	31.3	31.7		2	18.5	18.1	4.7
8	31.2	31.2	3.8	1	17.6	17.5	
7	31.0	31.7					
6	31.2	31.7	3.8	590	17.2	16.7	4.4
625	30.5	31.7		9	16.0	16.3	
4	30.9	31.4	3.8	8	15.2	16.3	4.6
3	31.0	31.5		7	14.3	15.7	
2	30.7	31.5	4.1	6	13.0	15.6	4.6
1	30.7	31.1					
620	30.9	31.7	4.0	585	11.9	14.6	
9	30.6	31.0		4	12.2	15.1	4.9
8	30.4	30.7	4.0	3	12.0	14.1	
7	30.0	30.8		2	11.1	14.7	4.8
6	30.3	29.9	3.8	1	10.2	13.1	
				580	10.9	13.3	4.9

and

$$I = RI_0 e^{-\alpha x}. \tag{3}$$

In Eq. (2), I_0 is the intensity of the radiation at a given wavelength with no cell in the path; I is the intensity with the cell in the path, and α_t is a total attenuation coefficient due to the absorption of the water plus the losses due to the interface reflections. In Eq. (3), I_0 and I are again the intensities with the cell out of and in position. R is a transmission factor containing the surface reflection losses; its value was found to be 0.916. To obtain this value, the transmission of light through the empty cells was measured over the spectral region; the value of this transmission representing the loss due to four air-glass surfaces was 0.850 ± 0.005 for both cells. From this observed value, the air-glass reflection coefficient was calculated to be 0.0398 and the index of refraction of the glass to be 1.4985. Using 1.3328 as the index of refraction for water, the glass-water reflection coefficient was calculated to be 0.0034. Using these data, one arrives at a value of 0.916 for the transmission factor R for two air-glass and two glass-water surfaces. α now is the absorption coefficient due to the water alone.

Simultaneous solution of Eqs. (2) and (3) leads to a relation for the correction term k where

$$k = \alpha_t - \alpha = -(\ln R/x) = 0.0883/x. \tag{4}$$

For the 60.0-cm cell, k equals 14.7×10^{-4} cm^{-1}; for the 132.0-cm cell, k equals 6.7×10^{-4} cm^{-1}.

Table I contains the absorption coefficient α for all three types of water in the region 790–580 mμ. For the distilled water and the artificial sea water, it is reported at 20-mμ intervals from 790–760 mμ, and at 1-mμ intervals from 760–580 mμ. For the heavy water, the coefficient is given for 2-mμ intervals from 790–760 mμ, at 1-mμ intervals from 760–640 mμ, and at 2-mμ intervals from 640–580 mμ. Table II contains the absorption coefficient for heavy water at 2-mμ intervals from 580–390 mμ, and for distilled water at 2-mμ intervals from 450–400 mμ.

TABLE II. Absorption coefficient α for heavy water (580–390 mμ) and for distilled water (450–400 mμ) ($\times 10^4$ cm^{-1}).

Wavelength (mμ)	Heavy water	Distilled water	Wavelength (mμ)	Heavy water	Distilled water
580	4.9	...	480	13.1	...
8	4.8		8	13.4	
6	5.2		6	13.3	
4	5.3		4	13.3	
2	5.2		2	13.5	
570	5.4	...	470	14.1	...
8	5.4		8	14.4	
6	5.3		6	14.5	
4	5.4		4	15.0	
2	5.5		2	14.9	
560	5.3	...	460	15.6	...
8	5.5		8	15.6	
6	5.6		6	15.7	
4	5.8		4	16.5	
2	5.8		2	17.7	
550	5.8	...	450	18.2	3.3
8	5.9		8	18.7	3.0
6	6.1		6	19.1	3.4
4	6.4		4	19.7	3.4
2	6.3		2	20.3	3.5
540	6.3	...	440	20.8	3.2
8	6.4		8	21.3	3.2
6	6.9		6	21.7	3.6
4	7.0		4	22.3	3.4
2	7.2		2	23.3	3.9
530	7.4	...	430	23.4	4.0
8	7.4		8	23.4	4.1
6	7.9		6	24.3	4.5
4	7.9		4	24.7	4.2
2	8.3		2	26.2	3.6
520	8.6	...	420	26.3	3.8
8	8.7		8	26.8	4.5
6	8.7		6	27.4	4.5
4	8.9		4	27.8	4.7
2	8.8		2	28.4	5.3
510	8.9	...	410	28.5	4.7
8	9.2		8	28.9	4.7
6	8.9		6	29.7	5.3
4	9.6		4	31.5	6.2
2	9.6		2	31.7	5.8
500	10.1	...	400	31.8	5.8
8	10.2		8	31.9	
6	10.2		6	32.5	
4	10.4		4	33.3	
2	10.9		2	33.7	
490	11.6	...	390	33.3	
8	11.7				
6	12.2				
4	12.2				
2	12.6				

DISCUSSION OF RESULTS

The absorption coefficients shown in Tables I and II represent the mean value of at least three observations in all cases except for the distilled water from 450–400 mμ; for many wavelengths the number of observations was greater.

Representative standard deviations from the mean absorption coefficient for a given wavelength were calculated for 30-mμ intervals throughout the spectral region. Table III shows the absolute standard deviation σ and the relative deviation in percentage for all three types of water in the region 790–580 mμ. Table IV contains similar deviations for the heavy water in the region 570–390 mμ.

From Tables III and IV, it is clear that the absolute deviation is fairly independent of the absolute value of the coefficient—for example, at 760 mμ, σ is the same for the distilled water and the heavy water, despite the great difference (a factor of 40) in the absolute value of the coefficient itself. On the other hand, σ decreases as one goes from the red toward the green; this is due to the greater lamp intensity in this region and to the need for less amplification resulting in an improved signal-to-noise ratio.

The relative deviation, indicated by the percentage, can be seen to increase with decreasing absorption. The large relative deviation in the heavy water is due to the generally small absorption it exhibits throughout the entire visible region.

TABLE III. Absolute and relative deviations from the mean value of the absorption coefficient for the three types of water from 790–580 mμ.

Wave-length (mμ)	Distilled water σ	Distilled water %	Artificial sea water σ	Artificial sea water %	Heavy water σ	Heavy water %
790	0.8[a]	0.4	2.5[a]	1.2	0.7[a]	9.5
760	0.5	0.2	1.1	0.4	0.5	7.2
730	1.1	0.5	1.4	0.6	0.3	5.1
700	0.4	0.6	0.4	0.6	0.6	12.8
670	0.2	0.5	0.3	0.7	0.2	5.7
640	0.2	0.6	0.2	0.6	0.2	6.3
610	0.2	0.7	0.2	0.7	0.4	10.0
580	0.3	2.8	0.3	2.2	0.3	6.1

[a] All values in these columns are times 10^4 cm^{-1}.

From these tables, it is not clear whether the deviations are due to the difference in the water samples or to errors in measurement. No extensive systematic investigation of this was made. Fortunately, there was one case when a single sample of heavy water was measured three times at 730 mμ. This showed a standard deviation of 0.2×10^{-4} cm^{-1} (it is recognized that three samples are not really enough to calculate a standard deviation), while five different samples showed a standard deviation of 0.3×10^{-4} cm^{-1}. If the differences between samples and the errors of measurement are treated as independent in this case, the deviation due to the differences between samples is

$$[(0.3)^2-(0.2)^2]^{\frac{1}{2}}=0.22(\times10^{-4}\text{ cm}^{-1}),$$

so that the two effects contribute about equally to the measured standard deviation. If this is true in general, the standard deviation due to the differences in the samples is $\sigma/\sqrt{2}$.

The absorption coefficients for the distilled water and the artificial sea water differ little from each other in the observed region. Both show a strong absorption in the region 790–730 mμ, that of the artificial sea water being slightly greater in the neighborhood of the maximum. The maximum is found to be at 745±3 mμ in both the distilled and artificial sea water, with some evidence of a smaller maximum in the latter at 760 mμ. This maximum, which is observed in both types of water, is undoubtedly the same as that reported by Curcio and Petty.[4] The value we report for the maximum absorption coefficient is about 4% higher than

that reported by them. McLennan et al.[7] in their investigation reported a diffuse absorption band from 635–599 mμ. We were able to find no maximum in this region. We did observe an almost step-like decrease in the absorption from 680–590 mμ very similar to that shown by McLennan.[7] This is also in general agreement with the findings of Hulburt.[3]

In the area where our investigation of distilled water coincided with that of Hulburt,[3] our results were generally 8%–10% higher. The reason for this discrepancy is not known.

TABLE IV. Absolute and relative deviations from the mean value of the absorption coefficient for heavy water from 570–390 mμ.

Wavelength (mμ)	σ ($\times10^4$ cm^{-1})	%
570	0.2	3.7
540	0.3	4.8
510	0.3	3.4
480	0.2	1.5
450	0.1	0.6
420	0.2	0.7
390	0.4	1.1

The spectrum of the heavy water shows a very small absorption throughout the visible region. No maximum was observed. There is a gradual decrease evident in the absorption from 790–700 mμ, reaching a broad minimum in the region 700–600 mμ. The absorption then gradually increases with decreasing wavelength until, in the neighborhood of 390 mμ, it has increased by a factor of ten over the minimum value. In absorption measurements of D_2O, H_2O, and HDO between 190 and 185 mμ, Barret and Mansell[8] found that the absorption increased with decreasing wavelength, but that at 185 mμ the absorption of the distilled water is larger than that of heavy water by a factor greater than 10.

CONCLUSION

Our investigation shows that for distilled water and artificial sea water the absorption maxima representing the higher harmonics of the fundamental vibrational frequency in the visible region observed—with the exception of the maximum at 745 mμ—are too weak to be observed. Our high-resolution examination of the observed regions leads us to conclude that there is no "optical window" in this portion of the spectrum. A comparison of the absorption spectra of the distilled water and the artificial sea water indicates that the variations in the observed spectra of ordinary sea water

[7] J. C. McLennan, R. Ruedy, and A. C. Burton, Proc. Roy. Soc. (London) **A120**, 296 (1928).
[8] J. Barret and A. L. Mansell, Nature **187**, 138 (1960).

are due to something other than the water itself. The spectrum of the heavy water, while showing no maximum in the visible region, suggests that this maximum might be found in the infrared as might be expected from the isotopic effect of the deuterium.

Further investigations of the absorption spectra of all three types of water in the ultraviolet would be of significant interest.

The author wishes to thank Dr. Karl F. Herzfeld, Dr. Shirleigh Silverman, and Mr. Jarus Quinn of the Catholic University of America for their assistance and advice.

10

Reprinted from *Opt. Soc. Am. J.* **41**(5):302–304 (1951)

The Near Infrared Absorption Spectrum of Liquid Water

Joseph A. Curcio and Charles C. Petty
Naval Research Laboratory, Washington, D. C.
(Received February 14, 1951)

The near infrared absorption spectrum of liquid water at 20°C has been reinvestigated using a PbS cell detector system. The total spectral range investigated was from 0.70 to 2.50μ. A curve is included which shows five prominent absorption bands at 0.76, 0.97, 1.19, 1.45, and 1.94μ; and a table gives experimental results of water absorption at 20°C.

THE near infrared absorption spectrum of water has been the subject of several previous investigations, notably those of Hulburt and Hutchinson,[1] and Collins.[2] Hulburt and Hutchinson did their work in conjunction with an investigation of the absorption by various salts in solution. Their water absorption data cover the wavelength range from 0.704 to 1.344 microns. In 1925 Collins published water absorption data from 0.70 to 2.10μ. He was primarily interested in the shift in position of the absorption bands as the water temperature was changed, rather than with the absorption coefficients as such. A thorough discussion of the problem has been given by Dorsey.[3]

However, since Collins, few results on this problem have been published. With the advent of new highly sensitive detector systems such as the PbS cells, which

[1] E. O. Hulburt and J. F. Hutchinson, Carnegie Institute of Washington Publication No. 260, 9 (1917).
[2] J. F. Collins, Phys. Rev. 26, 771 (1925).
[3] N. E. Dorsey, *Properties of Ordinary Water Substance* (Reinhold Publishing Corporation, New York, 1940).

143

make possible an extension of the wavelength rang of these measurements, it was thought worth while to reopen the problem.

In this investigation, measurements were made in the spectral region 0.70 to 2.50μ. There are five prominent water absorption bands in the near infrared which occur at 0.76, 0.97, 1.19, 1.45, and 1.94μ. A weaker band occurs at 0.845μ. Collins found the intensity of the bands to be enhanced when the water temperature was raised to 95°C. He further found that as the temperature of the water was increased, the position of maximum absorption was shifted to slightly shorter wavelengths. In the present investigation the water temperature was kept constant at 20°C.

A schematic diagram of the experimental arrangement is shown in Fig. 1. The dispersing unit was a Gaertner constant deviation spectrometer with quartz optics. The source was a 1000-watt G. E. projection lamp connected through a voltage stabilizer. The light was collimated and the beam stopped down before going through the water cell.

The detector system consisted of a PbS cell with power supply and amplifier. Radiation was chopped at 17 cycles per second and the photocell output was amplified, rectified, and recorded on a 5-milliampere Esterline-Angus recorder. RMS voltage output readings were also made on a Hewlett-Packard voltmeter. The output was kept below 1 volt RMS—it had been previously established that under these conditions the response of the system was linear.

The wavelength scale of the spectrometer was calibrated by means of a mercury arc which has suitable lines in the near infrared. Accuracy of the wavelength setting was limited by the division markings on the wavelength dial. At the lower wavelengths, from 0.7μ–1.5μ, the settings could be made with an accuracy of about 0.001μ, while for the longer wavelengths above 1.5μ the accuracy was about 0.002μ. Band widths passed by the spectrometer were kept as narrow as possible, and were in the range 0.003–0.006μ. Measurements were made at intervals of 0.005μ in the region 0.70 to 1.16μ, and at intervals of 0.01μ in all other spectral regions. For investigation the spectrum from 0.7μ to 2.5μ was broken up into five regions each containing one of the five prominent absorption bands mentioned above. These regions are defined in Table (I) which also gives the cell pathlength, spectrometer band width, and the absorption band position and intensity associated with each region.

Water used in the experiment was triple-distilled in an all-Pyrex system, and it appeared to be free of particles and air bubbles. Absorption data were obtained within 48 hours after final distillation and immediately after the absorption cells were filled. The cells were made of Pyrex tubing with Glyptal cemented Pyrex windows, and the lengths used varied with the region investigated; the two shortest cells consisted of quartz plates, separated by a lead gasket. Path lengths were measured

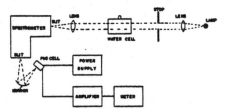

FIG. 1. Diagram of experimental arrangement.

by microscope and micrometer, and the accepted value for the path length is in each case the average of about 10 independent measurements. For all but the shortest cell, absorption measurements were made by comparing the transmissions of two water-filled cells of different length to obviate correction for reflection losses by the cell windows. A slider arrangement was provided whereby the thick and thin cells could be inserted easily into the collimated beam. In the case of the thinnest cell, where the pathlength was 0.022 cm, only one cell was used and suitable corrections were made for glass-air and glass-water reflection losses.

Scattered light in the spectrometer was reduced by means of filters. A Corning No. 2403 red filter, which did not transmit any light below 0.63μ was used between 0.70 and 0.90μ. A dyed Cellophane filter which did not transmit radiation below 0.88μ was used from 0.90 to 2.50μ.

For monochromatic radiation, the absorption coefficient, α, is defined by the relationship

$$I = I_0 e^{-\alpha x},$$

where I_0 is the original intensity of radiation and I the intensity after passing through x cm of water. Since the spectral transmission T is defined by

$$T = I/I_0,$$

the absorption coefficient may be written as

$$\alpha = (1/x) \ln(1/T).$$

Each absorption coefficient plotted in Fig. 2 is the average of at least three independent measurements. The independent measurements deviated by less than 0.002 cm⁻¹ from the average plotted value for each

TABLE I. Water absorption at 20°C.

Wavelength	Water path length (cm)	Spectrometer band width (μ)	Wavelength of max. absorption	Absorption coefficient at max. absorption (cm⁻¹)
0.70–0.90	12.972	0.006	0.76	0.026
0.90–1.15	1.993	0.003	0.97	0.46
1.15–1.35	0.987	0.003	1.19	1.05
1.35–1.80	0.090	0.003	1.45	26.0
1.80–2.50	0.022	0.004	1.94	114.0

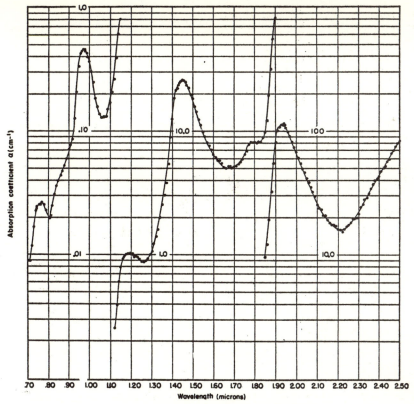

FIG. 2. Absorption coefficients of water at 20°C.

wavelength in the region from 0.70μ to 0.90μ. From 0.90μ to 2.5μ, the mean deviation for each plotted average value was less than 5 percent of the average value.

Figure 2 shows absorption coefficients of water at 20°C for the spectral region 0.70μ to 2.50μ. The regions of maximum absorption occur at 0.76, 0.97, 1.19, 1.45, and 1.94μ. A weaker band is evident by the shoulder on the curve at about 0.85μ. Collins found this band to be more pronounced and enhanced when the water temperature was raised to 95°C.

The absorption coefficients shown in Fig. 2 are lower by 5 percent than those obtained by Hulburt and Hutchinson for the wavelength interval 0.7μ to 1.34μ. The present measurements were made with a resolution of 0.003μ to 0.006μ, whereas those of Hulburt and Hutchinson were made with a resolution of 0.002μ. The present values agree to within 2 percent with those obtained by Collins except in the spectral interval 1.6μ to 1.9μ, where the present values of α are 10 percent lower. A shoulder on the 1.94μ absorption band was found at 1.80μ, which accounts for the lower α in this region. Collins' resolution was 0.005μ to 0.015μ. The present data show that beyond 2.1μ, α drops to a minimum at 2.2μ, then increases steadily out to 2.5μ, the long wavelength limit of the present measurements.

11

Reprinted from *C. R. Séances Acad. Sci.* [Paris] **240**:954–955 (Feb. 28, 1955)

SUR L'ABSORPTION DU RAYONNEMENT
ULTRAVIOLET PAR L'EAU DISTILIÉE

Jacqueline Lenoble
Bernard Saint-Guilly

Le coefficient d'absorption de l'eau distillée a été mesuré entre 4 000 et 2 200 Å par spectrophotométrie photographique, avec des tubes d'eau de 4 et 1 m de longueur.

L'eau pure présente dans le proche ultraviolet une transparence assez grande, que la présence d'impuretés à l'état de traces peut perturber rapidement ; la difficulté d'obtenir de l'eau de bonne qualité jointe à la difficulté des mesures de faibles absorptions explique le peu de résultats qui ont été donnés pour le domaine ultraviolet et la dispersion de ces résultats (on trouve pour la longueur d'onde 3 000 Å des valeurs du coefficient népérien d'absorption de 0,09 et 2,1 m^{-1}.) En dehors de quelques mesures limitées aux radiations inférieures à 3 000 Å ou supérieures à 3 600 Å, les deux séries de mesures les plus complètes sont celles de Sawyer ([1]) et celles de Dawson et Hulburt ([2]), dont nous comparerons plus loin les résultats aux nôtres.

Il nous a donc semblé utile, avant d'aborder l'étude de l'absorption de l'eau de mer, de faire une série de mesures sur l'eau distillée.

L'eau distillée, préparée juste avant les mesures, ne présente sans doute pas une pureté parfaite, mais est d'une bonne qualité, facile à préparer et bien constante. Un essai sur de l'eau bidistillée, préparée avec le plus grand soin, mais malheureusement conservée un certain temps en bouteilles, a donné, à la précision de nos mesures, des valeurs analogues à celles de l'eau distillée.

La méthode employée pour les mesures est la spectrophotométrie photographique ; la source est une lampe à hydrogène et le spectrographe à optique de quartz a une dispersion telle que le spectre entre 4 000 et 2 500 Å est étalé sur environ 10 cm. L'eau à étudier est placée sur un faisceau parallèle dans un grand tube de matière plastique fermé aux deux extrémités par des fenêtres de quartz ; les mesures ont été faites entre 4 000 et 2 650 Å avec un tube de 4 m de longueur et entre 3 500 et 2 200 Å avec un tube de 1 m.

([1]) W. R. SAWYER, *Conts. Can. Biol. and Fish.*, 7, 1931, p. 75.
([2]) L. H. DAWSON et E. O. HULBURT, *J. Opt. Soc. Amer.*, 24, 1934, p. 175.

Coefficient népérien de l'eau distillée (en m^{-1}).

λ(Å).	Tube 4 m.	Tube 1 m.	Moyenne.	Mesures de Sawyer.	Mesures de Dawson et Hulburt.
4000........ .	0,12	–	–	0,07	0,08
3950.........	0,13	–	–	–	–
3900.........	0,13	–	–	0,10	–
3850.........	0,13	–	–	–	–
3800.........	0,14	–	–	0,15	0,13
3750.........	0,14	–	–	–	–
3700.........	0,15	–	–	0,20	–
3650.........	0,15	–	–	–	–
3600.........	0,16	–	–	0,28	0.19
3550.........	0,17	–	–	–	–
3500.........	0,17	0,18	0,18	0,33·	–
3450.........	0,18	0,21	0,19	–	–
3400.........	0,19	0,21	0,20	0,38	0,28
3350.........	0,20	0,21	0,21	–	–
3300.........	0,21	0,23	0,22	0,46	–
3250.........	0,22	0,23	0,22	–	–
3200.........	0,23	0,23	0,23	0,58	0,43
3150.........	0,25	0,23	0,24	·	–
3100.........	0,26	0,25	0,26	0,84	–
3050.........	0,29	0,25	0,27	–	–
3000.........	0,31	0,25	0,28	–	0,64
2950.........	0,33	0,27	0,30	–	–
2900.........	0,35	0,32	0,34	–	–
2850.........	0,40	0,34	0,37	–	–
2800.........	0,43	0,39	0,41	–	0,77
2750.........	0,47	0,43	0,45	–	–
2700.........	0,47	0,55	0,51	–	–
2650.........	0,49	0,59	0,54	–	–
2600.........	–	0,62	–	–	0,92
2550.........	–	0,68	–	–	–
2500.........	–	0,75	–	–	–
2450.........	–	0,84	–	–	1,35
2400.........	–	0,93	–	–	–
2350.........	–	0,93	–	–	–
2300.........	–	0,98	–	–	–
2250.........	–	1,00	–	–	–
2200.........	–	1,14	–	–	–

Les valeurs du coefficient népérien d'absorption (en m^{-1}) données dans le tableau représentent la moyenne de cinq mesures avec chaque tube ; la précision à attendre est environ de 0,02 m^{-1} pour les mesures faites avec le grand tube et 0,08 m^{-1} pour les mesures avec le petit tube ; les cinq séries de mesures concordent à cette précision et dans la région commune aux deux tubes, l'accord entre eux est également raisonnable.

A titre de comparaison, nous avons porté également dans ce tableau les valeurs de Sawyer ([1]) et celles de Dawson et Hulburt ([2]) qui, tout en restant du même ordre de grandeur, semblent à partir de 3 700 Å nettement plus fortes que les nôtres.

Part IV

VOLUME SCATTERING FUNCTION:
INSTRUMENTS AND DATA

Editor's Comments
on Papers 12, 13, and 14

The volume scattering function (defined by R. W. Preisendorfer, Paper 3 Equation 20) is measured by means of a collimated beam of artificial light, and measurements can thus be made *in situ* or on contained samples. S. Q. Duntley, in his paper "Light in the Sea" (Paper 2), gives an excellent conceptual discussion of the volume scattering function and introduces optical methods for its measurement.

From the defining equation it can be seen that in order to accurately determine the volume scattering function it is necessary to have detailed information about the size of the scattering volume. The size of the scattering volume is delineated by the intersection of the beam of light and the beam of detectivity and if not controlled, will change in shape and size with the angle of observation. It is also necessary to determine accurate relative values, at every angle of measurement, of the radiant flux entering the sample volume from the source and leaving the sample volume toward the detector as scattered flux.

The methods that have been used to control and evaluate these factors can be classified into three main categories: methods that maintain a constant sample volume; methods that are based on accurate measurements of the important geometrical factors; and methods that employ a calibration procedure that eliminates the geometrical properties by cancellation. The method of constant sample volume has been employed extensively in the measurement of the scattering properties of contained samples and has been especially useful for the determination of the value of the scattering function at 90° (often called the Rayleigh Ratio), for distilled water and artificial sea water.

For *in situ* measurements, a constant sample volume has been achieved by means of the Waldrum stop that reduces the effective optical thickness of the water sample volume in direct proportion to its increase in length as the angle of observation moves away from 90°. N. G. Jerlov (Paper 14) has made direct use of the principle of the Waldrum stop in a simple but effective instrument and has also made use of the work of R. G. Beutell and A. W. Brewer (1949) to measure the total scattering coefficient of ocean water.

T. J. Petzold (Paper 12) makes use of accurate values of the geometrical constants of the optical system, some of which are shown in his Figures 2 and 5. In his work, the dimensions of all optical stops and critical distances were measured to four significant figures. The accuracy of scattering determinations from this equipment depends on these measurements and also on the high quality of the optical components used.

A method of measuring the volume scattering function in the atmosphere, based on the calibration procedure, was worked out by B. S. Pritchard and W. G. Elliott (1960). This method was adapted to the measurement of the volume scattering function of ocean water by J. E. Tyler (Paper 13).

The numerical value of the volume scattering function for small angles from the forward direction can be 3 to 4 orders of magnitude larger than its value at 10°, and another 2 to 3 orders of magnitude larger than its value at 90°. Because of this, and because of the special optical problems inherent in the measurement of the scattering function at small forward angles, it is necessary to use an instrument of special design. The basic design features for such an instrument are outlined in S. Q. Duntley's paper "Light in the Sea," and have been employed in the design of the instrument used by Petzold, Paper 12, to obtain measurements of the scattering function in the forward direction. Petzold's data for the scattering function of ocean water types is considered to be the most reliable data available at the present time.

REFERENCES

Beutell, R. G. and A. W. Brewer. 1949. Instruments for the measurement of the visual range. *J. Sci. Instrum.* **26**:357.

Pritchard, B. S. and W. G. Elliott. 1960. Two instruments for atmospheric optics measurements. *Opt. Soc. Am. J.* **50**(3):191.

VOLUME SCATTERING FUNCTIONS FOR SELECTED OCEAN WATERS

Theodore J. Petzold

Visibility Laboratory of the Scripps Institution of Oceanography
University of California, San Diego

LOW ANGLE SCATTERING METER

This instrument was designed and constructed in 1966. Its purpose is to determine the volume scattering function for small angles. Subsequent development and testing has led to modifications in both the photometric and optical systems and the instrument is considerably improved in its reliability and precision.

The optics were modified to better define the limits of the solid angle of the measurement, to reduce scattering within the instrument, and to permit operation of the instrument in air in order to determine the scattering contributed by the optical system. Even though care was taken to reduce the instrument's own internal scattering, it is still significant relative to the small angle forward scattering of clear waters, and its magnitude must still be known and accounted for if erroneous high values for the small angle forward scattering are to be avoided.

A schematic drawing of the optical system is shown in Fig. 1. The projector, which has a small point source of light at the focal point of a long focal-length lens, produces a beam of highly collimated light. (The projected beam has a 1/4 milliradian half-angle divergence in water.) After traversing the sample path, the light enters an identical long focal-length lens in the receiver and an image of the point source is formed at its focal point. The light which traverses the water and is neither absorbed nor scattered will fall within this small image. Light which is scattered will arrive at the image plane displaced from the axis at a distance proportional to the angle through which it has been scattered and to the focal length of the receiver lens. In the plane of this image lies a disk with four special field stops which can be sequentially indexed into position by remote control through the cable. The first field stop is a small hole which allows the light in the image of the source to reach the detector (a photomultiplier tube positioned behind the field stop). The other three field stops are annuluses. The annuluses have a circular opaque center surrounded first by a clear ring and then by an opaque area. The inner and outer radii of the annuluses determine the angular interval over which the scattered light is accepted and allowed to pass through to the photomultiplier tube. The nominal limits for the field stops in terms of angular acceptance are:

Stop No.	Limits (milliradians)
1	0 to 1
2	1 to 2
3	2 to 4
4	4 to 8

These limits were chosen to provide an adequate and roughly equal signal level for positions 2, 3, and 4. A calibrated neutral density filter in position 1 reduces the signal from the main beam to the same order of magnitude as the other signals. This reduces the dynamic range which the photometer must cover.

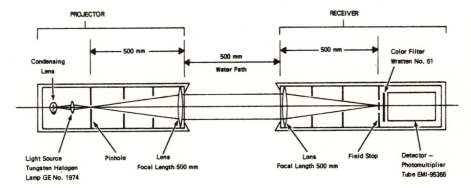

VERTICAL DIMENSIONS ARE EXAGGERATED

Figure 1.   Low Angle Scattering Meter Optical Schematic.

The volume scattering function, $\sigma(\theta)$, is calculated from:

$$\sigma(\theta) = \frac{P(\theta)}{P(0)} \cdot \frac{1}{\Omega \cdot l},$$

where

$\sigma(\theta)$ = the volume scattering at angle θ

$P(0)$ = the light flux entering the sample volume and traveling in the direction $\theta = 0°$

$P(\theta)$ = the light flux entering a small solid angle Ω about the angle θ at which the measurement is made

θ = the angle of scattering

Ω = the solid angle over which the measurement of $P(\theta)$ is made (steradians)

l = length of sample volume (meters).

The solid angle, Ω, is limited by the angles θ_1, θ_2 imposed by the annular field stops and is calculated, for these small angles, from $\Omega = 2\pi(\theta_2^2 - \theta_1^2)$, where θ_1 and θ_2 are in radians. The signal, v, from the photometer is linear with the light flux, P. If $v(\theta)$ is the signal obtained from $P(\theta)$, $v(0)$ is the signal obtained through a filter with transmission T_f of $P(0)$, and the sample path length is 0.500 meters, then

$$\frac{P(\theta)}{P(0)} = \frac{T_f \cdot v(\theta)}{v(0)}$$

and

$$\sigma(\theta) = 2T_f \cdot \frac{v(\theta)}{v(0)} \cdot \frac{1}{2\pi(\theta_2^2 - \theta_1^2)} .$$

The overall spectral response of the instrument, including the spectral output of the light source, the spectral transmission of the filter, and the spectral response of the detector, is shown in Fig. 2.

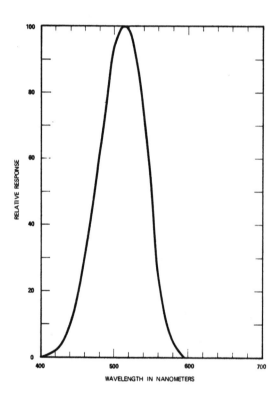

Figure 2.   Spectral Response of Low Angle Scattering Meter and General Angle Scattering Meter.

GENERAL ANGLE SCATTERING METER

This instrument was developed in 1969. Its purpose is to determine the volume scattering function between the limits of $\theta = 10°$ in the forward direction and $\theta = 170°$ in the backward direction. The optical system is depicted in Fig. 3.

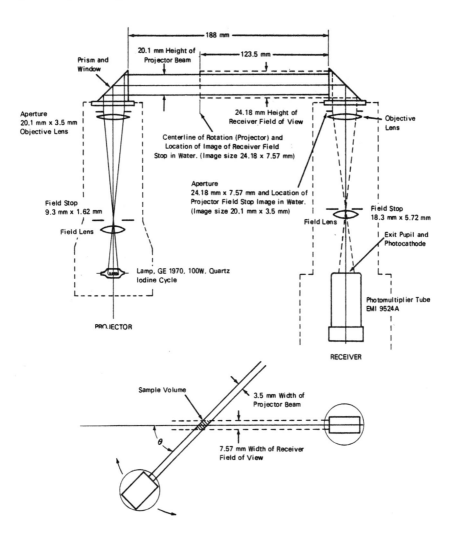

Figure 3.   General Angle Scattering Meter Optical System.

The projector rotates about the sample volume from $\theta = 0°$ through $\theta = 180°$. At $\theta = 0°$ a measurement is made of the total power in the projected beam. Measurement of the scattered light is made, as the projector rotates, between $\theta = 10°$ and $\theta = 170°$, and at 180 degrees the receiver entrance is blocked and the dark signal is recorded. Just beyond 180 degrees the projector beam is blocked and a recording is made of the background ambient light level. A calibrated attenuator which lies in the main beam path at $\theta = 0°$ keeps the on-axis measurement within the range of the photometer.

The instrument provides three analog voltage signals: depth, scan angle position, and photometer signal. Two methods of data collection are available. One can obtain either a continuous trace of photometer signal versus depth at any fixed angle between 10 and 170 degrees or a continuous trace of photometer signal versus scattering angle at a fixed depth. An x-y recorder is used to record the data.

The photometer spans almost six orders of magnitude with an output from 0 to 10 volts. This dynamic range is required to handle the large signal excursions encountered in this type of measurement. The photometer can be operated either in a linear or logarithmic mode. The logarithmic mode is not truly logarithmic, however, and it is necessary to use a calibration curve. No calibration curve is required when using the linear mode.

The volume scattering function in absolute units is obtained from the expression

$$\sigma(\theta) = \frac{P(\theta)}{P(0)} \cdot \frac{1}{\Omega \cdot 1},$$

where:

$P(\theta)$ = the power scattered into the solid angle Ω in the direction θ

$P(0)$ = the total power entering the sample volume

Ω = the solid angle over which $P(\theta)$ is collected and measured

1 = the distance through the sample volume in the direction of $P(0)$.

The length 1 is determined by the width of the receiver beam and the angle θ. Introducing the constants of the system we can calculate $\sigma(\theta)$ from

$$\sigma(\theta) = C_1 \cdot T_f \cdot \frac{C_2}{T_f P(0)} \cdot \frac{P(\theta)}{C_2} \cdot F_v(\theta) \sin\theta,$$

where:

C_1 = a constant of the system geometry and includes the solid angle Ω and the length 1 at $\theta = 90°$ ($C = 1/1 \cdot \Omega = 1.76 \times 10^4$ m^{-1} sr^{-1})

C_2 = a constant determined by the gain of the photometer

T_ℓ = the transmission of the attenuator located in the beam path at $\theta = 0°$ $(T_\ell = 5.62 \times 10^{-5})$

$F_v(\theta)$ = a volume correction factor which allows for the slight divergence of the receiver field of view beyond the midpoint of the sample volume

$\dfrac{T_\ell P(0)}{C_2}$ = the relative value obtained from the photometer at $0°$

$\dfrac{P(\theta)}{C_2}$ = the relative value obtained from the photometer at $\theta°$.

The light source used is a tungsten-halogen lamp operating at a color temperature of 2900 degrees Kelvin.   The detector is a photomultiplier tube with an S-11 response.   The spectral response of the instrument is shown in Fig. 2.

VOLUME SCATTERING FUNCTION CURVES

Figures 6, 10, and 14 show the volume scattering function, $\sigma(\theta)$, versus angle, θ, for three types of natural ocean waters.   The $\sigma(\theta)$ function is in absolute units (meter^{-1}, steradian^{-1}).   The symbols on the curves mark the data points used.   One data point at 0.0859 degrees (0.00149 radians) falls below the 0.1-degree edge of the format and is not shown.   For the clear waters, the signal level for this point was too low to be usable.   However, this point was used for the more turbid waters.

PROBABILITY CURVES

The curves in Figs. 7, 11, and 15 show graphically the function which is the ratio

$$P = \frac{2\pi \displaystyle\int_0^\theta \sigma(\theta) \cdot \sin\theta \cdot d\theta}{\cdot \; 2\pi \displaystyle\int_0^\pi \sigma(\theta) \cdot \sin\theta \cdot d\theta} = \frac{2\pi \displaystyle\int_0^\theta \sigma(\theta) \cdot \sin\theta \cdot d\theta}{s}.$$

The integral $2\pi \int_0^\theta \sigma(\theta) \cdot \sin\theta \cdot d\theta$ is that portion of the total scattering coefficient which lies between zero and the angle θ.   These ratios represent the ratio of the power scattered into angles less than θ relative to the total scattered power only for short path lengths, l, where $s \cdot l \ll 1$ and the approximation $1 - e^{-sl} = sl$ is valid.

SCATTERING DATA TABULATION

Figures 4, 5; 8, 9; 12, 13 show a listing of the values entered into the computer program and the corresponding $\sigma(\theta)$ values. Also listed are the total volume attenuation coefficient, α, at wavelength $\lambda = 530$ nm and the calculated values for the total scattering coefficient, s, the absorption coefficient, a, and the ratios s/α, a/α, and B/S.

The ratio B/S is calculated from:

$$B/S = \frac{2\pi \int_{90^\circ}^{180^\circ} \sigma(\theta) \cdot \sin\theta \cdot d\theta}{s} \ .$$

It is the ratio that shows what portion of the scattering coefficient occurs in the backward direction between 90 and 180 degrees.

The slope, m, is the log-log slope where

$$\log \sigma(\theta) = m \log \theta + \log C.$$

The slope m is determined from the last two data points from the low angle meter and the constant log C is found from the intercept of the line through these points at the $\theta = 1°$ ordinate. The logarithmic function is used to extrapolate to $\theta = 0°$ and to compute that part of the integral from 0 to 0.1 degrees:

$$2\pi \int_{0}^{0.1°} \sigma(\theta) \cdot \sin\theta \cdot d\theta \ .$$

Also shown are the ratios $\sigma(\theta)/s$ for $\theta = 20, 40, 45,$ and 90 degrees and the median angle where the probability function is one-half.

Following this tabulation is a listing of $\sigma(\theta)$ values, the integral $2\pi \int_{0}^{\theta} \sigma(\theta) \cdot \sin\theta \cdot d\theta$, and this integral divided by the total scattering coefficient s (normalized integral). The computer program computed 371 points. To reduce the volume, a sampling of 55 of these points is given for increments of angle $\Delta \log \theta = 0.1$ from $\theta = 0.1°$ to $\theta = 10°$ and increments of $\Delta\theta = 5°$ from $\theta = 10°$ to $\theta = 180°$.

DATA PRESENTATION

Data on the scattering properties of three very different types of ocean water are presented in the form of curves and tabulations. The measurements were taken with the instruments described.

In the computer-printed tabulation, the volume attenuation coefficient, α, the volume scattering co-efficient, s, and the volume absorption coefficient, a, all have the unit meter^{-1}.

Data from three locales are presented: (1) the "Tongue of the Ocean," Bahama Islands, (2) offshore southern California, (3) San Diego Harbor, California. The data for the Tongue of the Ocean are labeled "AUTEC-TEST-161." The approximate location and bottom depth for this station is given below.

Station No.	Latitude	Longitude	Depth
8	24°29' N	77°33' W	850 to 870 fathoms

The data for offshore southern California are labeled HAOCE. Station No. 11 was in the channel be-tween Catalina Island and the California coast at latitude 33°30.0' N and longitude 118°23' W. The data taken in San Diego Harbor are designated NUC.

Reference to the volume scattering functions obtained for various water types, shows the functions to differ in absolute level by more than an order of magnitude but to be very similar in form. The curves show a slight indication that the low angle forward scattering in the clearest water does not rise, rela-tively, as rapidly as in the other waters and that the backward scattering for the very turbid waters, again relatively, does not rise as sharply as for the others.

The assumption that the slope of the function continues unchanged from the last data point to $\theta = 0°$, when computing the colume scattering coefficient, s, is pure conjecture although there is no indication of any dramatic change in this area. If anything, the slope appears to be decreasing. From 3 to 15 percent of the total volume scattering obtained by integration lies at angles smaller than $\theta = 0.1°$. Allowing the slope to go to zero when $\theta = 0°$ with a parabolic fit would have affected the total s obtained very little. Omitting this part of the curve would have reduced the total s obtained by the 3 to 15 percent mentioned, the actual amount depending on the type of water. The probability curves show the importance of the low angle forward scattering. For the waters investigated, 18 to 28 percent of the total volume scattering co-efficient is included in the area between 0° and 1.0° and 58 to 75 percent is between 0° and 10°.

Under static conditions in the laboratory the low angle scattering meter has a repeatability of better than 1 percent. Under field conditions, time-varying changes in the water sample path cause small fluc-tuations in the output signal. Four readings were taken and averaged for each data point. The dark signal was recorded and the alignment tested before and after the instrument was used at each station. The data from the low angle measurements in the clearest water is the most susceptible to error. In very clear water the internal scattering and the dark signal of the low angle scattering meter are significant relative to the measurement. For this reason the scattering function for the lowest angle, $\theta = 0.086$ degrees, in clear water was not usable, and the data for 0.169 degrees in the clear water is possibly no better than ±30 percent. With the exception just noted, ±20 percent is a conservative estimate of the precision of the data obtained with the low angle scattering meter. The general angle scattering meter has a long his-tory of good stability and reliability. It has a readout precision of ±2 percent. Calibration errors and, again, noise caused by particle movement in the sample volume degrade the precision and an error band of ±5 percent would be reasonable.

The two scattering instruments used give data from which the volume scattering function can be computed. The results are on an absolute basis, not relative.

The low angle scattering meter was designed for use in fairly clear waters. The sample path length was longer than desirable for use in the turbid waters and it was necessary to make corrections to the data. When making measurements of the optical properties of water, the measurement path of the instrument used should be suited to the type of water being investigated, if possible.

VALIDATION EXPERIMENT

This experiment was designed to demonstrate if the two instruments used for the *in situ* scattering measurements do in fact respond to the scattering properties of the water and if they are, as they should be, insensitive to the absorption properties of the water.

The volume attenuation coefficient, α, is the sum of the total volume scattering coefficient, s, and the absorption coefficient, a. The volume attenuation coefficient, α, is easily determined with a beam transmissometer from $T = e^{-\alpha l}$. The scattering meters provide data from which the volume scattering function $\sigma(\theta)$ is obtained and from which the total scattering coefficient, s, is calculated by performing the summation indicated by $s = 2\pi \int_0^\pi \sigma(\theta) \cdot \sin\theta \cdot d\theta$. Since $\alpha = s + a$, any increase in scattering would cause an increase in s, Δs, and an equal increase in α, $\Delta\alpha$. Similarly, an increase in absorption would cause an increase in a, Δa, and an equal increase in α, $\Delta\alpha$, but no change in s.

In this experiment scattering material was introduced into clear water and the resultant change in α, as determined with a beam transmissometer, was compared with the change in s, as determined from measurements with the scattering instruments. Ideally, the increments would be equal. Then, an absorbing material was added and again the resulting changes in the measured values of α and s were compared. Ideally, in this case, the scattering meters would indicate no increase in s although the transmissometer would show an increase in α.

A test tank at the Visibility Laboratory was filled with fresh water which was pumped through a filter containing diatomaceous earth. The filtering was stopped and during the rest of the experiment the water was kept agitated and thoroughly mixed by a constantly running large diameter mixing propeller. The instruments described and an instrument for measuring total attenuation α were placed in the water and data recorded for three types of water: (1) clear filtered water, (2) clear water with artificially introduced scattering properties, and (3) water No. 2 with the addition of artificially introduced absorption properties.

Aluminum hydroxide and magnesium hydroxide are compounds which principally cause scattering when dispersed in water. A mixture of these two compounds was used for the scattering agent. The black organic dye Nigrosin was used to increase the absorption. When the scattering agent is added, the change in s, Δs, should ideally equal the change in α. $\Delta\alpha$. Similarly, an addition of the absorbing dye causes an increase in α but ideally should have no affect on s.

160

At the start of the test the beam transmissometer measurement indicated $\alpha = 0.102m^{-1}$ and the total s computed from the measurements made with the two scattering meters was $s = 0.009m^{-1}$, which would make $a = 0.093m^{-1}$. After adding enough scattering material to produce a significant change in the beam transmission (from 90.2 percent per meter to 50.4 percent per meter) and allowing time for the water to become thoroughly mixed, the measurements were repeated. From the beam transmissometer and from the scattering meters, $\alpha = 0.685m^{-1}$ and $s = 0.544m^{-1}$ were obtained. The addition of a material which is largely scattering in nature produced a change in α of $\Delta\alpha = 0.583m^{-1}$ and a change in s of $\Delta s = 0.535m^{-1}$. The change in the volume scattering coefficient calculated from the two sets of measurements with the two scattering meters was 91.8 percent of the change in the volume attenuation coefficient! These results would indicate that the volume absorption coefficient was changed by $\Delta a = 0.048m^{-1}$, which is 8.2 percent of the change in α and 9.0 percent of the calculated change in s.

To determine how sensitive the scattering meters were to absorption, the black dye was added to the water again in a large enough amount to produce a significant change in the beam transmission (from 50.4 percent per meter to 27.3 percent per meter). A third set of measurements gave $\alpha = 1.340m^{-1}$ and s was calculated to be $0.573m^{-1}$. This time, the addition of an absorbing dye caused a change in the volume attenuation coefficient of $\Delta\alpha = 0.655m^{-1}$ and a change in the calculated volume scattering coefficient of $\Delta s = 0.029m^{-1}$. This change in s is only 4.4 percent of the change in α. The corresponding change in the volume absorption coefficient would be $\Delta a = 0.626m^{-1}$ or 96.6 percent of the change in α.

The computed α, a, and s values are tabulated below. The absorption coefficient a was obtained by subtraction: $a = \alpha - s$.

Water Type	α	s	a	$\Delta\alpha$	Δs	Δa	$\dfrac{\Delta s}{\Delta\alpha}$	$\dfrac{\Delta a}{\Delta\alpha}$
Filtered water	0.102	0.009	0.093					
				0.583	0.535	0.048	0.918	0.082
Water with scattering agent added	0.685	0.544	0.141					
				0.655	0.029	0.626	0.044	0.956
Water with absorbing agent added	1.340	0.573	0.767					

The water was not absolutely stable during the time of the test. Constant monitoring with the beam transmissometer showed a very slow drift in the direction of increasing α. This could account for the small increase in absorption, $\Delta a / \Delta\alpha \times 100 = 8.2$ percent, when the scattering agent was added, as well as the even smaller increase in scattering, $\Delta s / \Delta\alpha \times 100 = 4.4$ percent, when the dye was added. It also is likely that the scattering material used was not ideal and did absorb light to a small extent, which is reflected in the measurements. An earlier test similar to this one, for which the changes in scattering and absorption were much smaller, gave even better, although nearly identical, results.

The scattering meters can detect a difference in scattering of less than 2 percent, but in the data which had noise related to the water scattering and the sample volume size, the precision was degraded to probably no better than ±5 percent under the conditions of this test. Considering that two sets of measurements were used for each determination of Δa, Δs, and Δa, the results are unexpectedly good.

Adding a quantity of material to the water which is highly scattering and low in absorption allowed the change in the volume scattering coefficient to be determined with a beam transmissometer. Assuming no change in absorption, the change in a as determined with the beam transmissometer was equal to the change in the volume scattering coefficient. Thus the performance of the scattering meters was tested by comparison with a simple measurement taken with an instrument of proven reliability. The beam transmissometer was also used to give a measure of the change in a when a large amount of absorbing black dye was added to the water to evaluate the insensitivity of the scattering meters to absorption.

This entire test was performed to determine the validity of the *in situ* measurements obtained with the scattering meters and the computation of the total volume scattering coefficient from these measurements. The results indicate that: (1) the instruments are capable of providing accurate data on the scattering properties of the water, (2) the total volume scattering coefficient, s, can be computed from these data, and (3) the scattering instruments are relatively insensitive to absorption.

ACKNOWLEDGMENTS

Many people contributed in different ways to the work which is summarized in this report. Particular credit must be given to several Visibility Laboratory staff members: R. W. Austin for his encouragement and insight; Wayne Wilson, who voluntarily worked many hours lending analytical assistance and performing all of the computer work; and Don Webb for his painstaking work and assistance in developing the optical systems used in the instruments. Special mention must also be made of Stevens P. Tucker, Naval Postgraduate School, Monterey, California, whose dedicated energy and oceanographic know-how overcame many obstacles and enabled the work at the Tongue of the Ocean to be accomplished.

Development of the scattering meters and the work at the Tongue of the Ocean was supported by the Naval Air Development Center, Warminister, Pennsylvania 18974, Contract No. N62269-71-C-0676.

NUC — 05OCT71 — STATION 2040

	DATA READ IN			ITERATED DATA	
	ANGLE (DEG)	SIGMA		ANGLE (DEG)	SIGMA
1	8.5900E-02	9.9700E 03	0	8.5900E-02	4.0024E 03
2				1.0000E-01	3.2620E 03
3	1.6900E-01	4.0100E 03	0	1.6900E-01	1.6098E 03
4	3.3800E-01	1.5400E 03	0	3.3800E-01	6.1823E 02
5	5.7300E-01	6.7500E 02	10	5.7300E-01	3.0791E 02
6	1.7200E 00	9.2000E 01	10	1.7200E 00	5.4684E 01
7	5.7300E 00	7.5000E 00	10	5.7300E 00	5.9699E 00
8	1.0000E 01	2.3200E 00	0	1.0000E 01	2.1107E 00
9	1.5000E 01	9.6000E-01	0	1.5000E 01	9.0405E-01
10	2.0000E 01	4.6500E-01	0	2.0000E 01	4.4523E-01
11	2.5000E 01	2.8300E-01	0	2.5000E 01	2.7335E-01
12	3.0000E 01	1.6600E-01	0	3.0000E 01	1.6128E-01
13	4.0000E 01	8.0900E-02	0	4.0000E 01	7.9133E-02
14	5.0000E 01	4.4700E-02	0	5.0000E 01	4.3884E-02
15	6.0000E 01	2.5900E-02	0	6.0000E 01	2.5483E-02
16	7.0000E 01	1.6800E-02	0	7.0000E 01	1.6550E-02
17	8.0000E 01	1.1400E-02	0	8.0000E 01	1.1239E-02
18	9.0000E 01	8.5300E-03	0	9.0000E 01	8.4110E-03
19	1.0000E 02	6.7900E-03	0	1.0000E 02	6.6940E-03
20	1.1000E 02	5.9800E-03	0	1.1000E 02	5.8912E-03
21	1.2000E 02	5.6400E-03	0	1.2000E 02	5.5491E-03
22	1.3000E 02	5.2500E-03	0	1.3000E 02	5.1541E-03
23	1.4000E 02	4.9300E-03	0	1.4000E 02	4.8223E-03
24	1.5000E 02	4.7700E-03	0	1.5000E 02	4.6344E-03
25	1.6000E 02	5.3700E-03	0	1.6000E 02	5.1417E-03
26	1.7000E 02	6.1000E-03	0	1.7000E 02	5.5497E-03
27				1.8000E 02	5.6857E-03

ITERATIONS=   7
ITERATION CHANGE IN S/ALPHA LESS THAN 0.10 PERCENT

ALPHA=	2.190	S/ALPHA=	0.833
S=	1.824	A/ALPHA=	0.167
A=	0.366	B/S=	0.020

SIGMA(0.1 DEGREES)=	3.2620E 03	
SLOPE=	-1.346	
S UP TO 0.1 DEGREES=	9.5452E-02	NORMALIZED= 5.23221E-02

RATIO OF SIGMA(THETA) TO S

THETA(DEG)	20.0	40.0	45.0	90.0
RATIO	2.4405E-01	4.3377E-02	3.2108E-02	4.6105E-03

	MU	RADIANS	DEGREES
MEDIAN	0.9967	0.8169E-01	4.680

Figure 4

163

NUC - 05OCT71 - STATION 2040

ANGLE(RAD)	ANGLE(DEG)	SIGMA	INTEGRAL	NORM. INTEGRAL	
1.7453E-03	1.0000E-01	3.2620E 03	9.5452E-02	5.2322E-02	1
2.1972E-03	1.2589E-01	2.3974E 03	1.1099E-01	6.0837E-02	11
2.7662E-03	1.5849E-01	1.7566E 03	1.2905E-01	7.0740E-02	21
3.4824E-03	1.9953E-01	1.2745E 03	1.4994E-01	8.2190E-02	31
4.3841E-03	2.5119E-01	9.2603E 02	1.7395E-01	9.5349E-02	41
5.5192E-03	3.1623E-01	6.7637E 02	2.0166E-01	1.1054E-01	51
6.9483E-03	3.9811E-01	5.0267E 02	2.3400E-01	1.2827E-01	61
8.7474E-03	5.0119E-01	3.7048E 02	2.7204E-01	1.4912E-01	71
1.1012E-02	6.3096E-01	2.6762E 02	3.1606E-01	1.7325E-01	81
1.3864E-02	7.9433E-01	1.8974E 02	3.6591E-01	2.0057E-01	91
1.7453E-02	1.0000E 00	1.3288E 02	4.2157E-01	2.3109E-01	101
2.1972E-02	1.2589E 00	9.1912E 01	4.8297E-01	2.6474E-01	111
2.7662E-02	1.5849E 00	6.2795E 01	5.4985E-01	3.0140E-01	121
3.4824E-02	1.9953E 00	4.1708E 01	6.2142E-01	3.4063E-01	131
4.3841E-02	2.5119E 00	2.7365E 01	6.9604E-01	3.8153E-01	141
5.5192E-02	3.1623E 00	1.7926E 01	7.7356E-01	4.2402E-01	151
6.9483E-02	3.9811E 00	1.1723E 01	8.5395E-01	4.6809E-01	161
8.7473E-02	5.0119E 00	7.6549E 00	9.3718E-01	5.1371E-01	171
1.1012E-01	6.3096E 00	5.0388E 00	1.0234E 00	5.6096E-01	181
1.3864E-01	7.9433E 00	3.3017E 00	1.1134E 00	6.1032E-01	191
1.7453E-01	1.0000E 01	2.1107E 00	1.2057E 00	6.6088E-01	201
2.6180E-01	1.5000E 01	9.0405E-01	1.3676E 00	7.4963E-01	206
3.4907E-01	2.0000E 01	4.4523E-01	1.4702E 00	8.0587E-01	211
4.3633E-01	2.5000E 01	2.7335E-01	1.5437E 00	8.4617E-01	216
5.2360E-01	3.0000E 01	1.6128E-01	1.5966E 00	8.7515E-01	221
6.1086E-01	3.5000E 01	1.1087E-01	1.6356E 00	8.9658E-01	226
6.9813E-01	4.0000E 01	7.9133E-02	1.6671E 00	9.1381E-01	231
7.8540E-01	4.5000E 01	5.8575E-02	1.6921E 00	9.2752E-01	236
8.7266E-01	5.0000E 01	4.3884E-02	1.7128E 00	9.3887E-01	241
9.5993E-01	5.5000E 01	3.2882E-02	1.7291E 00	9.4780E-01	246
1.0472E 00	6.0000E 01	2.5483E-02	1.7427E 00	9.5524E-01	251
1.1345E 00	6.5000E 01	2.0408E-02	1.7535E 00	9.6120E-01	256
1.2217E 00	7.0000E 01	1.6550E-02	1.7631E 00	9.6642E-01	261
1.3090E 00	7.5000E 01	1.3446E-02	1.7706E 00	9.7058E-01	266
1.3963E 00	8.0000E 01	1.1239E-02	1.7774E 00	9.7430E-01	271
1.4835E 00	8.5000E 01	9.6372E-03	1.7829E 00	9.7728E-01	276
1.5708E 00	9.0000E 01	8.4110E-03	1.7880E 00	9.8009E-01	281
1.6581E 00	9.5000E 01	7.3961E-03	1.7921E 00	9.8234E-01	286
1.7453E 00	1.0000E 02	6.6940E-03	1.7961E 00	9.8455E-01	291
1.8326E 00	1.0500E 02	6.2195E-03	1.7994E 00	9.8632E-01	296
1.9199E 00	1.1000E 02	5.8912E-03	1.8027E 00	9.8817E-01	301
2.0071E 00	1.1500E 02	5.7291E-03	1.8055E 00	9.8967E-01	306
2.0944E 00	1.2000E 02	5.5491E-03	1.8084E 00	9.9129E-01	311
2.1817E 00	1.2500E 02	5.3427E-03	1.8107E 00	9.9255E-01	316
2.2689E 00	1.3000E 02	5.1541E-03	1.8132E 00	9.9392E-01	321
2.3562E 00	1.3500E 02	4.9670E-03	1.8151E 00	9.9492E-01	326
2.4435E 00	1.4000E 02	4.8223E-03	1.8171E 00	9.9603E-01	331
2.5307E 00	1.4500E 02	4.6347E-03	1.8184E 00	9.9678E-01	336
2.6180E 00	1.5000E 02	4.6344E-03	1.8200E 00	9.9764E-01	341
2.7053E 00	1.5500E 02	4.8998E-03	1.8210E 00	9.9819E-01	346
2.7925E 00	1.6000E 02	5.1417E-03	1.8223E 00	9.9888E-01	351
2.8798E 00	1.6500E 02	5.3587E-03	1.8229E 00	9.9924E-01	356
2.9671E 00	1.7000E 02	5.5497E-03	1.8238E 00	9.9971E-01	361
3.0543E 00	1.7500E 02	5.6177E-03	1.8240E 00	9.9981E-01	366
3.1416E 00	1.8000E 02	5.6857E-03	1.8243E 00	1.0000E 00	371

Figure 5

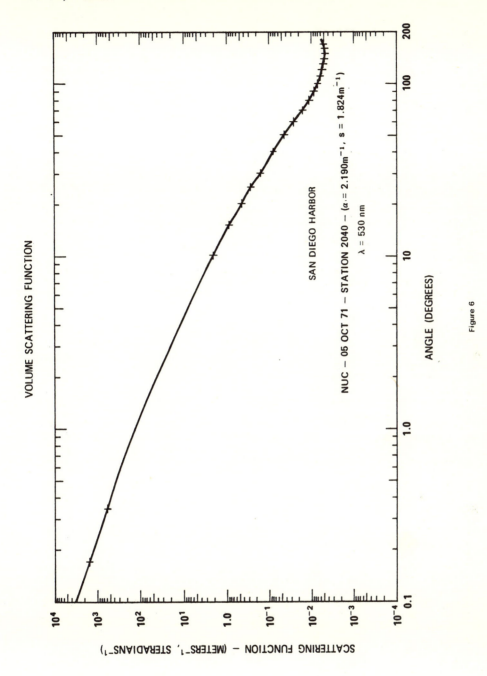

VOLUME SCATTERING FUNCTION

SAN DIEGO HARBOR

NUC — 05 OCT 71 — STATION 2040 — (α = 2.190m^{-1}, s = 1.824m^{-1})

λ = 530 nm

ANGLE (DEGREES)

SCATTERING FUNCTION — (METERS^{-1}, STERADIANS^{-1})

Figure 6

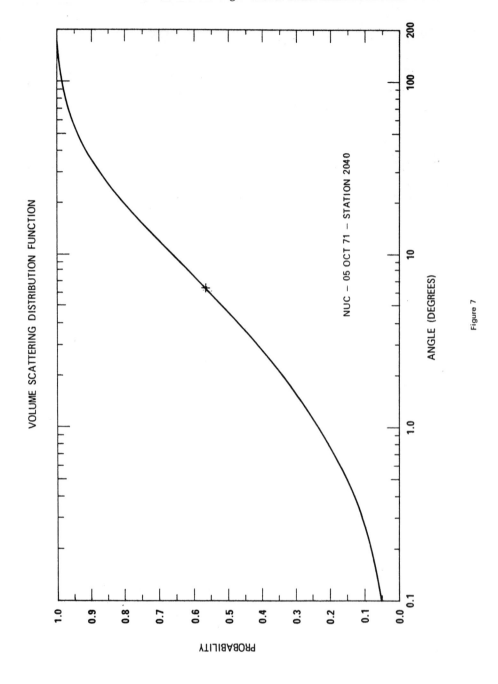

VOLUME SCATTERING DISTRIBUTION FUNCTION

NUC – 05 OCT 71 – STATION 2040

ANGLE (DEGREES)

PROBABILITY

Figure 7

AUTEC — TEST 161 — 13JUL71 — STATION 8

| | DATA READ IN | | | ITERATED DATA | |
	ANGLE (DEG)	SIGMA		ANGLE (DEG)	SIGMA
1				1.0000E-01	5.3182E 01
2	1.6900E-01	2.9000E 01	0	1.6900E-01	2.8464E 01
3	3.3800E-01	1.2700E 01	0	3.3800E-01	1.2465E 01
4	5.7300E-01	6.0000E 00	10	5.7300E-01	5.9045E 00
5	1.7200E 00	1.0300E 00	10	1.7200E 00	1.0191E 00
6	5.7300E 00	1.1400E-01	10	5.7300E 00	1.1347E-01
7	1.0000E 01	4.1700E-02	0	1.0000E 01	4.1620E-02
8	1.5000E 01	2.0400E-02	0	1.5000E 01	2.0375E-02
9	2.0000E 01	1.1000E-02	0	2.0000E 01	1.0990E-02
10	2.5000E 01	6.1700E-03	0	2.5000E 01	6.1656E-03
11	3.0000E 01	3.8900E-03	0	3.0000E 01	3.8877E-03
12	4.0000E 01	1.9000E-03	0	4.0000E 01	1.8991E-03
13	5.0000E 01	1.0200E-03	0	5.0000E 01	1.0196E-03
14	6.0000E 01	6.0300E-04	0	6.0000E 01	6.0280E-04
15	7.0000E 01	4.0700E-04	0	7.0000E 01	4.0688E-04
16	8.0000E 01	3.0200E-04	0	8.0000E 01	3.0191E-04
17	9.0000E 01	2.4600E-04	0	9.0000E 01	2.4593E-04
18	1.0000E 02	2.2400E-04	0	1.0000E 02	2.2394E-04
19	1.1000E 02	2.2400E-04	0	1.1000E 02	2.2393E-04
20	1.2000E 02	2.3400E-04	0	1.2000E 02	2.3392E-04
21	1.3000E 02	2.6300E-04	0	1.3000E 02	2.6290E-04
22	1.4000E 02	2.7500E-04	0	1.4000E 02	2.7488E-04
23	1.5000E 02	3.0900E-04	0	1.5000E 02	3.0882E-04
24	1.6000E 02	3.6300E-04	0	1.6000E 02	3.6268E-04
25	1.7000E 02	4.6800E-04	0	1.7000E 02	4.6710E-04
26				1.8000E 02	5.0190E-04

ITERATIONS=   3
ITERATION CHANGE IN S/ALPHA LESS THAN 0.10 PERCENT

ALPHA=	0.151	S/ALPHA=	0.247
S=	0.037	A/ALPHA=	0.753
A=	0.114	B/S=	0.044

SIGMA(0.1 DEGREES)=   5.3182E 01
SLOPE=   -1.191
S UP TO 0.1 DEGREES=   1.2585E-03   NORMALIZED= 3.37509E-02

RATIO OF SIGMA(THETA) TO S

THETA(DEG)	20.0	40.0	45.0	90.0
RATIO	2.9474E-01	5.0931E-02	3.6785E-02	6.5953E-03

	MU	RADIANS	DEGREES
MEDIAN	0.9941	0.1091	6.252

Figure 8

```
AUTEC - TEST 161 - 13JUL71 - STATION 8
  ANGLE(RAD)     ANGLE(DEG.)      SIGMA        INTEGRAL     NORM. INTEGRAL
  1.7453E-03     1.0000E-01     5.3182E 01    1.2585E-03     3.3751E-02      1
  2.1972E-03     1.2589E-01     4.0424E 01    1.5161E-03     4.0660E-02     11
  2.7662E-03     1.5849E-01     3.0727E 01    1.8265E-03     4.8983E-02     21
  3.4824E-03     1.9953E-01     2.3735E 01    2.2029E-03     5.9077E-02     31
  4.3841E-03     2.5119E-01     1.8141E 01    2.6627E-03     7.1407E-02     41
  5.5192E-03     3.1623E-01     1.3598E 01    3.2140E-03     8.6194E-02     51
  6.9483E-03     3.9811E-01     9.9536E 00    3.8614E-03     1.0355E-01     61
  8.7474E-03     5.0119E-01     7.1793E 00    4.6060E-03     1.2352E-01     71
  1.1012E-02     6.3096E-01     5.1100E 00    5.4518E-03     1.4621E-01     81
  1.3864E-02     7.9433E-01     3.5911E 00    6.3992E-03     1.7161E-01     91
  1.7453E-02     1.0000E 00     2.4976E 00    7.4489E-03     1.9976E-01    101
  2.1972E-02     1.2589E 00     1.7191E 00    8.5998E-03     2.3063E-01    111
  2.7662E-02     1.5849E 00     1.1710E 00    9.8486E-03     2.6412E-01    121
  3.4824E-02     1.9953E 00     7.7576E-01    1.1182E-02     2.9987E-01    131
  4.3841E-02     2.5119E 00     5.0866E-01    1.2569E-02     3.3707E-01    141
  5.5192E-02     3.1623E 00     3.3399E-01    1.4011E-02     3.7574E-01    151
  6.9483E-02     3.9811E 00     2.1960E-01    1.5512E-02     4.1601E-01    161
  8.7473E-02     5.0119E 00     1.4459E-01    1.7078E-02     4.5798E-01    171
  1.1012E-01     6.3096E 00     9.5219E-02    1.8711E-02     5.0178E-01    181
  1.3864E-01     7.9433E 00     6.2816E-02    2.0414E-02     5.4746E-01    191
  1.7453E-01     1.0000E 01     4.1620E-02    2.2196E-02     5.9525E-01    201
  2.6180E-01     1.5000E 01     2.0375E-02    2.5612E-02     6.8686E-01    206
  3.4907E-01     2.0000E 01     1.0990E-02    2.8075E-02     7.5292E-01    211
  4.3633E-01     2.5000E 01     6.1656E-03    2.9789E-02     7.9888E-01    216
  5.2360E-01     3.0000E 01     3.8877E-03    3.1021E-02     8.3190E-01    221
  6.1086E-01     3.5000E 01     2.6802E-03    3.1972E-02     8.5741E-01    226
  6.9813E-01     4.0000E 01     1.8991E-03    3.2723E-02     8.7756E-01    231
  7.8540E-01     4.5000E 01     1.3717E-03    3.3321E-02     8.9360E-01    236
  8.7266E-01     5.0000E 01     1.0196E-03    3.3798E-02     9.0638E-01    241
  9.5993E-01     5.5000E 01     7.6833E-04    3.4183E-02     9.1672E-01    246
  1.0472E 00     6.0000E 01     6.0280E-04    3.4496E-02     9.2511E-01    251
  1.1345E 00     6.5000E 01     4.8832E-04    3.4761E-02     9.3220E-01    256
  1.2217E 00     7.0000E 01     4.0688E-04    3.4985E-02     9.3822E-01    261
  1.3090E 00     7.5000E 01     3.4571E-04    3.5182E-02     9.4350E-01    266
  1.3963E 00     8.0000E 01     3.0191E-04    3.5353E-02     9.4809E-01    271
  1.4835E 00     8.5000E 01     2.6810E-04    3.5509E-02     9.5226E-01    276
  1.5708E 00     9.0000E 01     2.4593E-04    3.5648E-02     9.5599E-01    281
  1.6581E 00     9.5000E 01     2.3152E-04    3.5779E-02     9.5952E-01    286
  1.7453E 00     1.0000E 02     2.2394E-04    3.5902E-02     9.6280E-01    291
  1.8326E 00     1.0500E 02     2.2254E-04    3.6022E-02     9.6603E-01    296
  1.9199E 00     1.1000E 02     2.2393E-04    3.6137E-02     9.6913E-01    301
  2.0071E 00     1.1500E 02     2.2651E-04    3.6253E-02     9.7221E-01    306
  2.0944E 00     1.2000E 02     2.3392E-04    3.6363E-02     9.7518E-01    311
  2.1817E 00     1.2500E 02     2.5050E-04    3.6476E-02     9.7822E-01    316
  2.2689E 00     1.3000E 02     2.6290E-04    3.6587E-02     9.8118E-01    321
  2.3562E 00     1.3500E 02     2.6615E-04    3.6695E-02     9.8407E-01    326
  2.4435E 00     1.4000E 02     2.7488E-04    3.6794E-02     9.8672E-01    331
  2.5307E 00     1.4500E 02     2.8957E-04    3.6889E-02     9.8928E-01    336
  2.6180E 00     1.5000E 02     3.0882E-04    3.6975E-02     9.9160E-01    341
  2.7053E 00     1.5500E 02     3.3044E-04    3.7057E-02     9.9380E-01    346
  2.7925E 00     1.6000E 02     3.6268E-04    3.7128E-02     9.9570E-01    351
  2.8798E 00     1.6500E 02     4.0732E-04    3.7193E-02     9.9743E-01    356
  2.9671E 00     1.7000E 02     4.6710E-04    3.7243E-02     9.9877E-01    361
  3.0543E 00     1.7500E 02     4.8450E-04    3.7278E-02     9.9972E-01    366
  3.1416E 00     1.8000E 02     5.0190E-04    3.7289E-02     1.0000E 00    371
```

Figure 9

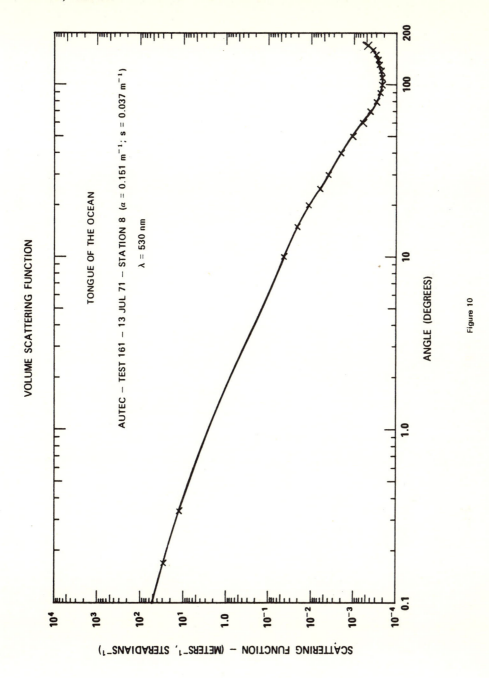

VOLUME SCATTERING FUNCTION

TONGUE OF THE OCEAN

AUTEC – TEST 161 – 13 JUL 71 – STATION 8 ($a = 0.151$ m^{-1}; $s = 0.037$ m^{-1})

$\lambda = 530$ nm

ANGLE (DEGREES)

SCATTERING FUNCTION – (METERS^{-1}, STERADIANS^{-1})

Figure 10

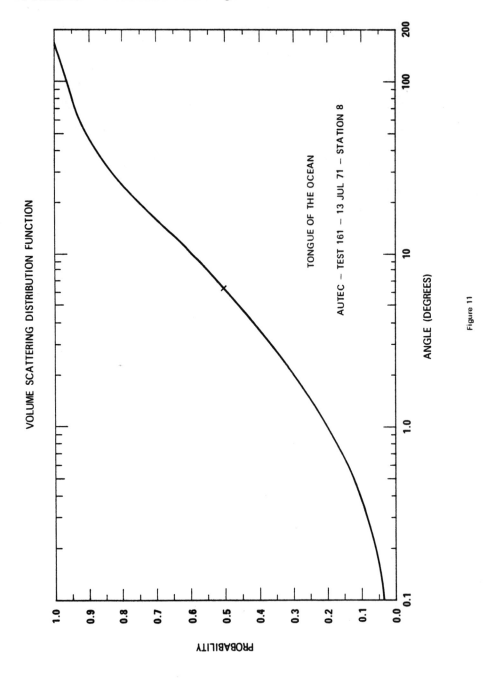

VOLUME SCATTERING DISTRIBUTION FUNCTION

PROBABILITY

ANGLE (DEGREES)

TONGUE OF THE OCEAN

AUTEC – TEST 161 – 13 JUL 71 – STATION 8

Figure 11

HAOCE - 05AUG71 - STATION 11

	DATA READ IN			ITERATED DATA	
	ANGLE (DEG)	SIGMA		ANGLE (DEG)	SIGMA
1				1.0000E-01	6.5329E 02
2	1.6900E-01	3.2400E 02	0	1.6900E-01	2.9035E 02
3	3.3800E-01	1.1100E 02	0	3.3800E-01	9.9472E 01
4	5.7300E-01	4.7500E 01	10	5.7300E-01	4.3225E 01
5	1.7200E 00	6.7000E 00	10	1.7200E 00	6.2941E 00
6	5.7300E 00	6.5000E-01	10	5.7300E 00	6.3242E-01
7	1.0000E 01	2.1800E-01	0	1.0000E 01	2.1554E-01
8	1.5000E 01	9.3500E-02	0	1.5000E 01	9.2828E-02
9	2.0000E 01	4.4500E-02	0	2.0000E 01	4.4268E-02
10	2.5000E 01	2.4000E-02	0	2.5000E 01	2.3900E-02
11	3.0000E 01	1.4500E-02	0	3.0000E 01	1.4450E-02
12	4.0000E 01	6.0300E-03	0	4.0000E 01	6.0140E-03
13	5.0000E 01	3.0000E-03	0	5.0000E 01	2.9934E-03
14	6.0000E 01	1.7400E-03	0	6.0000E 01	1.7366E-03
15	7.0000E 01	1.0960E-03	0	7.0000E 01	1.0940E-03
16	8.0000E 01	7.2500E-04	0	8.0000E 01	7.2376E-04
17	9.0000E 01	5.2500E-04	0	9.0000E 01	5.2412E-04
18	1.0000E 02	4.3700E-04	0	1.0000E 02	4.3626E-04
19	1.1000E 02	4.0800E-04	0	1.1000E 02	4.0727E-04
20	1.2000E 02	3.9800E-04	0	1.2000E 02	3.9723E-04
21	1.3000E 02	4.0800E-04	0	1.3000E 02	4.0710E-04
22	1.4000E 02	4.4700E-04	0	1.4000E 02	4.4582E-04
23	1.5000E 02	5.2500E-04	0	1.5000E 02	5.2319E-04
24	1.6000E 02	6.7000E-04	0	1.6000E 02	6.6651E-04
25	1.7000E 02	9.5000E-04	0	1.7000E 02	9.3927E-04
26				1.8000E 02	1.0302E-03

ITERATIONS=   4
ITERATION CHANGE IN S/ALPHA LESS THAN 0.10 PERCENT

```
       ALPHA=   0.398        S/ALPHA=   0.551
           S=   0.219        A/ALPHA=   0.449
           A=   0.179            B/S=   0.013


       SIGMA( 0.1 DEGREES)=     6.5329E 02
       SLOPE=                      -1.545
       S UP TO 0.1 DEGREES=     2.7506E-02        NORMALIZED= 1.25411E-01
```

RATIO OF SIGMA(THETA) TO S

THETA(DEG)	20.0	40.0	45.0	90.0
RATIO	2.0184E-01	2.7421E-02	1.8892E-02	2.3897E-03

	MU	RADIANS	DEGREES
MEDIAN	0.9990	0.4423E-01	2.534

Figure 12

HAOCE - 05AUG71 - STATION 11

ANGLE(RAD)	ANGLE(DEG)	SIGMA	INTEGRAL	NORM. INTEGRAL	
1.7453E-03	1.0000E-01	6.5329E 02	2.7506E-02	1.2541E-01	1
2.1972E-03	1.2589E-01	4.5768E 02	3.0541E-02	1.3925E-01	11
2.7662E-03	1.5849E-01	3.2064E 02	3.3911E-02	1.5461E-01	21
3.4824E-03	1.9953E-01	2.2517E 02	3.7657E-02	1.7169E-01	31
4.3841E-03	2.5119E-01	1.5788E 02	4.1824E-02	1.9069E-01	41
5.5192E-03	3.1623E-01	1.1038E 02	4.6449E-02	2.1178E-01	51
6.9483E-03	3.9811E-01	7.7309E 01	5.1575E-02	2.3515E-01	61
8.7474E-03	5.0119E-01	5.3705E 01	5.7248E-02	2.6102E-01	71
1.1012E-02	6.3096E-01	3.6749E 01	6.3454E-02	2.8931E-01	81
1.3864E-02	7.9433E-01	2.4805E 01	7.0127E-02	3.1974E-01	91
1.7453E-02	1.0000E 00	1.6623E 01	7.7241E-02	3.5217E-01	101
2.1972E-02	1.2589E 00	1.1060E 01	8.4769E-02	3.8650E-01	111
2.7662E-02	1.5849E 00	7.3058E 00	9.2677E-02	4.2256E-01	121
3.4824E-02	1.9953E 00	4.7505E 00	1.0090E-01	4.6006E-01	131
4.3841E-02	2.5119E 00	3.0669E 00	1.0933E-01	4.9849E-01	141
5.5192E-02	3.1623E 00	1.9772E 00	1.1795E-01	5.3778E-01	151
6.9483E-02	3.9811E 00	1.2728E 00	1.2675E-01	5.7789E-01	161
8.7473E-02	5.0119E 00	8.1825E-01	1.3571E-01	6.1877E-01	171
1.1012E-01	6.3096E 00	5.2847E-01	1.4485E-01	6.6042E-01	181
1.3864E-01	7.9433E 00	3.4020E-01	1.5420E-01	7.0307E-01	191
1.7453E-01	1.0000E 01	2.1554E-01	1.6365E-01	7.4617E-01	201
2.6180E-01	1.5000E 01	9.2828E-02	1.8032E-01	8.2214E-01	206
3.4907E-01	2.0000E 01	4.4268E-02	1.9081E-01	8.7000E-01	211
4.3633E-01	2.5000E 01	2.3900E-02	1.9759E-01	9.0089E-01	216
5.2360E-01	3.0000E 01	1.4450E-02	2.0229E-01	9.2233E-01	221
6.1086E-01	3.5000E 01	9.0629E-03	2.0565E-01	9.3767E-01	226
6.9813E-01	4.0000E 01	6.0140E-03	2.0812E-01	9.4889E-01	231
7.8540E-01	4.5000E 01	4.1435E-03	2.0996E-01	9.5730E-01	236
8.7266E-01	5.0000E 01	2.9934E-03	2.1138E-01	9.6379E-01	241
9.5993E-01	5.5000E 01	2.2525E-03	2.1251E-01	9.6893E-01	246
1.0472E 00	6.0000E 01	1.7366E-03	2.1343E-01	9.7310E-01	251
1.1345E 00	6.5000E 01	1.3689E-03	2.1418E-01	9.7652E-01	256
1.2217E 00	7.0000E 01	1.0940E-03	2.1480E-01	9.7934E-01	261
1.3090E 00	7.5000E 01	8.7821E-04	2.1531E-01	9.8168E-01	266
1.3963E 00	8.0000E 01	7.2376E-04	2.1573E-01	9.8362E-01	271
1.4835E 00	8.5000E 01	6.0355E-04	2.1609E-01	9.8526E-01	276
1.5708E 00	9.0000E 01	5.2412E-04	2.1640E-01	9.8666E-01	281
1.6581E 00	9.5000E 01	4.7034E-04	2.1667E-01	9.8789E-01	286
1.7453E 00	1.0000E 02	4.3626E-04	2.1692E-01	9.8901E-01	291
1.8326E 00	1.0500E 02	4.1890E-04	2.1714E-01	9.9006E-01	296
1.9199E 00	1.1000E 02	4.0727E-04	2.1736E-01	9.9104E-01	301
2.0071E 00	1.1500E 02	3.9941E-04	2.1756E-01	9.9197E-01	306
2.0944E 00	1.2000E 02	3.9723E-04	2.1776E-01	9.9285E-01	311
2.1817E 00	1.2500E 02	3.9841E-04	2.1794E-01	9.9369E-01	316
2.2689E 00	1.3000E 02	4.0710E-04	2.1812E-01	9.9449E-01	321
2.3562E 00	1.3500E 02	4.2193E-04	2.1828E-01	9.9525E-01	326
2.4435E 00	1.4000E 02	4.4582E-04	2.1844E-01	9.9598E-01	331
2.5307E 00	1.4500E 02	4.7749E-04	2.1860E-01	9.9668E-01	336
2.6180E 00	1.5000E 02	5.2319E-04	2.1874E-01	9.9735E-01	341
2.7053E 00	1.5500E 02	5.8236E-04	2.1888E-01	9.9798E-01	346
2.7925E 00	1.6000E 02	6.6651E-04	2.1901E-01	9.9858E-01	351
2.8798E 00	1.6500E 02	7.8226E-04	2.1913E-01	9.9912E-01	356
2.9671E 00	1.7000E 02	9.3927E-04	2.1923E-01	9.9958E-01	361
3.0543E 00	1.7500E 02	9.8473E-04	2.1930E-01	9.9989E-01	366
3.1416E 00	1.8000E 02	1.0302E-03	2.1933E-01	1.0000E 00	371

Figure 13

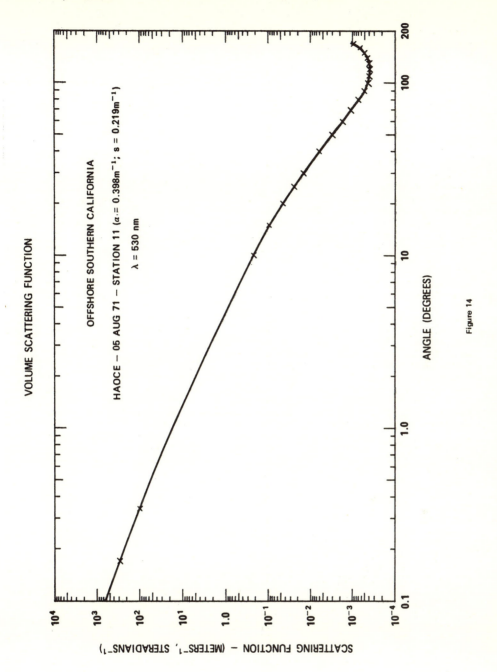

Figure 14

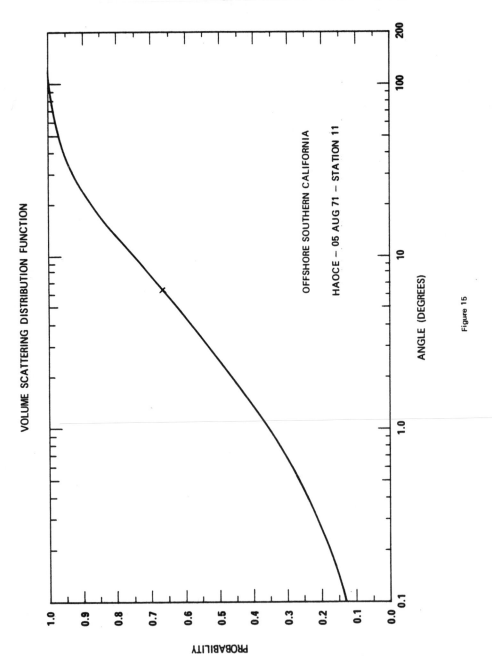

OFFSHORE SOUTHERN CALIFORNIA

HAOCE – 05 AUG 71 – STATION 11

Figure 15

13

Reprinted from *Appl. Opt.* **2**(3):245–248 (1963)

Design Theory for a Submersible Scattering Meter

John E. Tyler

This paper extends the theoretical analysis of an atmospheric polar nephelometer by Pritchard and Elliott[1] and applies it to the problem of measuring the volume scattering function of ocean water. Procedures for obtaining the necessary absolute calibrations are also treated theoretically. The analysis yields an equation for the volume scattering function, $\sigma(\theta)$, for each angle θ which requires for its solution a detailed knowledge of the optical properties of a diffusing plastic calibration plate, a somewhat tedious calibration procedure to allow for dependence on the sample volume and on the irradiance input to this volume, and a knowledge of the transmittances of the water used for calibration and the water being measured. The analysis is the foundation for an in situ scattering meter which is currently under construction. No data are given.

Introduction

The quantitative determination of the volume scattering function is based on Eq. (1):

$$\sigma(\theta) = \frac{dJ(\theta)}{H\ dv},\qquad(1)$$

where H is the irradiance at the sample volume, dv, produced by a beam of radiation and dJ is the increment of intensity scattered out of the beam at an angle θ from the forward direction of the beam.

The classical experiment for determining the volume scattering function of a medium is to project a beam of radiation through the medium and measure the intensity of the scattered radiation at a selected point of the beam and at various angles of observation, θ, by means of a flux detector of known field of view.

Experimental measurements of the volume scattering function of liquids have been made in the laboratory on samples contained in glass cells (see, for example, ref. 2) and in situ by means of instruments that could be immersed in the medium (see, for example, Tyler[3] and Jerlov[4]).

It will be appreciated that in making these measurements the sample volume, dv, depends in a nonlinear way on the angle of observation, θ, and that there are considerable practical difficulties involved in obtaining comparable measurements of dJ and H.

The author is at the Visibility Laboratory, Scripps Institution of Oceanography, University of California, La Jolla, California.

Received 11 December 1961.

This research has been supported by the National Science Foundation, Earth Sciences Division and by the U. S. Navy Electronics Laboratory.

In 1960 Pritchard and Elliott[1] published the theory of their polar nephelometer for scattering measurements in the atmosphere. This definitive paper develops a calibration technique that avoids these difficulties to a large extent.

The purpose of this paper is to extend the theoretical analysis of Pritchard and Elliott[1] and to apply it to the *in situ* measurement of the volume scattering function of ocean water. An underwater calibration procedure is also worked out which does not require the calibration plate to have cosine emitting properties and which does not depend on MgO as a standard for absolute calibration.

The steps in this analysis will be as follows:

1. Derivation of the volume calibration equation by use of a calibration screen.

2. Derivation of the measurement equation for an unknown sample.

3. Derivation of the equation for $\sigma(\theta)$.

4. Determination of the input and output functions of the calibration screen.

1. Derivation of the Volume Calibration Equation by Use of a Calibration Screen

Relative values of the sample volume are determined by means of a diffusing plastic calibration screen (noncosine) which is driven along the axis of the detection system and through the volume defined by the intersection of the two beams (a beam of light and a beam of detection). The reading of the phototube will be directly proportional to the sectional area of the volume at each point, x, along the axis. A recording of this output as a function of x is therefore directly

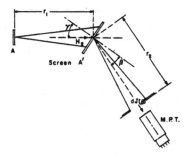

Fig. 1. Geometry of measurement.

proportional to a plot of the cross-sectional area vs the thickness of the volume and the integral of this curve will be directly proportional to the sample volume, dv, at the fixed observation angle θ.

In Fig. 1 the output radiant intensity of an element dA' at the center of the screen in the direction of the phototube will be:

$$dJ(\beta) = N(\beta) \, dA', \qquad (2)$$

where β is any suitable angle between the normal to the calibration screen and the phototube axis, and $N(\beta)$ is the inherent radiance of the screen in the direction β.

The flux reaching the phototube will be

$$dP(\beta) = C_1 \, dJ(\beta)e^{-\alpha r_1}, \qquad (3)$$

where α_1 is the total attenuation coefficient of the water in which the calibration is being conducted, and C_1 is the proportionality constant. C_1 has the dimensions of solid angle. Its magnitude is set largely by the solid angle of acceptance of the multiplier phototube but in detail it is also the coupling factor between dP and dJ and its magnitude can change within the solid angle of acceptance of the multiplier phototube. This effect is also present during the measurement of a water sample, and in Eq. (22) the effect cancels out.

The reading produced by this flux on the recorder will be

$$dK(\beta) = C_2 \, dP(\beta),$$

where C_2 is no more than the proportionality constant of the recorder and circuit, and is assumed to be constant for all values of $dP(\beta)$. From Eqs. (2), (3), and (4)

$$dK(\beta) = C_1 \, C_2 e^{-\alpha r_1} N(\beta) \, dA'. \qquad (5)$$

The radiance of a noncosine calibration screen can be described as follows:

$$N(\beta) = N(0) f(\beta), \qquad (6)$$

where $f(\beta)$ must be obtained experimentally.

The inherent radiance of the screen when viewed normally is

$$N(0) = \frac{H_2 T(\gamma,0)}{\Omega}, \qquad (7)$$

where H_2 is the irradiance arriving on the screen, $T(\gamma,0)^*$ is the transmittance of the screen in the direction of its normal (i.e., $\beta = 0$) when irradiated at an angle γ to its normal, and Ω is the solid angle of acceptance.

If H_1 is the irradiance arriving on the screen when $\gamma = 0$, then

$$H_2 \frac{A'}{\cos\gamma} = H_1 A'$$

or

$$H_2 = H_1 \cos\gamma \qquad (8)$$

and

$$N(0) = \frac{H_1 \cos\gamma \, T(\gamma,0)}{\Omega}. \qquad (9)$$

From Eqs. (9) and (6) therefore

$$N(\beta) = \frac{H_1 \cos\gamma \, T(\gamma,0)}{\Omega} f(\beta). \qquad (10)$$

Under water, H_1 will depend on the properties of the water. We note therefore that

$$H_1 = H_0 \, e^{-\alpha r_1}, \qquad (11)$$

where H_0 is the irradiance that would appear at A' if the water had zero attenuation. (H_0 varies only with lamp output, which can be controlled.) Using Eq. (11), Eq. (10) becomes

$$N(\beta) = \frac{H_0 e^{-\alpha r_1} \cos\gamma \, T(\gamma,0)}{\Omega} f(\beta). \qquad (12)$$

If we now substitute Eq. (12) into Eq. (5) we obtain

$$dK(\beta) = C_1 C_2 \, e^{-\alpha(r_1+r_2)} f(\beta) \, dA' \frac{H_0 \cos\gamma \, T(\gamma,0)}{\Omega} \qquad (13)$$

and

$$K(\beta) = T(\gamma,0)e^{-\alpha(r_1+r_2)} \frac{f(\beta) \cos\gamma}{\Omega} C_2 \int_{A'} C_1 H_0 \, dA'. \qquad (14)$$

In Eq. (14) if H_0 is held constant $K(\beta)$ is directly proportional to that area of the calibration plate which is both irradiated and measured. The volume is therefore proportional to the integral,

$$\int K(\beta) \, dx.$$

Thus the equation

$$\int K(\beta) \, dx = T(\gamma,0)e^{-\alpha(r_1+r_2)} \frac{f(\beta) \cos\gamma}{\Omega} C_2 \int_{x} \int_{A'} C_1 H_0 \, dv \qquad (15)$$

---

* Note that the transmittance term $T(\gamma,\beta)$ must be replaced by a reflectance term, $R(\gamma,\beta)$ in the backward quadrant. Also note that the $\theta = \gamma + \beta$ so that $T(\gamma,0)$ is the same as $T(\theta)$.

is the volume calibration equation for any fixed angular setting θ.

2. Derivation of the Measurement Equation

In measuring the scattered flux from this same volume of an unknown hydrosol at this same angle θ, the intensity of the flux will be, by definition,

$$dJ(\theta) = \sigma(\theta) H_1 \, dv \qquad (16)$$

and the flux received by the phototube will be

$$dP(\theta) = C_1 \, dJ(\theta) \, e^{-\alpha_2 r_2}, \qquad (17)$$

where α_2 is the total attenuation coefficient of the unknown hydrosol.

The reading obtained will be

$$dK(\theta) = C_2 \, dP(\theta) \qquad (18)$$

or

$$dK(\theta) = C_1 C_2 e^{-\alpha_2 r_2} \, \sigma(\theta) H_1 \, dv. \qquad (19)$$

Using Eq. (11), this becomes

$$dK(\theta) = C_1 C_2 e^{-\alpha_2 r_2} \sigma(\theta) H_0 e^{-\alpha_2 r_1} \, dv \qquad (20)$$

and

$$K(\theta) = \sigma(\theta) e^{-\alpha_2 (r_1 + r_2)} C_2 \int C_1 H_0 \, dv. \qquad (21)$$

3. Derivation of the Equation for $\sigma(\theta)$

The equation for $\sigma(\theta)$ can now be obtained by dividing Eq. (21) by Eq. (15).

$$\frac{K(\theta)}{K(\beta) \, dz} = \frac{\sigma(\theta) e^{-\alpha_2(r_1+r_2)} \, \Omega}{T(\gamma,\theta) \, f(\beta) \cos\gamma \, e^{-\alpha_1(r_1+r_2)}} \cdot \frac{C_2 \int C_1 H_0 \, dv}{C_2 \int C_1 H_0 \, dv} \qquad (22)$$

which reduces to

$$\sigma(0) = \frac{K(\theta)}{\int K(\beta) \, dz} \cdot \frac{T(\gamma,0) \, f(\beta) \cos\gamma}{\Omega} \cdot \frac{e^{-\alpha_2(r_1+r_2)}}{e^{-\alpha_1(r_1+r_2)}}. \qquad (23)$$

Since $e^{-\alpha_1(r_1+r_2)}$ is the transmittance (T_1) of a path length $(r_1 + r_2)$ of the water used for calibration, and $e^{-\alpha_2(r_1+r_2)}$ is the transmittance (T_2) of an identical path length of unknown water, Eq. (23) can be written

$$\sigma(\theta) = \frac{K(\theta)}{\int K(\beta) \, dz} \cdot \frac{T(\gamma,0) \, f(\beta) \cos\gamma}{\Omega} \cdot \frac{T_1}{T_2}. \qquad (24)$$

4. Determination of the Input and Output Functions of the Calibration Plate

The remaining undetermined functions of Eq. (24) are $T(\gamma,0)$ and $f(\beta)$.

It will be remembered that $T(\gamma,0)$ represents the transmittance of the calibration plate when viewed normally and irradiated at an angle from its normal.

Relative values of $T(\gamma,0)$ can be obtained experimentally by fixing the calibration plate at one or more angles, β, to the axis of the phototube system and scanning through all angles of θ. A typical recording

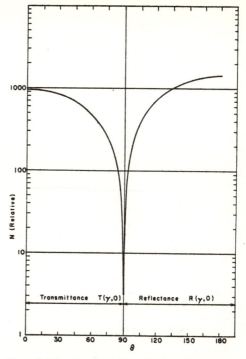

Fig. 2. Relative transmitted (0° to 90°) and reflected (90° to 180°) radiance from a typical white diffusing plastic screen.

at $\beta = 0$ is illustrated in Fig. 2. Experimental values in the region near $\theta = 90°$ are subject to greater error due to the magnitude of $dN/d\theta$ but since the change in sample volume with θ is very small in this same region, interpolation of the relative values of sample volume through 90° should not lead to large inaccuracies.

Alternatively, the value of β can be manipulated to obtain the relative volume over the whole range of θ.

In either case it becomes necessary to determine the true transmittance at some one point on the relative curve of transmittance vs θ. This should be done under water and is accomplished as follows:

The axis of the phototube system is made coincident with that of the irradiating system with $\theta = 0$ and the phototube is moved up to the position of the calibration plate as in Fig. 3(a). A diffusing plate is placed at the exit pupil of the irradiating system and the distance from this plate to the entrance pupil of the phototube system is made equal to r_1 (which for convenience is also equal to r_2).

The flux accepted by the phototube under these circumstances will be (following the pattern of the previous argument):

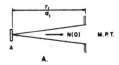

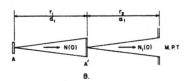

Fig. 3.   Geometry of calibration.

$$P = C_1 N(0) A\, e^{-\alpha_1 r_1}, \qquad (25)$$

where $N(0)$ is now the inherent radiance of the diffusing plate. With the diffusing screen still at A the calibration screen is next placed at A' and the phototube is moved back a distance $r_2 = r_1$ from the calibration screen, as in Fig. 3(b).   The flux accepted by the phototube is now

$$P_1 = C_1 N_1(0) A'\, e^{-\alpha_1 r_1}, \qquad (26)$$

where $N_1(0)$ is the radiance emitted by the calibration plate in the direction of its normal.   Combining (26) and (25) gives

$$\frac{P_1 e^{+\alpha_1 r_1}}{P} = \frac{N_1(0)\, A'}{N(0) e^{-\alpha_1 r_1}\, A}. \qquad (27)$$

If all of the geometric factors are kept the same in the measurements specified by Eqs. (25) and (26), it follows that

$$\frac{P_1 e^{+\alpha_1 r_1}}{P} = \frac{N_1(0)}{N(0) e^{-\alpha_1 r_1}} = T(0,0) \qquad (28)$$

which is the true transmittance of the calibration screen for normal irradiance and normal viewing (i.e., for $\gamma = 0$, $\beta = 0$) and is the desired calibration point on the curve of relative transmittance vs θ.

The function $f(\beta)$ represents the angular variation of the radiance output of the calibration screen for a specified setting of γ.   Values for $f(\beta)$ are needed only when $\beta > 0$ for when $\beta = 0$, $f(\beta) = 1$.   Values of $f(\beta)$ can be obtained by fixing the calibration plate relative to the axis of the irradiating beam and scanning through the desired range of β.   When $\beta = 0$ the flux accepted by the phototube will be

$$P(0) = C_1 N(0) A'\, e^{-\alpha_1 r_1}. \qquad (29)$$

When the phototube is turned at an angle β to the normal of the calibration screen, the flux accepted will be

$$P(\beta) = C_1 N(\beta) A'\, \cos\beta\, e^{-\alpha_1 r_1} \qquad (30)$$

Dividing Eq. (30) by (20) gives,

$$\frac{(P\beta)}{P(0)} = \frac{N(\beta)}{N(0)} \cos\beta$$

or, since

$$N(\beta)/N(0) = f(\beta),$$

$$\frac{P(\beta)}{P(0)} = f(\beta) \cos(\beta). \qquad (31)$$

Equation (31) makes it possible to determine the appropriate values of $f(\beta)$ experimentally.   It is probable that $f(\beta)$ will vary with γ, and values of $f(\beta)$ should therefore be determined for the specific experimental conditions to which they apply.

References

1. B. S. Pritchard and W. G. Elliott, J. Opt. Soc. Am. **50**, 191 (1960).
2. K. A. Stacy, *Light-Scattering in Physical Chemistry* (Academic Press, New York, 1956).
3. J. E. Tyler, J. Opt. Soc. Am. **48**, 354 (1958).
4. N. G. Jerlov, Medd. Oceanog. Inst. Göteborg **30**, Ser. B., Band 8, No. 11 (1961).

14

Reprinted from pp. 9, 10–14 of *Göteborgs K. Vetenskapsakad. Vitterh. Samh. Handl.* Ser.
B, **6** (*Medd. Oceanogr. Inst. Göteborg 30*):1–40 (1961)

OPTICAL MEASUREMENTS IN
THE EASTERN NORTH ATLANTIC

("DISCOVERY II" EXPEDITION
OF AUGUST AND SEPTEMBER 1959)

BY

N. G. JERLOV

[*Editor's Note:* In the original, material precedes this excerpt

Fig. 4. The *in situ* scattering meter in starting position.

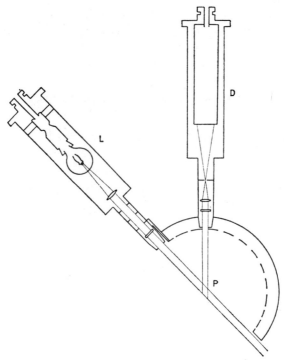

Fig. 5. Plan of the *in situ* scattering meter.

In Situ Scattering Meter

The essential parts of the meter are the lamp unit L (Fig. 4 and 5) and the detector unit D which is provided with a photomultiplier tube. The collimated beam from the lamp, which has a filament of 2×2 mm, intersects with the beam of detectivity thus defining the scattering volume element.

The meter operates in the following way. The detector unit is fixed in a vertical position facing downwards and the lamp unit is mounted close to it with an initial setting of 10° between the optical axis of the two units. When released by means of a simple mechanism the lamp unit falls slowly — checked by a paddlewheel revolving in the water — and rotates around the centre P of the scattering volume element. The rotation brings in succession 12 stops ($\Theta=10°$, 20°, 30°, 45° (15°) 165°) in front of the detector. The width of the stops is proportional to sin Θ so that the volume of the scattering element is constant for all settings. The three stops for 10°, 20° and 30° were provided with 2 mm NG 5 neutral glasses with a reduction factor of 25 %.

The electric unit was the same as used in connection with the irradiance meter and is described in Section 4. The records were taken on a Speedomax. The sensitivity of the detector was reduced once during the rotation, namely when the deflection dropped to zero between 45° and 30°.

For scattering values at low angles due regard must be paid to the slight divergence of the two beams. The correction for this was evaluated theoretically as well as experimentally with the instrument submerged in water. It was found to be important for $\Theta = 10°$ only.

Observations

The experiments were conducted at night in the surface stratum of the sea. In order to avoid disturbing stray light, most artificial light sources on board were put out.

The records for different localities show no palpable deviations, and therefore the results of 31 records for blue light (465 mμ) have been condensed into one series of relative values of $\beta(\Theta)$. With a view to the theoretical interpretation of scattering it is preferred to represent in Fig. 6 the factor $\beta(\Theta)/b$ where b is the total scattering coefficient defined by the equation:

$$b = 2\pi \int_0^\pi \beta(\Theta) \sin \Theta \, d\Theta \qquad (2)$$

The integration necessitates a rather delicate extrapolation of $\beta(\Theta)$ from 10° to 0° and another from 165° to 180° which is readily made. The extension of the curve from 10° to 0° as suggested in Fig. 6 is subject to uncertainty which may cause some error in b. It must be considered as an urgent task to furnish experimental evidence for scattering at low angles (Cf KOZLYANINOV, 1957).

The curve in Fig. 6 gives evidence of a very pronounced forward scattering. The back-scatter field in equation (2), *i. e.* the integral from $\pi/2$ to π makes up only 0.6 % of b. The scattering function attains a flat minimum at about 100° and shows an appreciable increase toward the greatest angles.

Total Scattering Coefficient

The factor $\beta(\Theta)/b$ is independent of instrumental sensitivity and will yield $\beta(\Theta)$ in absolute units if the total scattering coefficient b is directly measured. BEUTELL and BREWER (1949) introduced a

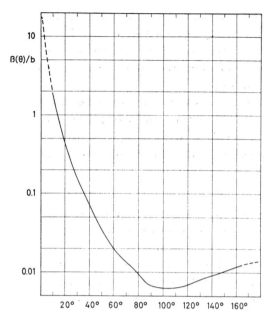

Fig. 6. The ratio between the volume scattering function, $\beta(\Theta)$, and the total scattering coefficient, b.

method for determining absolute values of b. Their meter though designed for visual use in air may serve as an effective tool in oceanographic optics provided that the theory is modified to hold good for scattering in an absorbing medium like the sea.

The oceanographic meter in principal operates as follows (Fig. 7). A small light source L emits light as a perfect diffuse radiator, *i. e.* its intensity is $I \sin \Theta$. A radiance detector is placed at O with its optical axis parallel to the surface of L and facing a light trap T. The volume element at R is irradiated by

$$E_R = \frac{I \sin \Theta}{h^2 \operatorname{cosec}^2 \Theta} e^{-ch \operatorname{cosec} \theta} = \frac{I \sin^3 \Theta}{h^2} e^{-ch \operatorname{cosec} \theta} \tag{3}$$

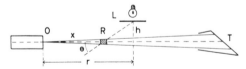

Fig. 7. Plan of the meter measuring the total scattering coefficient, b, according to the principle of BEUTELL and BREWER (1949).

The attenuation coefficient c is the sum of the absorption coefficient a and the scattering coefficient b.

$$c = a + b \tag{4}$$

The intensity scattered by the volume element dv in the direction RO is:

$$dI_0 = E_R\,\beta(\Theta)\,dv = E_R\,\beta(\Theta)\,x^2dx\,d\omega \tag{5}$$

The radiance of dv recorded at O will be

$$dB = \frac{dI_0}{x^2d\omega}\,e^{-cx} \tag{6}$$

Considering that $x = r - h \cot\Theta$ and $dx = h\,\mathrm{cosec}^2\Theta d\Theta$ it is obtained from equations (3), (5) and (6),

$$dB = \frac{I}{h}\,\beta(\Theta)\,\sin\Theta\,e^{-cr\,-ch\,(\mathrm{cosec}\,\theta\,-\cot\,\theta)}\,d\Theta \tag{7}$$

The instrument is so designed that $h \ll r$; hence integration can be performed from $\Theta = 0$ to $\Theta = \pi$. In the forward-scatter field, $0 - \dfrac{\pi}{2}$, the expression $h(\mathrm{cosec}\,\Theta - \cot\Theta)$ goes from 0 to h and in the back-scatter field, $\dfrac{\pi}{2} - \pi$, from h to ∞. We have seen, however, that the back-scatter field in ocean water contributes less than one percent of the total integral. Therefore the attenuation term $e^{-ch\,(\mathrm{cosec}\,\theta\,-\cot\,\theta)}$ can be neglected and the radiance is determined with sufficient accuracy for all practical purposes by the equation:

$$B = \frac{I}{h}\,e^{-cr}\int_0^{\pi}\beta(\Theta)\,\sin\Theta\,d\Theta \tag{8}$$

or from equation (2)

$$B = \frac{I\,b}{2\,\pi\,h}\,e^{-cr} \tag{9}$$

which with the exception of the attenuation term e^{-cr} is the same form as the function derived by BEUTELL and BREWER.

An application of this formula requires a calibration of the instrument in order to find I. This was accomplished by submerging the meter in the Göteborg institute tank which contained water with an attenuation coefficient c_1 comparable to that of ocean water. A perfect diffuse reflector Q was placed at a distance r_1 from the source L so that the line connecting L and Q is perpendicular to their plane surfaces. The irradiance at Q is

$$\frac{I}{a_1{}^2} e^{-c_1 r_1}$$

and the radiance of the reflector in any direction

$$B_1 = \frac{I}{\pi a_1{}^2} e^{-c_1 r_1} \tag{10}$$

From equations (9) and (10) it is finally found that

$$b = \frac{2h \, B}{a_1{}^2 B_1} e^{cr - c_1 r_1} \tag{11}$$

which contains only measurable quantities. It may be added that the optical distance cr is small for blue light and ocean water. It is advisable to make $cr = c_1 r_1$ so that the attenuation term vanishes.

By comparing records of the $\beta(\Theta)$-meter and those of the b-meter with both instruments submerged in the tank, the b-value pertaining to the experimental results in Fig. 6 was found to be 0.037.

[Editor's Note: Material has been omitted at this point.]

REFERENCES

Beutell, R. G., and A. W. Brewer. 1949. Instruments for the measurement of the visual range. *J. Sci. Instrum.* **26:**357.
Kozlyaninov, M. V. 1957. New instrument for measuring optical properties of sea water. *Tr. inst. Okeanol. Akad. Nauk SSSR* **25:**134.

Part V

RADIANCE DISTRIBUTION:
INSTRUMENTS AND DATA

Editor's Comments
on Papers 15 Through 21

Radiance distribution is measured in the natural light field underwater. It is a complex and difficult measurement to make because useful data depend on the stability of the natural lighting as well as on the water mass, for relatively long periods of time.

The measurement of radiance distribution has intrigued many oceanographers. The early measurements of H. Pettersson, N. G. Johnson (Jerlov) and G. Liljequist, and the theoretical work of L. V. Whitney and H. H. Poole have already been discussed. In 1940, Y. Takenouti published radiance measurements in the vertical plane for several spectral band widths. During the *Calipso* Expedition (1956) J. LeNoble obtained monochromatic ultraviolet radiance measurements in the zenith and nadir directions and in one azimuth direction.

During the period 1950 to 1960, major research programs to determine the details of underwater radiance distribution as a function of depth were being conducted by Dr. S. Q. Duntley and coworkers at the Scripps Institution of Oceanography, by Prof. T. Sasaki and coworkers at the Tokyo University of Fisheries, and by Prof. N. G. Jerlov and coworkers at the Oceanographic Institute in Göteborg.

These programs made use of remarkably different approaches. Duntley's group undertook to build a complex submersible instrument that could be oriented in both azimuth and tilt angle by remote control. Sasaki's group made their measurements while submerged in the submersible observation chamber "Kuroshio," the hatch of which was rotated to obtain horizontal radiance scans. Jerlov's group installed an east-west guide wire in the waters of Gullmar Fjord and made multiple lowerings of a simple instrument that was adjusted to a new tilt angle for each lowering. Sasaki and coworkers published their results in a series of seven or more papers between 1955 and 1961. Data obtained by Sasaki et al. (1957) clearly show the edge of the Snell circle at about 50° (the theoretical position for calm water and $n = 1.33$ is 48.6°). The loss of contrast at the edge of the Snell circle as a result of increasing depth is clearly shown, and the elongation of the lower hemisphere radiance distribution due to the low sun zenith angle (42.3°) is also evident in his data. In 1958 Sasaki et al. described a new instrument for measuring radiance distribution and reported measurements of radiance in the horizontal plane. These data demonstrated the effect of sun altitude on the horizontal distribution of the radiance. In 1961 Prof. Sasaki, presented a paper at the 10th Pacific Science Congress in Hawaii (reproduced here as Paper 15). He discussed his program of research and presented new radiance distribution data for the vertical plane including the sun's direction. Jerlov and M. Fukuda (1960) published their experimental results and included a theoretical computation for radiance distribution, based on the independent treatment of sunlight, skylight, and scattered light. Their graphical data show the shift in position of the maximum radiance vector toward the zenith direction as depth is increased. Their data also show the edge of the Snell circle, in measurements near the air-water interface.

The work of Duntley's group was reported in several papers. Details of the instrument were described by Duntley et al. (1955). The results of radiance measurements obtained at Lake Pend Oreille (Idaho, USA) were reported in papers by J. E. Tyler. The phenomena of the Snell circle and its change in contrast with increasing depth were critically examined by Tyler under ideal conditions of overcast sky and flat-calm water surface (Paper 16).

In 1960 Tyler published the extensive numerical data for radiance distribution under clear, sunny sky conditions and under totally overcast conditions. The data had been obtained during the field work at Lake Pend Oreille in 1957. A selection from this work appears here as Paper 19.

These investigations by Sasaki, Tyler, and Jerlov are in full agreement with the predictions of Whitney and Poole and others. They foresaw that the distribution of radiance under water would be strongly influenced by the relative strength of scattering and absorbing phenomena, and that at some depth in homogeneous water, the radiance distribution would reach an asymptotic shape, called asymptotic radiance distribution.

R. W. Preisendorfer, Paper 20, has used phenomological theory to derive a general equation that demonstrates the limiting form of radiance distribution in water having specified inherent and apparent optical properties. His result formally confirms, in greater detail, the earlier work of W. R. G. Atkins, L. V. Whitney, H. H. Poole, and others who argued that the asymptotic shape of the radiance distribution was controlled by scattering, which increased the obliquity of the distribution, and absorption, which decreased the obliquity.

To be fully useful for the computation of the apparent optical properties of an ocean water type, radiance distribution data must be obtained under stable surface lighting conditions. Scanning instruments such as those used by Jerlov, Sasaki, and Tyler are slow and tedious to operate. With these instruments it is consequently necessary to confine the data taking to periods when the altitude of the sun is not changing rapidly, a limitation that controls either the number of data points obtainable at a single depth or the number of depths that can be scanned.

In order to avoid this limitation R. C. Smith et al. constructed the photographic radiance meter described in Paper 17. This instrument was successfully used by Smith to obtain radiance distribution data in the Mediterranean Sea, Paper 18.

The natural light field underwater exhibits polarization that varies in magnitude and orientation with the angle of observation, relative to the sun's direction. This phenomenon was investigated in considerable detail by T. H. Waterman (1954, 1955, 1956) and by A. Ivanoff (1956). Waterman's interest stemmed from his work on the orientation of marine animals, whereas Ivanoff was more interested in the use of underwater polarization to describe and differentiate ocean masses. In

1958 these two scientists cooperated in a detailed study of underwater polarization and its variability with depth and wavelength (Paper 21).

REFERENCES

Duntley, S. G., R. J. Uhl, R. W. Austin, A. R. Boileau, and J. E. Tyler. 1955. An underwater photometer. *Opt. Soc. Am. J.* **45:**904(A)

Ivanoff, A. 1956. Facteur de Polarisation du résidu sous-marin de lumière du jour. *Ann. Geophys. (Paris)* **12**(1):45.

————. 1956. Au sujet du facteur de polarisation du résidu sous-marin de lumière du jour, sur la ouest de la Corse. *Comptes rendus des seances de l'Academie des Sciences* **243:**1345, 1430.

————. 1974. Polarization measurements in the sea. In N. G. Jerlov and E. Steemann Nielsen, eds., *Optical Aspects of Oceanography*. Academic Press, Inc., London, pp. 151–175.

Jerlov, N. G., and M. Fukuda. 1960. Radiance distribution in the upper layers of the sea. *Tellus* **12**(3):348.

Lenoble, Jacqueline. 1958. Resultats Scientifiques des Campagnes de la "Calypso" Fascicule III. Mason et Cie, Editeurs, Paris, 40 pp.

Sasaki, T., S. Watanabe, G. Oshiba, and N. Okami. 1958. Measurements of angular distribution of submarine daylight by means of a new instrument. *Oceanogr. Soc. Jpn. J.* **14**(2):47.

Takenouti, Y. 1940. Angular distribution of submarine solar radiations and the effect of altitutde of the sun upon the vertical extinction coefficient. *Jpn. Soc. Sci. Fish. Bull.* **8**(5):213.

Waterman, T. H. 1954. Polarization patterns in submarine daylight. *Science* **120**(3127):927.

———— and W. E. Westell. 1956. Quantitative effect of the sun's position on submarine light polarization. *Sears Found. J. Mar. Res.* **15**(2):149.

Waterman, T. H. 1955. Papers in Marine Biology and Oceanography, Contribution No. 763. Woods Hole Oceanog. Inst., Woods Hole, Mass., pp. 426–434.

15

Reprinted by permission from pp. 19–24 of *Physical Aspects of Light in The Sea: A Symposium*, J. E. Tyler, ed., Honolulu, Hawaii: University of Hawaii Press, 1964, 64 pp.

ON THE INSTRUMENTS FOR MEASURING ANGULAR DISTRIBUTIONS OF UNDERWATER DAYLIGHT INTENSITY

TADAYOSHI SASAKI

Tokyo University of Fisheries, Tokyo, Japan

THE STUDIES ON THE ANGULAR distribution of daylight penetrating into water were begun by Pettersson, Jerlov, Liljequist, *et al.* Many physical oceanographers took an interest in these studies, measured the angular distribution by various instruments constructed according to their own ideas, and published reports.

With the development of instruments, detailed data on the angular distribution of underwater daylight as a function of depth have been obtained. The data serve to explain the theory of radiative transfer through a hydrosol and have enabled measurement of the complicated radiance distribution of upper ocean layers.

In 1951, in collaboration with Inoue, I constructed the undersea observation chamber *Kuroshio.* After determining the position of the sun, relative to the point where measurements were carried out, the measurement of the angular distribution of the underwater daylight was performed, using this chamber. The measurement was made with a light receiver having a resolving power of 10° and containing a photomultiplier tube (RCA 931-A). The studies of angular distribution by means of *Kuroshio* were carried out for three years; however, for better control in selecting points, time, and frequency of measurements, a remote control instrument suspended from a ship was needed. Accordingly, we attempted, and in 1958 succeeded, in the construction of an instrument for measuring the horizontal angular distribution of underwater daylight. In 1959 we accomplished the construction of a remote control instrument for both vertical and horizontal angular distribution of underwater daylight. In order to determine the sun's azimuth in the measurement of the vertical plane, a method was first adopted to turn the horizontal light receiver in the direction of maximum in-

tensity of horizontal underwater daylight. Later, this method was modified to include a bearing indicator.

We then constructed a new, improved instrument to increase the accuracy of measurement, and repeated measurements have confirmed that sufficient accuracy is possible.

The underwater unit suspended by the cable is shown in Figure 1 and it is diagrammed in Figures 2 and 3. The underwater unit is low-

FIG. 1. Sketch of the light path.

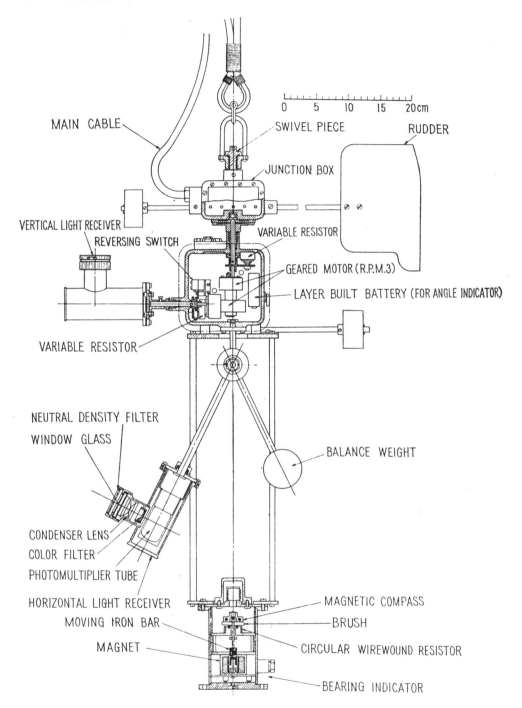

MAIN CABLE

SWIVEL PIECE

RUDDER

0   5   10   15   20cm

JUNCTION BOX

VERTICAL LIGHT RECEIVER

REVERSING SWITCH

VARIABLE RESISTOR

GEARED MOTOR (R.P.M.3)

LAYER BUILT BATTERY (FOR ANGLE INDICATOR)

VARIABLE RESISTOR

NEUTRAL DENSITY FILTER

WINDOW GLASS

BALANCE WEIGHT

CONDENSER LENS

COLOR FILTER

PHOTOMULTIPLIER TUBE

HORIZONTAL LIGHT RECEIVER

MOVING IRON BAR

MAGNET

MAGNETIC COMPASS

BRUSH

CIRCULAR WIREWOUND RESISTOR

BEARING INDICATOR

FIG. 2. General arrangement of the underwater unit.

191

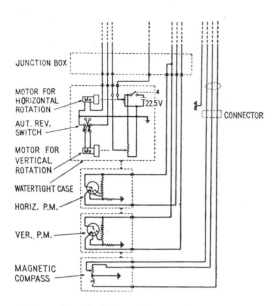

JUNCTION BOX

MOTOR FOR
HORIZONTAL
ROTATION

AUT. REV.
SWITCH

MOTOR FOR
VERTICAL
ROTATION

WATERTIGHT CASE

HORIZ. P.M.

VER. P.M.

MAGNETIC
COMPASS

22.5V

CONNECTOR

FIG. 3. Electrical circuit diagram of the underwater unit.

tical light receiver will be perpendicular to the vertical plane with respect to the sun. If the horizontal light receiver is manually turned 90° from the former position, then the intensity in the 0° vertical plane will be measured by the vertical light receiver.

With increased solar altitude, turbidity of the water, cloudiness of the sky, and depth to be measured, the determination of the maximal intensity of horizontal light by the horizontal light receiver becomes gradually more difficult. If the horizontal light receiver is inclined to the direction of the refracted light of the sun, the accuracy of the maximal intensity measurement will be increased slightly; however, the difficulty described above will remain. In this case the plane being scanned by the vertical light receiver is determined by the bearing indicator. This indicator consists of a magnetic compass, a circular wire-wound resistor, an electric magnet, and other mechanisms in a watertight housing. The magnetic compass consists of two magnets, each 50 mm long and 5 mm in diameter. The jewelled compass bearing is supported by a hardened steel pivot to the vertical shaft attached to the iron bar of the electric magnet. A long brush and a shorter one are attached to the lower part of the compass and the circular wire-wound resistor and a metal ring are below the brushes. The iron bar is supported by a coil spring and the compass points freely in N-S direction. When current is passed through the coil, the iron bar is drawn up and the two brushes contact the resistor and the metal ring respectively. The value of the resistor changes according to the bearing. The housing is filled with transformer oil in order to dampen the compass.

ered into the sea by a main cable through a swivel piece having a rustproof ball bearing. The rudder attached to the junction box stabilizes the unit within the horizontal plane in order to prevent reaction when the photometer rotates horizontally. If the geared motor, connected by the vertical shaft to the junction box, rotates, the watertight housing will rotate horizontally and involve the horizontal light receiver, the vertical light receiver, and the bearing indicator. The rotary angle, limited within 340°, will be indicated on shipboard by means of the variable resistor connected to the vertical shaft. When measuring the angular distribution in a horizontal plane, the horizontal light receiver must, of course, be directed to the horizontal.

The vertical light receiver is rotated by another motor. The rotation, limited within 340° by a reversing switch, is reversed automatically and the rotary angle is indicated on shipboard in the same way as that of the horizontal rotary angle.

When the optical axis of the horizontal light receiver and the rotary shaft of the vertical light receiver are placed in the vertical plane, then when the maximal light intensity is received by the horizontal light receiver, the ver-

Each light receiver consists of condenser lenses, a green filter, a diaphragm with a small hole, and a photomultiplier tube in a cylindrical watertight housing fitted with a glass window. A suitable neutral density filter may be attached to the window to decrease the light intensity when it is too great. By means of the lenses and a diaphragm inserted between the glass window and the photomultiplier tube, a resolving power of 3.8° is obtained in water. The center of gravity of the spectral sensitivity curve of the photomultiplier tube (green filter, Matsuda VG-1) is about 520 μ.

The measurement is carried out by supplying to the phototube 800 volts, 1000 volts, and 1200 volts from a stabilized high voltage power supply. In the three steps of voltage, the changing ratio of the radiance to the photocurrent is corrected; therefore, all measurements can be compared relatively.

Figure 3 shows the electrical circuit diagram of the underwater unit.

The sky photometer, used to measure the angular distribution of the sky light, is shown in Figure 4. The main parts of the sky photometer are a scaled disk for bearing indication kept in the horizontal plane by gimbals and balance weight, a photocell with a resolving power of 10° stabilized by a cylindrical hood, a magnetic compass attached to the outer case, and a microammeter. The adjustment of the incident light intensity is done by changing suitable neutral density filters.

Figures 5 and 6 show the electrical circuit diagrams of the operating unit and the bearing indicator, both located on shipboard. Figure 7 shows the angular distribution in the vertical plane with respect to the sun measured at a depth of about 20 cm below the surface off Ito City, Shizuoka Prefecture, measured under a clear sky in June, 1960. In this figure the curve at zenith angle —90° is level; however, that at +90° clearly indicates the maximum. J. E. Tyler in 1958 and N. G. Jerlov in 1960 published reports on the peak appearing at +90°. Jerlov clearly explained by theoretical calculation that the peak is due to the scattered light from below the surface totally reflected at the surface. It is considered that the levelling

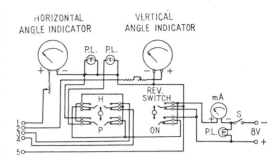

FIG. 5. Electrical circuit diagram of the operating unit on shipboard.

off at —90° may be due to the surface effect.

Figure 8 shows the angular distribution in the vertical plane with respect to the sun at depths to 40 m below the surface measured off Ito City in July 1961. The sky at the time of measurement was covered with thin clouds; however, the position of the sun could be accurately ascertained. The water temperature at the surface was 26.5°C and decreased gradually with increase of depth. The temperature at 20 m below the surface was 19.7°C and that at 50 m was 15.2°C. The vertical distribution of the turbidity in the region decreased gradually at depths to 10 m below the surface and at depths more than 10 m became almost constant and yielded similar distributions of turbidity. The curve of the angular distribution at a depth of 1 m below the surface decreased rapidly at +60° and —40° and for about 20°, centering at +80° and —60° respectively; the curve lies under the curves of 3 m and 5 m depth. This tendency is similar to that measured by Tyler under overcast sky conditions.

Figure 9 shows the angular distributions at depths to 15 m below the surface in the ver-

FIG. 4. The sky photometer.

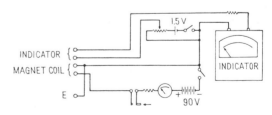

FIG. 6. Electrical circuit diagram of the bearing indicator on shipboard.

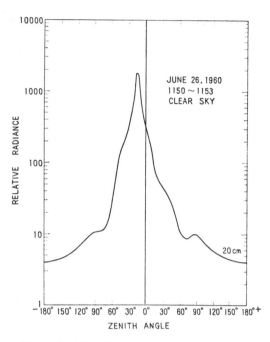

FIG. 7. Zenith angle.

the sky light is unpolarized and that the sea surface is flat. Using Fresnel's formula, calculate the sky light penetrating into the sea, taking into consideration the correction for the decrease of the solid angle. The relation between this result and refracting angle is plotted on the dotted line of Figure 9. As clearly shown, the maximum appearing at −47° in the air moves to about −33° in sea water, the minimum at −20° in the air moves to about −15° in sea water, and the minimum appearing at −58° in the air does not appear in sea water. This may occur because of the rapid decrease of the penetrating light at incidental angles above 58°. In the calculated distribution of sky light below the surface, the maximum and minimum just below the surface are not as distinct as in those at a depth of 1 m below. This is due to the fact that the resolving power of the underwater light receiver is 3.8°, while that of the sky photometer is 10°.

It has been demonstrated that the radiance distribution in the near-surface layer of the sea

tical plane with respect to the sun off Usami, Shizuoka Prefecture, measured in July 1961 under a cloudy sky. The water temperature at the surface was 26.4°C and that at the depth of 3 m was 27.8°C. With continued increase of depth the temperature decreased gradually; that at a depth of 10 m was 23.3°C, and that at 20 m was 18.8°C. In observing the vertical distribution of turbidity we found that the layer from 3 m to 10 m below the surface was homogeneous and had less turbidity than those at depths above and below this layer. In the curve of the angular distributions at 1 m below the surface the maximum appears at zenith angle of −30° and the minimum appears at −15°. Both maximum and minimum appear in the curves at 3 m and 5 m, but not as distinctly as those at 1 m. This phenomenon becomes indistinct with increase of depth and disappears completely at the depth of 10 m. The radiance distribution of the sky light is shown in the upper portion of Figure 9. The minima appear at −20° and −58° in the distribution curve of the sky light and the maximum appears at −47°. To show how the distribution of the sky light changes in the sea water, assume that

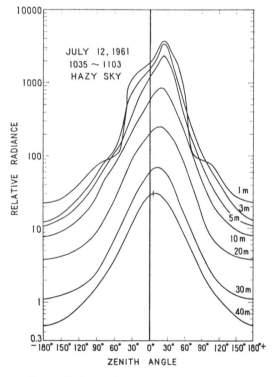

FIG. 8. Zenith angle.

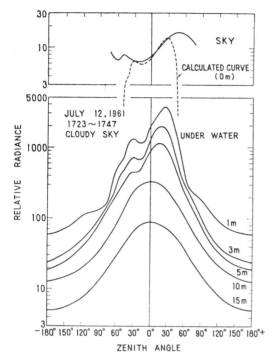

FIG. 9. Zenith angle.

depends considerably on the radiance distribution in the sky. Accordingly, my collaborators and I will attempt to improve instruments capable of automatically measuring the radiance distribution of underwater light and sky light simultaneously. If we succeed we will continue our studies with radiance distribution.

16

Reprinted from Sears Found. J. Mar. Res. 16(2):96–99 (1958)

COMPARISON OF LIGHT DISTRIBUTION ABOVE AND BELOW WATER[1]

By

JOHN E. TYLER

Scripps Institution of Oceanography, University of California
La Jolla, California

ABSTRACT

Recent measurements of natural light distribution in homogeneous lake water have revealed interesting and heretofore unrecorded optical effects in the near-surface layers. These effects are graphically described and explained.

From 1935 to 1945, studies of the angular distribution of natural light underwater revealed many interesting features of the underwater light field (1, 2, 4, 5, 7, 8). Concurrently, instrumentation in this field advanced with the development of the "shadowing-screen" photometer, described by Pettersson (1) and by Johnson and Liljequist (2), and of the Gershun (3) tube photometer, which directly limited the solid angle of acceptance. However, the work was performed at depths below five meters, and the instruments were limited to a low resolving power (15 to 20°). Consequently, no information was obtained on the distribution of light in the near-surface layers.

It is clear that, from a point just below the calm surface of a large body of water, the directional distribution of natural light overhead will be a compressed version of the scene above the surface. This is in keeping with Snell's law of refraction. As the depth of observation is increased, the light field will change from this complex structure; image detail will be lost and the shape of the radiance distribution solid will become smooth; with increasing depth the brightest spot in the field will move toward the zenith and the shape of the radiance distribution solid will become more symmetrical around the vertical axis. Finally, at some unknown depth, the radiance distribution solid becomes substantially fixed in shape.

Because of the exponential nature of the attenuation of light in water, these changes in light distribution will take place most rapidly in the surface layers. Thus the upper layers are at once the most interesting, the most complex, and the most difficult to explore.

[1] Contribution from Scripps Institution of Oceanography, University of California, New Series. This work has been supported by the Bureau of Ships and by the Bureau of Aeronautics under Contract NObs-72092.

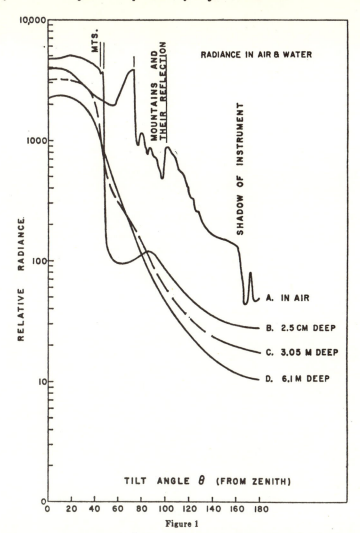

Figure 1

During recent field work at Lake Pend Oreille, Idaho, details of some of the optical effects that occur in these upper layers were explored with an underwater photometer having a resolving power of 6.6°; this underwater photometer was first described in 1955 (6). Measurements of the radiance distribution in a single vertical plane were made on an overcast day,[2] first in air and then from different depths beneath the surface. The fully corrected data are shown in Fig. 1.

[2] The sun was not visible. Its position was approximately 45° west of the instrument heading and at an altitude of 61.0°.

Curve A shows results of observations taken in air. From $\theta = 0°$ to $\theta = 74°$ the curve reports the radiance of the overcast sky; at $\theta = 74°$ the dark mountainous skyline is recorded, and this is followed by detail representing patches of dark rock or snow. From $\theta = 90°$ to $\theta = 102°$ the reflection of the mountains is recorded and this is then followed by reflection of the sky disturbed by slight wave action. At 162° the instrument's own shadow image is recorded.

Curve B shows the radiance distribution in the same direction as that in curve A but from a depth of 2.5 cm below the water surface. In the zenith direction, B shows higher radiance than A because the space light contributed by 2.5 cm of water is greater than the absorption of this same path. (It is also possible that some change took place in the zenith overcast during the short time interval between curves A and B.) Optical compression of the sky tends to keep the radiance at a high level for the first 40°. The skyline itself is very much compressed and can no longer be resolved, but the edge of the Snell circle of light appears sharp and distinct at 48°. Beyond 48° the instrument measures the radiance of the short path to the surface plus a longer reflected path that looks down into the water at an angle equal to $(180° - \theta)$. This accounts for the changes in slope of the curve between 48 and 88°. Beyond 88° the path radiance is due entirely to space light from one direction only, which decreases steadily as the tilt angle increases to 180°.

Curve C shows the radiance distribution in the same direction as curves A and B but from a depth of 3.05 m. Some detail can still be observed. The steep slope between 40 and 48° is the remaining radiance change at the edge of the Snell circle of light. The dark band at 60° exhibited in B is nearly obscured by space light; in C, between 50 and 80°, it can just be seen together with its brighter rim.

Curve D shows the radiance distribution in the same direction as the other curves but from a depth of 6.1 m. At this depth all fine detail is lost, leaving only the gross changes in radiance between zenith and nadir settings.

Between 90 and 180° the underwater curves for various depths can be approximately predicted from the equation

$$N_D = N_0\, e^{-CD},$$

where N_D is the radiance at depth D, N_0 is the radiance at depth 0 and $C = f(\theta)$. Below 6.1 m the coefficient C tends to become independent of θ, approaching an asymptotic value which is characteristic of the hydrosol.

REFERENCES

(1) PETTERSSON, HANS
  1938.   Measurements of the angular distribution of submarine light.   Rapp.
          Cons. Explor. Mer, *108* (2): 9.
(2) JOHNSON N. G. AND G. LILJEQUIST
  1938.   On the angular distributions of submarine daylight and on the total sub-
          marine illumination.   Svenska hydrogr.-biol. Komm. Skr., Hydrogr. *14:* 3.
(3) GERSHUN, A.   (Translated by PARRY MOON)
  1939.   The Light-Field. J. Math. Phys., *18:* 51–151.
(4) POOLE, H. N.
  1945.   The angular distribution of submarine daylight in deep water.   Sci. Proc.
          R. Dublin Soc., *24* (4): 29–42.
(5) WHITNEY, L. V.
  1941.   The angular distribution of characteristic diffuse light in natural waters.
          J. mar. Res., *4* (2): 122–131; J. opt. Soc. Amer., *31:* 714–722.
(6) DUNTLEY, S. Q., *et al.*
  1955.   An underwater photometer.   J. opt. Soc. Amer., *45:* 904(A).
(7) TAKENOUTI, YOSITADA
  1940.   Angular distribution of submarine solar radiations and the effect of altitude
          of the sun upon the vertical extinctions coefficient.   Bull. Jap. Soc. sci.
          Fish., *8:* 213–219.
(8) JERLOV, N. G. (JOHNSON)
  1951.   Optical studies of ocean waters.   Reports of the Swedish Deep-sea Expedi-
          tion, 1947–1948.   *3*, Physics and Chemistry #1: 1–59.

An Oceanographic Radiance Distribution Camera System

R. C. Smith, R. W. Austin, and J. E. Tyler

An oceanographic optical instrument has been designed and constructed to record the radiance distribution of natural radiant energy underwater. The instrument contains two cameras placed back-to-back, each equipped with a fisheye (180° field of view) lens and is fabricated so that film can be exposed by remote control. The instrument is designed to operate at depths up to 100 m. Values of underwater radiance can be obtained from the exposed films by means of photographic photometry. Radiance distributions obtained in natural waters will provide basic information needed for the study and solution of several problems in optical oceanography. First, the radiance distribution is an important input for the study of the interaction of electromagnetic radiation with the sea. From the distribution of radiance in the natural light fields in the ocean, many of the important optical properties which relate to radiative transfer processes in the ocean can be calculated. Second, from these optical properties one can compute the magnitude of the deterioration of image contrast of submerged objects and thus furnish information for the study and solution of underwater visibility problems, which include problems in underwater television and photography. Finally, since radiant energy is critical to the beginning of the marine food chain through photosynthetic plankton, radiance distribution measurements will provide information of fundamental importance to the problem of primary productivity in natural waters.

I. Radiance Distribution and Its Usefulness

A. Introduction

Radiance is the power per unit area per unit solid angle incident on a point from a specific direction. A radiance distribution is the totality of radiance values for every direction about the point.

The general form of the distribution of radiance underwater was explored by Jerlov,[1] Whitney,[2,3] Takenouti,[4] and others. Additional incomplete data giving the relative magnitude of radiance in a few directions were obtained by Jerlov and Fukuda[5] and Sasaki *et al.*[6–8]

The importance of complete radiance distribution data was recognized by early workers in optical oceanography, but it was not until 1967 that the first sufficiently complete sets of data were obtained in lake water by Tyler.[9] Computations from this data were described by Tyler *et al.*[10] and by Tyler and Shaules.[11]

The instrumentation used by Tyler to obtain the 1957 data was unsuitable for general oceanographic research. His work, however, confirmed the usefulness of radiance distribution data and stimulated

The authors are with the Visibility Laboratory, Scripps Institution of Oceanography, University of California at San Diego, San Diego, California 92152.

Received 4 February 1970.

further efforts to devise a simple method for its measurement. The equipment described below is the outgrowth of those efforts.

B. Radiative Transfer Theory and Optical Properties of the Ocean

Many of the important optical properties which relate to radiative transfer processes in the ocean can be calculated from radiance distribution data on natural light fields underwater. Although numerous measurements have been made of some single optical property of the ocean, almost without exception important related properties have not been determined. As a consequence, most of these data have been of limited usefulness. The radiance distribution cameras described herein surmount this problem to a large extent by making it possible to record the essential radiance distribution data rapidly and completely. The usefulness of radiance distribution data for the description and study of underwater light fields is outlined in Fig. 1.[12]

On the chart shown in Fig. 1, $N(Z,\theta,\phi)$ represents radiance distribution data taken at various depths, Z, where θ is the zenith angle and ϕ the azimuth of a radiance value. From the radiance distribution data, the optical properties shown on the chart can be determined.

The irradiance, H, is the power per unit area arriving at a point on a surface. $H(-)$ is the downwelling irradiance; i.e., it is the power per unit area measured

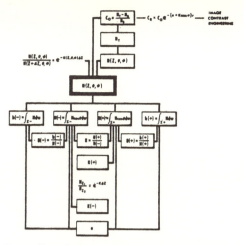

Fig. 1. Chart illustrating optical properties which can be derived from radiance distribution data. The properties are defined and discussed in the text.

by a horizontally oriented cosine collector facing upward. Similarly, $H(+)$ is the upwelling irradiance. The rate of change of horizontal irradiance with depth provides the essential information for calculating other optical properties of the hydrosols.

With appropriate spectral consideration the horizontal irradiances have frequently been useful in problems dealing with visual tasks, camera systems, and photosynthesis.

On the chart in Fig. 1, $h(-)$ and $h(+)$ are the downwelling and upwelling scalar irradiances, respectively, which when summed give the total scalar irradiance. The total scalar irradiance, when divided by the velocity of light in the medium, gives the total amount of energy per unit volume at a given point. To date, accurate measurements of scalar irradiance have not been made in natural waters.

The distribution functions, $D(\pm)$, provide simple means for characterizing the depth dependence of the shape of the radiance distributions. In addition, these functions play important roles in the equations of applied radiative transfer theory, particularly in those equations which link the inherent and apparent optical properties of the medium.

The K functions, shown in the chart, are defined as the logarithmic depth derivatives of the irradiance functions. Thus,

$$K(Z) = (-1/H)(dH/dZ), \qquad (1)$$

or by solving for H one obtains the relationship

$$H(Z_2)/H(Z_1) = \exp[-K(Z_2 - Z_1)], \qquad (2)$$

where K has units of reciprocal length, and Z is the

depth at which H is measured. Physically, the K functions are the quantities that specify the individual depth dependence of irradiance functions. Historically, the K functions were derived from the experimental fact that, in general, radiant energy decreases exponentially with depth. These functions have been used in connection with both theoretical and experimental aspects of radiant energy in natural waters.

Preisendorfer has shown, by a method which makes use of the divergence relation of the radiant energy field, that the volume absorption function a, can be computed without the requirement of previous knowledge of the volume attenuation function, α, or the volume scattering function, s. This may be done by dividing the rate of change with depth of the net upwelling irradiance by the scalar irradiance, that is,

$$a(Z) = -1/h(Z)[d\bar{H}(Z,+)/dZ], \qquad (3)$$

where

$$\bar{H}(Z+) = H(Z,+) - H(Z,-) \qquad (4)$$

is the net upwelling irradiance, and

$$h(Z) = h(Z,-) + h(Z,+) \qquad (5)$$

is the total scalar irradiance.

The Scripps Spectroradiometer[13] was developed to measure irradiance as a function of wavelength, and measurements have been made in a variety of lake[14] and ocean[15] waters. The measurements of spectral irradiance, made at the same time as radiance distribution measurements, complement and aid in interpreting the radiance distribution data.

C. Image Contrast Engineering

The upper part of Fig. 1 summarizes the application of radiance measurements to the problems of image contrast engineering. For an object to be visually detectable, it must differ photometrically from its background in the direction of view of the sensor. Whatever difference does occur constitutes an optical signal. This optical signal at the object can be described by its inherent contrast[16] defined as

$$C_o = (N_T - N_B)/N_B, \qquad (6)$$

where N_T is the inherent radiance of the object and N_B is the inherent radiance of the background against which the object is seen. Thus the first major step in any detection problem is the assessment of the magnitude of the inherent contrast. In computing the inherent contrast of a submerged object, the background radiance, N_B, can be obtained directly from the radiance distribution. The inherent radiance of the object, N_T, depends on the distribution of the flux incident upon the object and the angular reflectance of the object. Thus knowledge of the gonioreflecting properties of the object and the radiance distribution of the light field in which the object is immersed, leads directly to the determination of inherent contrast for any path of sight.

Once the inherent contrast has been determined, the equation of contrast reduction,

$$C_R = C_o \exp[-(\alpha + K \cos\theta)r], \qquad (7)$$

can then be applied to problems of the visibility of submerged objects. Here C_R is the apparent contrast seen by a sensor at a distance, r, along a path, α is the volume attenuation function (the quantity measured by a transmissometer), and K is the diffuse attenuation function (for radiance) which can be calculated from the radiance distribution data by the equation defining $K(Z,\theta,\phi)$ in Fig. 1. The θ, as used in Fig. 1 and Eq. (7), is the angle between the vertical and the line of sight. Using Eq. (7), the apparent contrast can be calculated along any path of sight once the inherent contrast of the object has been determined. Thus it can be seen that the key to the problem of determining the contrast and detectability of submerged objects is a knowledge of the natural radiance distribution in the water in question.

D. Biological Applications

Radiance distribution information also provides new data and information which should be of interest to biologists. Scalar irradiance is a measure of the total energy per unit volume of space at a given point. It thus gives a measure of total energy incident from all directions on an organism in the water. The volume absorption function, a, refers to the conversion of radiant energy into other forms of energy (heat, chemical, another wavelength of radiant energy, etc.). It gives the actual energy absorbed by the medium (water plus organism) and is independent of the attenuation of energy due to scattering (without change of wavelength). Therefore, radiance distribution measurements, correlated with appropriate biological data can provide new information on the problem of energy utilization by photosynthetic organisms.

II. Fisheye Lens and Camera

A. Properties of the Fisheye Lens

The Nikkor 8 mm, $f/8$ fisheye lens, manufactured by Nippon Kogaku K.K. (Japan Optical Industry Company), has been chosen to measure the radiance distribution in the underwater environment. The properties of this lens which make it useful for this application, as well as its limitations, have been investigated in detail. This lens is not an extension of a wide-angle lens[17] and does not exhibit the usual distortions associated with such a lens. However, the fisheye lens has an inherent distortion which should not be considered an aberration, but rather the result of projecting a hemisphere onto a plane. Referring to Fig. 2, let the zenith angle of an incident ray from an infinite object be θ and the coordinates of the image of this ray on the plane of the film be (r^1,ϕ^1). The projection of the Nikkor $f/8$ fisheye lens conforms to an equidistant projection to within 5%[18]; that is,

$$r^1 = f\theta, \qquad (8)$$

where f is the lens focal length. Properties of the fish-

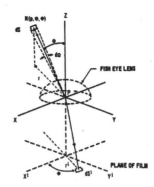

Fig. 2. Schematic diagram showing projection geometry of fisheye lens for recording radiance distributions on film.

eye lens which are of interest follow as a consequence of this projection.

In Fig. 2, $d\Omega$ is an element of solid angle in object space and dS^1 is the corresponding area in the image plane, and for an equidistant projection the following relation holds:

$$d\Omega = f^{-2}(\sin\theta/\theta)dS^1 \qquad (9)$$

The field radiance, N, is the energy per unit time per unit area per unit solid angle incident on a point P (projection origin of the lens) from the direction (θ,ϕ). The exposure given to a film, as used in sensitometry, is proportional to the amount of radiant energy per unit area incident on the film. Miyamoto has shown, using the above expressions, that the irradiance received by the film during the time of exposure, E, has a one-to-one correspondence to the field radiance, N, as shown in Eq. (10),

$$E = N(n/n^1)^2(\pi/4F^2)(\sin\theta/\theta). \qquad (10)$$

Here n and n^1 are the indices of refraction of object and image space, respectively, and F is the f/number of the optical system. Equation (10) shows that for a uniform radiance distribution, the irradiance of the film during exposure is relatively uniform, following a $\sin\theta/\theta$ relation, compared to ordinary lenses which obey the $\cos^4\theta$ law. This relative uniformity, the known one-to-one correspondence between the image exposure of the film and the incident radiance distribution, and the hemispherical field of view, are the reasons why this particular lens is ideally suited for the purpose of measuring radiance distributions. Thus, by the methods of photographic photometry, the density of an area in a film negative can be related to the exposure of that area, and this can, in turn, be related to the desired value of field radiance.

The Nikkor $f/8$ fisheye lens has one other feature which is necessary for our purposes, a built-in, six-position filter wheel. Neutral density filters have been mounted in this wheel to accommodate the six to eight order of magnitude range in radiance levels encountered when the instrument is near the surface under sunny conditions. No simple photographic film is available with a latitude sufficient to cover this range of radiance values. It is therefore necessary to expose a number of frames of film (one to four), using different neutral density filters which span the required range for each radiance distribution position. Thus a radiance distribution at a given depth is recorded on a number of film frames, each frame with an exposed area covering roughly two orders of magnitude of useful radiance information. To reduce the data from the films it is then necessary to sift the useful information, (i.e., correctly exposed), from each frame and then combine this information to form one complete radiance distribution.

One other comment is in order at this point. Mees[19] has stated that "... hardly any type of measurement contains so many pitfalls for the unwary as photographic photometry." These pitfalls, and the attendant lack of precision and inaccuracies, must be balanced against the chief advantage of photographic photometry; namely, that it can record a vast amount of information quickly on a single piece of film. Since a radiance distribution inherently represents a large amount of information, even with modest resolution and a vast amount of information with high resolution (approximately 50,000 bits of information with a 1° resolution), the acquisition of such data strongly indicates a photographic approach.

B. Tests of the Fisheye Lens

Our preliminary tests of the fisheye lens indicate that it conforms to an equidistant projection to within 2% between 10° and 90°, and that it varies by as much as 5% only in the 0–10° (zenith) region. Similar tests have been published by Crawley.[20] We have also measured the $\sin\theta/\theta$ dependence [Eq. (10)] and have found it to be correct to within 5%.

For use underwater, the fisheye lenses have been protected with acrylic plastic domes. These domes and the change of refractive index from an air to a water environment affect the projection properties of the lenses. Thus, final testing of the lens and its protective underwater housing must be made underwater.

These tests are of a calibration nature; that is, once the projection properties of a lens plus its protective dome have been measured in water, any departure from the theoretical projection is known and is a fixed property of the system. The measured, rather than the theoretical, projection properties will then be used to analyze the data, and these properties will not degrade the accuracy of the results.

On the other hand, there are characteristics of the lens system which are not fixed but vary with the environmental radiance distribution. A major concern in considering the fisheye lens for radiance measurements is the problem of flare light.[21] For the purposes of the intended measurement, it is important that the non-image-forming light due to flare be small compared with the image-forming light from the sought for radiance distribution. If this is not the case, we must be able to detect and correct for flare. For the discussion to follow we will speak of error due to flare as the ratio, at a point on the film, of the non-image-forming irradiance due to flare light to the image-forming irradiance coming directly from the object.

A Nikkor fisheye lens was tested for the purpose of making quantitative estimates of the effects of flare. For our purpose it was sufficient to consider flare light of the fisheye lens for a typical underwater radiance distribution. Actually, an atypical situation was studied by choosing the most extreme underwater lighting conditions likely to be encountered in order to determine the upper limit of the error that may be expected due to flare light. The environmental scene chosen for these flare studies was the underwater radiance distribution of the upper hemisphere on a clear, sunny day at a shallow depth.[22]

The Sky Simulator of the Visibility Laboratory proved to be a valuable facility for the flare light study. This lighting simulator is a 2.44-m hemisphere containing 196 individually controllable luminous panels which can accurately simulate different distributions of natural lighting. A number of techniques, which yielded similar results, were used to evaluate the error due to flare; the most straightforward being simply to photograph a simulated radiance distribution having symmetry about the vertical axis. To do this, one azimuthal sector of the simulator was unilluminated during exposure, i.e., left dark. Thus, a photograph of this radiance distribution taken with the fisheye lens showed the circular projected image of the hemispherical radiance distribution with a pie-shaped wedge corresponding to the black sector. If there were no flare, this wedge would be completely unexposed. On the other hand, any density found in the negative in the unexposed area is a measure of the flare of this lens for the given lighting distribution.

Results of these tests indicate that when the camera is pointed toward the zenith the error due to flare, even under the most extreme underwater environmental conditions, is less than a few percent between the zenith (0°) and 80° and that it becomes greater than this only in the 80° to 90° zonal region. A camera pointed toward the nadir records the radiance distribution from 90° to 180° and, since the radiance distribution in the lower hemisphere has a range of radiances no greater than two orders of magnitude (current data indicate one order of magnitude), the accumulated flare light in photographs of the lower hemisphere is always considerably less than 0.1%. Since the underwater radiance distribution is essentially smooth in the 80° to 100° region, simultaneous photographs of the upper (0° to 90°) and lower (90° to 180°) hemispheres automatically reveal the presence of any excessive flare in that region and allow an interpolation over this region. The fact that we are dealing with a smooth distribution also allows interpolation of the data over flare spots of ghost images with a minimum loss of accuracy.

From the results of these flare measurements we conclude: that even under the most unfavorable environmental conditions the error due to flare is less than a few percent; that the flare light for an underwater scene is distributed in such a way that simultaneous photographs of the upper and lower hemisphere detect its presence; that appropriate correction can be made, if necessary; and that under many typical environmental conditions (greater depth and/or overcast sky) the error due to flare becomes less than a few percent and, thus, does not degrade the accuracy of the measurement.

C. Camera Tests

For the purpose of photographic photometry, the focal plane shutter of the Nikon camera has the disadvantage of causing an uneven negative density if the curtain speed is not constant.

Experiments to measure the camera shutter speeds and speed variation across the focal plane have been made as a function of camera temperature from 5°C to 25°C. Results of a typical measurement are given in Fig. 3 which shows the shutter speed, normalized to 1.0 at the center of the focal plane, plotted against distance along the focal plane. It is seen that the greatest variations in shutter speed are at the edges of the focal plane where the greatest accelerations of the shutter curtain occur. This is fortunate, for our purposes, since the fisheye lens exposes only a 24-mm circle in the center of the focal plane and does not make use of the edges. These experiments have shown that, except for the slowest (1 sec) and fastest (1/1000 sec) shutter speeds, the shutter speed variation over the region of interest is less than ±3%.

These measurements also provide the actual shutter speed, as opposed to the rated speed, of the camera as a

function of temperature. It was found that, while the actual speed may be significantly different from the rated speed, the actual speed is reproducible to within ±2% over the range of expected temperature variations. This shutter speed measurement is considered a calibration of the camera and is checked for possible changes before and after data are taken in the field.

D. Summary

Above we discussed why the fisheye lens is ideal for measuring underwater radiance distributions and also some of its limitations for such measurements. This, of course, does not exhaust the discussion of the difficulties and limitations inherent in this technique of measuring underwater radiance, nor does it provide sufficient information for a full discussion of error.

Sources of error due to photographic photometry are well known to this laboratory, and their magnitudes can be estimated and to some extent controlled.[19,23,24] It is estimated that a goal of 25% absolute accuracy and 10% relative precision for the radiance measurements can be consistently attained. Analysis of our preliminary data tends to confirm these estimates. It should be noted from Fig. 1 that most of the optical quantities which are calculated from radiance values are ratios, and, consequently, their values are not seriously affected by the anticipated magnitude of absolute errors.

III.   Instrument Package

To successfully obtain radiance distribution data as a function of depth in natural underwater environments it is necessary to have reliable seaworthy equipment, capable of satisfying a number of requirements. The fisheye lens is the central feature of our instrumentation. The instrument contains two motor operated Nikon F cameras, each equipped with a fisheye lens. The two cameras are mounted with one oriented toward the zenith, the other toward the nadir. The upper and lower hemispheres are photographed simultaneously. This method provides a means of detecting and correcting for flare light, should it appear. Furthermore, simultaneous photographs ensure that the upper and lower hemisphere data were obtained under identical lighting conditions, thus simplifying data interpretation. Each camera lens is covered with a polished acrylic plastic hemisphere, which is centered with respect to the projection origin (P in Fig. 2) of the lens. This provides a coupling of the lens to the water, which yields undistorted recording of the radiant energy field.

Each lens–camera system is housed separately in a pressure case consisting of a cylindrical aluminum case on which is mounted the plastic hemisphere. The case is assembled with quick release bands to facilitate the process of changing film. Each housing is capable of withstanding the pressures at a depth of 100 m. The maximum depth for data taking is, of course, variable and is determined by the natural lighting, the sensitivity of the film, the maximum exposure time permitted, and the clarity of the water.

Under normal operating conditions, the maximum ex-

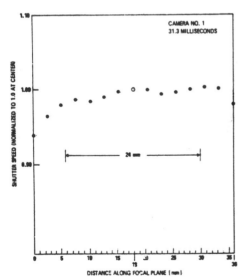

Fig. 3.   Camera shutter speed vs distance along the focal plane.

posure range determined by the light field is estimated to be roughly six to eight orders of magnitude. In order to cover this range it is necessary to have some means of determining the irradiance level at the depth of operation in order to set the camera for correct exposure time. Since radiance distribution data as a function of depth is the objective, it is also important to have some method of relating the radiance values, measured at the various depths, to each other.

In addition to the two radiance camera systems, the instrument has a third housing containing two photocells constructed so as to measure upwelling and downwelling irradiance.[25,26] These are used to determine correct camera exposure settings. These photocells also provide the information required to relate the data taken at the various depths and can be used to correct for any reciprocity law failure of the film. Provisions are incorporated in the instrument so that the magnitude of these irradiances, their ratio to each other, or their ratios to an irradiance meter mounted on deck can be recorded. The spectral sensitivity of these irradiance collectors are matched as closely as possible to the spectral sensitivity of the film and filter combination being used in the cameras. This spectral sensitivity of the radiance camera system is chosen after giving due consideration to the spectral selectivity of the natural water[27] under investigation.

The orientation and depth of the instrument package are important considerations. The azimuthal orientation of films obtained at different depths at the same site can be determined quite accurately, even at the greater depths, from the position of maximum radiance on the film. The direction of the vertical can be determined from the film, since the position of the nadir is unique by always being the position of minimum radiance. In some cases the location of the nadir on the film by this method will not be sufficiently accurate to orient the data. To ensure that the instrument package is level when an exposure is made (and therefore that the zenith and nadir are at the centers of the circular formats of the films), two tilt-sensing electrolytic transducers oriented at right angles to each other are used. These are arranged in series with the shutter switch and prevent operation of the shutter until the cameras are level. The depth of the instrument is indicated by means of an accurate depth transducer.

The upwelling and downwelling irradiance collectors, the two-axis, tilt-sensing device, the depth transducer, and various electronics are housed in a third pressure case. Figure 4 is a photograph of the combined underwater unit consisting of the electronics unit in center foreground with its downwelling irradiance collector visible, the upper camera unit in the left background, and the bottom of the lower camera unit in the right background. From an optical point of view, the instrument package should be arranged so that the up and down photographs are taken from as close to the same depth as possible. In the photograph the projection origins of the two fisheye lenses are separated by 29 cm.

The first field work with this instrument was in favorable environments, where the problem of instrument

Fig. 4.   Photograph of the underwater unit of the radiance distribution camera system.

orientation and stability was not of major concern. However, the instrument was designed so that it can be used in more demanding ocean environments. In order to obtain maximum stability in rough seas, the instrument package should be long and thin in the vertical direction. The modular form of the instrument package allows the pressure housings to be arranged one above the other and to be gimballed so as to minimize tilt due to currents and cable drag should this prove necessary or desirable. This alternate configuration may necessitate a significant optical correction of the data to account for the increased separation of the two lenses and does not allow use of the irradiance collectors.

The underwater instrument is remotely operated via a fourteen-conductor electrical cable from a deck control panel. This panel provides the necessary information inputs and command functions for the operation of the camera system. Figure 5 is a block diagram of the complete camera system. It shows the flow of information and commands to and from the deck control unit, the underwater electronics unit, and the two underwater cameras.

The exposure control commands are encoded into frequencies at the control panel and transmitted to the underwater electronics unit where they are decoded by a system of frequency selective reeds. The decoded command information is then transmitted to the two cameras where motors operating geneva mechanisms advance the shutter speed controls, the iris diaphragm controls, or, in the case of the upper camera, a neutral density filter wheel. Potentiometers attached to the control shafts return analog voltage position informa-

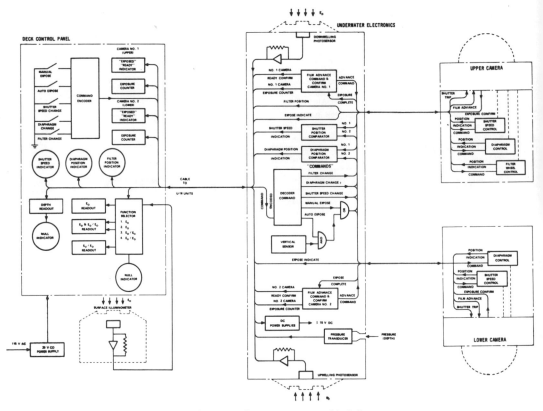

Fig. 5.   Underwater radiance camera system block diagram.

tion to the underwater electronics unit, indicating the condition of each control shaft in the two cameras. The two analog voltages indicating shutter speed are compared, and if they are equal, one of these analog signals is sent to the surface confirming that the two shutters are set to the same speed and indicating what this speed is. Should there be a lack of correspondence between the two voltages, the comparator will not pass a signal, and the lack of signal will indicate a malfunction. The information on the iris diaphragm position is handled in a similar manner. The filter position information from the upper camera is transmitted directly to the surface.

The exposure command can be given in an automatic mode, in which case the camera shutters will be tripped when the radiance distribution camera system is within about 1° of vertical, as sensed by the orthogonal pair of electrolytic level sensors. The requirement that the cameras be vertical can be overridden by a manual command. When the camera shutter is tripped, confirmation of the exposure is obtained from the closure of the flash synchronization contacts, and this confirmation is transmitted to the surface where a counter for

each camera is advanced showing the number of frames which have been exposed. After this confirmation of exposure is obtained, the motorized camera backs advance the film to the next frame.

The deck control panel also provides a continuous indication of camera depth by means of a manually balanced potentiometric measurement of the pressure transducer voltage. Similar measurement capability is provided for the various irradiances and irradiance ratios from which correct exposure, diffuse attenuation coefficient, and water reflectance can be determined.

Power for operation of the camera system can be obtained from batteries for remote field use or any convenient source of 30 V dc. Internal regulators supply critical voltage sensitive components so that the regulation characteristics of the primary source are unimportant.

IV.   Data Analysis

A.   General Technique for Data Analysis

As discussed in Sec. II. A, the radiance distribution information is recorded on film via the fisheye lens. That is, the field radiance from a given direction is represented uniquely by a density on the film at a particular position. Thus, to recover the radiance distribution information, the density of the film at each position on the film must be measured. The density of the developed film is measured with the use of a densitometer using standard techniques.[19,23,24] Since the (r^1, ϕ^1) coordinates (Fig. 2) of the negative correspond to the (θ, ϕ) coordinates of the radiance distribution, the densitometer must have some provison for accurately scanning the negative and a facility for recording the film density as a function of position. In addition, since the aperture and magnification of the densitometer determine the area dS^1 of the scanned film, and since dS^1 is related to the object space solid angle, $d\Omega$, by Eq. (9), it is the densitometer which determines the resolution of the radiance measurement. This resolution should be small enough so that detailed scanning of the sunspot can be made, but not so small that the sensitivity of the densitometer measurement is sacrificed. It is useful to choose and vary the resolution so that the complete information on the film is utilized, but the number of bits of information remains the minimum required for a specific calculation.

B.   Preliminary Radiance Distribution Data

The radiance distribution camera system described above has been used successfully in lake waters and from aborad ship on a number of stations in the Gulf of California. The instrument has proven to be reliable and seaworthy.

Data which have been manually processed show satisfactory agreement with theoretical expectations[22] and with available radiance distribution data with which it can be compared.[9] In addition, our preliminary analysis has given additional insight into more efficient and effective methods for reduction and analysis of large amounts of radiance data. The techniques of data analysis, the radiance distribution data and ancillary oceanographic measurements, and the quantities and conclusions derived from these data will be discussed in a separate publication.

References

1. N. G. Johnson and G. Liljequist, Svenska Hydrograf. Biol. domm. Skrifter, Ny Ser. Hydrograf. 14, 3 (1938).

2. L. V. Whitney, J. Marine Res. 4, 122 (1941).

3. L. V. Whitney, J. Opt. Soc. Amer. 31, 714 (1941).

4. Y. Takenouti, Bull. Japan. Soc. Sci. Fish. 8, 213 (1940).

5. N. G. Jerlov and M. Fukuda, Tellus 12, 348 (1960).

6. T. Sasaki, S. Watanabe, G. Oshiba, and N. Okami, Records Oceanog. Works Japan, Special No. 4, 197 (1960).

7. T. Sasaki, S. Watanabe, G. Oshiba, and N. Okami, J. Oceanog. Soc. Japan 14, 1 (1958).

8. T. Sasaki, in Physical Aspects of Light in the Sea—A Symposium, (Univ. of Hawaii Press, Honolulu, 1964), pp. 19–24.

9. J. E. Tyler, Univ. Calif. Bull. S.I.O. 7, 363 (1960).

10. J. E. Tyler, W. H. Richardson, and R. W. Holmes, J. Geophys. Res. 64, 667 (1959).

11. J. E. Tyler and A. Shaules, Appl. Opt. 3, 105 (1964).

12. J. E. Tyler, J. Quant. Spectrosc. Radiative Transfer 8, 338 (1968).

13. J. E. Tyler and R. C. Smith, J. Opt. Soc. Amer. 56, 1390 (1966).

14. R. C. Smith and J. E. Tyler, J. Opt. Soc. Amer. 57, 589 (1967).

15. J. E. Tyler and R. C. Smith, J. Opt. Soc. Amer. 57, 595 (1967).

16. S. Q. Duntley, The Sea (Wiley–Interscience, New York, 1962), Vol. 1, pp. 452–455.

17. K. Miyamoto, J. Opt. Soc. Amer. 54, 1060 (1964).

18. K. Miyamoto (private communication).

19. C. E. K. Mees, The Theory of Photographic Process (Macmillan, New York, 1954).

20. G. W. Crawley, Brit. J. Phot. 100 (January 1965).

21. L. A. Jones and H. R. Condit, J. Opt. Soc. Amer. 38, 123 (1948).

22. J. E. Tyler and R. W. Preisendorfer, The Sea (Wiley–Interscience, New York, 1962), Vol. 1, Chap. 4.

23. T. H. James and G. C. Higgins, Fundamentals of Photographic Theory (Morgan and Morgan, New York, 1960).

24. B. Dudley, Handbook of Photography (McGraw-Hill, New York, 1939) Chap. 7.

25. R. W. Austin and R. W. Loudermilk, Soc. Photo. Opt. Instrum. Eng., Seminar Proc. 12, 123 (1968).

26. R. C. Smith, J. Marine Res. 27, 341 (1969).

27. J. E. Tyler, Limnol. Oceanog. 4, 102 (1959).

18

Reprinted from pp. 95–119 of *Optical Aspects of Oceanography*, N. G. Jerlov and E. Steeman Nielsen, eds., London: Academic Press Inc. (London) Ltd., 1974, 494 pp.

Structure of Solar Radiation in the Upper Layers of the Sea*

RAYMOND C. SMITH

Visibility Laboratory, Scripps Institution of Oceanography, University of California, San Diego, La Jolla, California, U.S.A.

I. INTRODUCTION

The sun's electromagnetic energy which penetrates into the upper layers of the sea is a basic pre-requisite for life in the oceans, is characterized by fundamental physical processes, and is essential to many activities of man. Solar radiation supplies the energy for ocean ecosystems by conversion to chemical energy through photosynthesis and to heat energy by absorption. The sun's radiant energy is selectively absorbed and scattered as it penetrates the upper layers of the sea, and these basic processes distinctively alter the structure of the radiant energy field. From a description of the underwater radiant energy field, many of the inherent properties of the water and information on basic physical processes can be obtained. This knowledge and information is a requirement for the study and solution of many problems of interest to man.

To describe the time rate of flow of radiant energy, one must specify its magnitude (the square of the electric field vector), its polarization (direction of oscillation of the electric field vector), its wavelength (frequency of oscillation of the electric field vector), and its direction of propagation. The spectral characteristics (Tyler and Smith, 1970) and polarization (Lundgren, 1971) of underwater radiant energy are

* This work was supported by a grant from the National Science Foundation, NSF–GA–19738.

reviewed elsewhere. Radiance, the energy flux per unit solid angle per unit area normal to the direction of propagation incident on a point, specifies the remaining characteristics of radiant energy. A radiance distribution, the totality of radiance values for every direction about the point, gives a complete description of the geometrical structure of the radiant energy field. The significant features of underwater radiance distributions, along with techniques for their experimental measurement and theoretical analysis, are the principal subjects of this review.

The structure of the radiance distribution in the upper layers of the sea is dependent upon factors which modify the solar radiation in the earth's atmosphere, conditions of the air-sea interface and optical properties of the sea water. The radiance distribution above the sea surface depends primarily upon the altitude of the sun, the scattering-absorbing properties of atmospheric molecules and particles, the meteorologic conditions and the radiant energy reflected back from the sea surface. Recent measurements and a review of earlier measurements of the solar constant have been made by Thekaekara et al. (1969, 1970) and Arvesen et al. (1969). Sky radiance distributions have been reported for a variety of sun angles and meteorological conditions by Gordon et al. (1966a,b,c). Monte Carlo methods have been used to study the reflected and transmitted radiation in the atmosphere over a wide range of conditions and this work has been summarized by Plass and Kattawar (1971). The effect of wind stress on the surface roughness of the air-sea interface has been investigated by a number of authors (including Duntley, 1954; Cox and Munk, 1954; Cox, 1958; Wu, 1969) and discussed in a review article by Ursell (1956). The relationship of sea surface slopes to the underwater radiance distribution has been discussed by Cox and Munk (1955) and Gordon (1969). Primary emphasis in this chapter will be on the radiance distribution of solar energy as a function of depth below the sea surface and how this radiance distribution is altered and characterized by the optical properties of sea water.

II. UNDERWATER RADIANCE DISTRIBUTIONS

Underwater radiance distributions have been obtained by means of a radiance distribution camera system (Smith et al., 1970). This instrument contains two cameras placed back to back, each equipped with a 180° field of view ("fisheye") lens, and it is fabricated so that film can be exposed by remote control. The projection geometry of the 180° field of view lens is shown in Fig. 1, where the projection origin of the lens (origin of the x—y coordinates in the figure) is at a

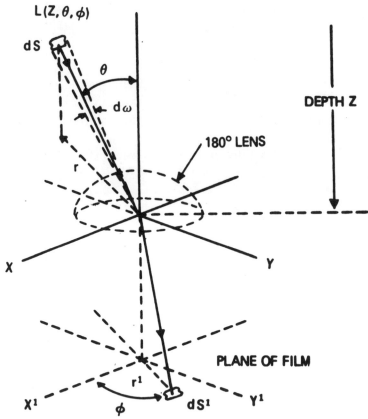

Fig. 1. Schematic diagram showing the projection geometry of the 180° field of view (fisheye) lens for obtaining radiance distributions.

depth z underwater. The radiance, $L(z, \theta, \varphi)$, is the energy per unit time per unit area per unit solid angle incident on the origin from the direction (θ, φ). Referring to Fig. 1, θ is the zenith angle of an incident ray from an infinite object and (r', φ) are the coordinates of the image of this ray on the plane of the film. As described by Miyamoto (1964) and discussed in detail by Smith et al. (1970), this lens has the property of projecting a hemisphere onto a plane such that:

$$r' = f\theta, \tag{1}$$

where f is the lens focal length. In addition, the irradiance received by the film during the time of exposure, $E'(r', \varphi)$, has a one-to-one corres-pondence to the field radiance given by:

$$E'(r', \varphi) = L(z, \theta, \varphi)\left(\frac{n}{n'}\right)^2\left(\frac{\pi}{4F^2}\right)\left(\frac{\sin \theta}{\theta}\right). \tag{2}$$

Here n and n' are the indices of refraction of object and image space, respectively, and F is the f/number of the optical system. By the methods of photographic photometry, the density of an area on a film negative can be related to the exposure of that area, and this can, in turn, be related to the desired field radiance.

Before considering quantitative data, the significant qualitative features of an underwater radiance distribution will be illustrated by means of a positive print, obtained from a data film negative, taken by the radiance distribution camera system. Figure 2 illustrates the upper

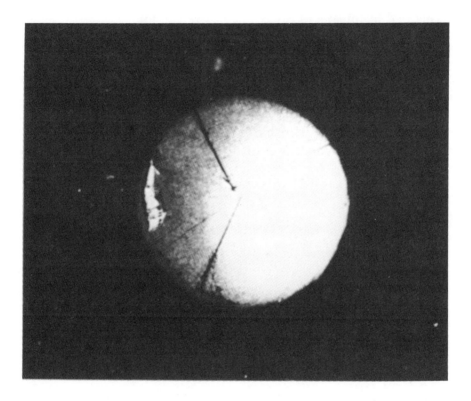

Fig. 2. Positive photographic print of a downwelling radiance distribution, obtained from a data film negative, taken by the radiance distribution camera system. In this photograph the zenith ($\theta = 0°$) is in the centre, the horizontal directions ($\theta = 90°$) form the outside circumference of the circular format, and the intermediate zenith angles are directly proportional to the radial distance from the centre (eqn. 1). The Snell circle is seen here as the brighter, inner circle extending to the critical angle ($\theta = 48\cdot6°$). Dark, radial lines are the ship's boom and the instrument's electrical and suspensions cables. Other information: depth 1·75 m, zenith angle of the sun 29° 30', refracted angle of the sun 21° 45', clear skies (less than 0·1 cloud cover), glassy-calm sea surface, Beaufort Wind Scale No. 0, instrument suspended 6 m from side of ship (R/V Helland-Hansen), 60 nm bandwidth (full width at half maximum) spectral response approximately centered at the wavelength of maximum transmittance of the water under investigation (λ max. $\approx$ 473 nm obtained using Wratten No. 48 filter), Mediterranean Sea (Sta. D2) 8 July 1971, 38° 22' N 07° 12' E.

hemisphere of a radiance distribution at a depth of 1·75 m. As noted in the caption to Fig. 2, the environmental conditions were ideal when the exposure for this photograph was taken. In this figure the zenith is at the centre, the horizontal directions form the outside circumference of the circular format, and intermediate angles from the zenith are proportional to the radial distance from the centre (eq. 1).

The role of refraction in structuring the distribution of radiance underwater is illustrated in Fig. 2 where many principal features can be explained by means of Snell's law of refraction:

$$m \sin i = n \sin j. \tag{3}$$

Figure 3 schematically depicts the application of this law to the experimental situation that existed when the photograph shown in Fig. 2 was obtained. Consider an observer (or his radiance instrument) at a depth of 1·75 m underwater viewing the upper hemisphere in a plane containing the observer, the zenith and the sun (the vertical plane of

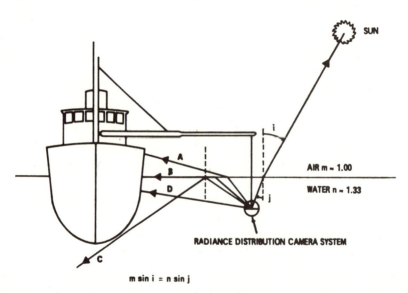

Fig. 3. Schematic diagram depicting the application of Snell's Law to the experimental situation that existed when the exposure for fig. 2 was obtained. See text for details.

the sun). To the right of the zenith (to the right of the centre in Fig. 2) the refracted image of the sun is seen as a bright spot with a diffuse glow surrounding it. To the opposite side of the zenith from the sun the refracted image of the ship above its water line can be seen along a path of sight such as "A". As the angle of view is increased from the zenith, the water line of the ship (and the horizon at other azimuth angles) is seen at the critical angle, j_c, defined as:

$$\sin j_c = \frac{m}{n} \sin 90°. \qquad (4)$$

The critical angle is approximately $48\cdot6°$ for the sea-air interface and is shown as path of sight "B" in Fig. 3. Thus the total 180° sky hemisphere above water is compressed by refraction into a $97\cdot2°$ cone underwater. This is clearly seen in Fig. 2 as the brighter inner circle containing the refracted images of the sun, sky and ship above its water line.

Total internal reflection from the surface occurs, e.g. path "C", when the direction of sight is increased to a zenith angle greater than $48\cdot6°$. As the direction of sight is further increased from the zenith, the path of sight "C" eventually strikes the bottom of the ship. From this angle, until the water line of the ship is viewed directly, the reflected (thus inverted) image of the ship below its water line is seen. Finally, for paths of sight from the water line downward, the ship is seen along a direct underwater path "D". The dark shadow along the lefthand circumference of Fig. 2 is due to the reflected and direct image of the ship below the water line.

A small maximum can appear in near surface radiance distributions at zenith angles of about 70° to 90° which depends upon the relative absorbing and scattering properties of the water. Consider paths of sight for zenith angles of $48\cdot6°$ to 90° and assume the ship was not present. The radiance as viewed along a path such as "C" is composed of backscattered radiant energy which is internally reflected from the sea-air interface plus radiant energy which is forward scattered between the surface and the observer. Both the upwelling internally reflected and the downwelling component of the radiance depend upon the absorptance and scatterance along the path of sight considered. The internally reflected component will decrease as the zenith angle is increased due to increased absorption along the longer path length to the surface. The forward scattered radiant energy, on the other hand, may increase as the path length to the surface is increased until the gain due to forward scatterance is balanced by increased absorptance. If such an increase occurs, a small maximum will appear at sun zenith angle a few degrees less than 90° (Tyler, 1958).

Quantitative values of the relative radiance versus zenith angle in the vertical plane of the sun are shown in Fig. 4. This downwelling data was obtained from a microdensitometer scan of the same negative used to produce the positive print shown in Fig. 2, where the down-

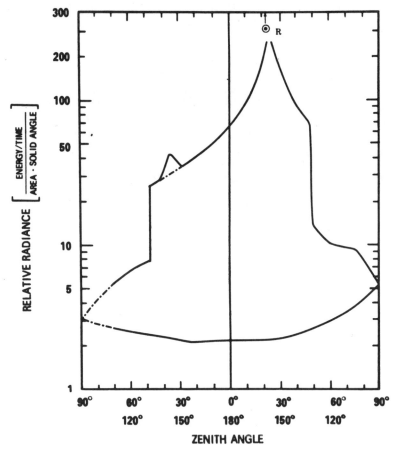

Fig. 4. Relative radiance versus zenith angle for up- and downwelling radiance in the vertical plane of the sun. The data for the upper curve were obtained from the same negative that was used to produce the positive print shown in Fig. 2. The upper curve gives the downwelling radiance from the zenith to the horizontal (90°), the lower curve gives the upwelling radiance from the horizontal to the nadir (180°). Dashed portions of the curves are extrapolations of the data to eliminate perturbations in the radiance distribution due to the ship. $\odot_R$ indicates the refracted angle of the sun. The angular resolution of the radiance data in this figure is 7·2°.

welling radiance is given by the upper curve from the zenith to the horizontal toward and away from the sun. Principal features discussed above can be seen in this curve. The greatest radiance values occur in the direction of the refracted image of the sun, $\odot_R$. Since a relatively coarse scanning resolution of 7·2° was used to obtain this curve, the sharp detail, such as the small sun's image, will be somewhat smoothed by the scanning process. The steep drop in radiance values at 48·6° on both sides of the zenith marks the Snell circle. The small maximum in the curve on the side away from the sun is light reflected from the ship above its water line. The maximum in the radiance on the side of the sun between 70° and 90° is due to the interplay of the internally reflected and forward scattered radiant energy. In this particular example, the perturbation of the radiance distribution due to the ship obscures a possible similar maximum on the side opposite the sun.

Considerable care was taken to minimize perturbations in the underwater radiance distribution caused by the ship and instrument cables. However, as with any underwater radiant energy measurement, such perturbations are very difficult to eliminate completely. It is an advantage of the photographic technique of measuring underwater radiance that one can directly see and identify the source of any perturbation on the film, as well as quantitatively measure its magnitude. Thus, by directly identifying sources of perturbation and by making use of the inherent symmetry about the vertical plane of the sun in underwater radiance distributions, the data can usually be accurately corrected for ship and cable perturbations.

Upwelling radiance, which is composed only of backscattered radiant energy, was obtained with a downward facing camera and 180° field of view lens. Upwelling radiance, obtained simultaneously with the corresponding downwelling radiance, is shown as the lower curve in Fig. 4. This curve goes from the horizontal (90°) on the side away from the sun, through the nadir (180°), to the horizontal on the side of the sun. Figure 4 should not be confused with a polar plot of radiance. There is no "origin" within the curves of this figure. It is merely a plot of relative radiance versus zenith angle with the curve from 90° to 180° folded back underneath the 0° to 90° curve. In this way the radiance distribution is plotted on a logarithmic scale, which allows the full range of radiance values to be displayed accurately, and yet the curves maintain the heuristic attributes of a polar plot.

This technique of plotting radiance allows a subtle feature of the radiance distribution to be readily observed. Note that the minimum radiance value is not at the nadir but at the refracted anti-solar point. This is because, in relatively clear water near the surface, the instrument casts and then "sees" its own shadow.

Figure 5 is a photograph showing this aureole effect (Minnaert, 1954). The various slopes in the water's surface cast a streak of light or shade behind it; all these streaks run parallel to the line from the refracted image of the sun through the projection origin of the observing lens. The instrument then records these lines as meeting perspectively in the refracted anti-solar point, that is, in the shadow image of the instrument. Thus, the radiance as seen by the instrument has a minimum at its own shadow image surrounded by a circle of slightly increased radiance. This effect is of little practical consequence but is nevertheless a pleasing phenomena to observe and is a further example of how our instruments, however subtly, perturb the environment we are attempting to study.

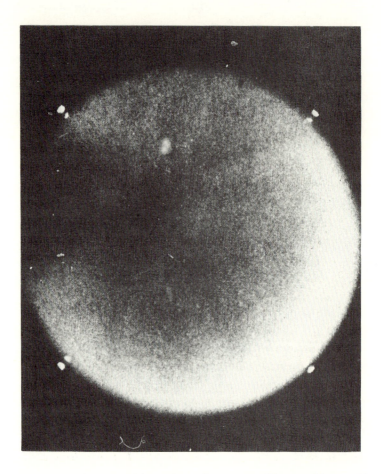

Fig. 5. Positive photographic print of an upwelling radiance distribution, obtained from a data film negative, taken by the radiance distribution camera system. View of upwelling radiance at 1·75 m showing the aureole effect. See text for details.

Figure 6 shows a positive photographic print of, and Fig. 7 shows the quantitative data for, a radiance distribution obtained at a depth of 20·4 m and with a sun zenith angle of 75° 30′. The remaining oceanographic conditions are the same as for Fig. 2. In Figs. 6 and 7 the refracted angle of the sun lies very close to the edge of the Snell circle and the diffuse glow surrounding the sun's image spreads out beyond this circle. This glow is broader than that seen at the shallower depth, because we are viewing the narrow angle forward scatterance of the sun over a longer path length. The Snell circle is still obvious but its edge is more diffuse, again because of the increase scatterance along a longer path at this greater depth. The dark shadow near the zenith is

Fig. 6. Positive photographic print of a downwelling radiance distribution. Other information: depth 20·4 m, zenith angle of the sun 75° 30′, refracted angle of the sun 46° 30′, remaining conditions are the same as for Fig. 2.

the image of the bottom of the ship and is seen as a dip in the radiance curve in Fig. 7. This figure indicates a slight increase in the radiance just prior to reaching the Snell circle, and an increase in brightness can also be seen at all azimuth angles near the Snell circle in the photograph shown in Fig. 6. This increased radiance is probably due to the distribution of atmospheric light.

The figures discussed above, obtained under ideal environmental conditions, illustrate that the structure of solar radiant energy just below a calm surface of the sea is primarily dependent upon the radiance distribution above the surface and Snell's law of refraction. A cloudy atmosphere, which obscures the sun, will decrease and diffuse the underwater radiance peak in the direction of the refracted angle of the sun. An increased sea state will produce a glitter pattern about the

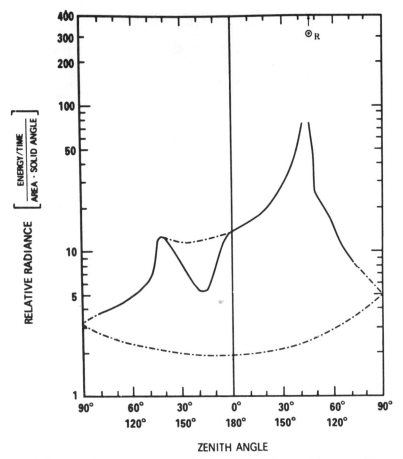

Fig. 7. Relative radiance versus zenith angle for downwelling radiance in the vertical plane of the sun. Data obtained from the same negative used to produce Fig. 6. Upwelling (dashed) curve estimated.

sun's image and diffuse the edge of the Snell circle. Even with clear skies and a calm surface, the radiance distribution will lose its sharp structure as the absorptance, the scatterance, or the depth of observation is increased.

The relative radiance versus zenith angle for three depths is shown in Fig. 8. Moderately rough seas produced a glitter pattern about the sun and diffused the Snell circle, as shown by the 5 m curve. With increasing optical depth this image detail is lost and the shape of the radiance

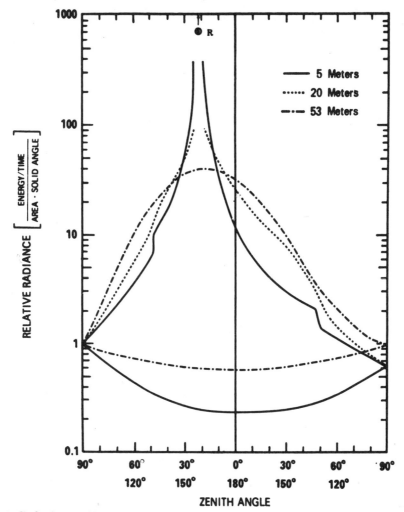

Fig. 8. Relative radiance versus zenith angle for up- and downwelling radiance in the vertical plane of the sun. The curves have been arbitrarily normalized to unity at $\theta = 90°$ so that they may be readily compared. Other information: zenith angle of the sun 28°, refracted angle of the sun 21° 30′, clear skies (less than 0·1 cloud cover), moderate breeze (Beaufort No. 4) with gusts to 25 knots, wind chopped moderate sea (1 to 1½ m waves) in lee of island, instrument suspended 4 m from stern of ship (R/V Ellen B. Scripps), 60 nm bandwidth (FWHM) spectral response approximately centered at the wavelength of maximum transmittance of the water under investigation (λ max. $\approx$ 497 nm obtained using Wratten No. 64 filter), Gulf of California (Fresnel II Cruise—Sta. 2) 16 March 1971, 25° 26′ N 111° 08′ W. The upwelling curve for 20·4 m is very nearly the same as the upwelling curve for 4·6 m and is not shown.

distribution becomes increasingly smooth, as shown by the 20 and 53 m curves. In addition, with increasing optical depth, the point of maximum radiance moves toward the zenith away from the direct refracted angle of the sun. It is postulated (Whitney, 1941) that the radiance distribution at great optical depths is symmetrical about the vertical axis and that its shape is dependent only upon the scatterance and absorptance of the water.

Preisendorfer (1959) has shown that the asymptotic radiance hypothesis is equivalent to the statement that the direction (θ, φ) and depth (z) dependence of the radiance distribution multiplicatively uncouple at great depths. Duntley (1963), extrapolating from Tyler's (1960) Lake Pend Oreille data, estimated that an asymptotic distribution was not reached until the depth was greater than 20 attenuation lengths $(1/c)$. To be precise, the depth at which an asymptotic distribution is reached depends primarily on the scattering length $(1/b)$. The absorption length $(1/a)$ then determines how many, if any, photons remain when asymptotic depths have been reached. For normal ocean water types, in the spectral region of maximum transmittance, the evidence indicates that the incident energy density will be reduced by a factor of at least a million or more before an asymptotic radiance distribution can be obtained. This is particularly true for waters where the absorption is greater than the scattering and for spectral regions where the the ratio of absorption to scattering is greater than unity. Thus, unless the radiance distribution input to the ocean is relatively diffuse due to cloudy skies and a rough sea surface or the water has a low ratio of absorption to scattering, the simplification which the asymptotic radiance hypothesis provides in the theoretical analysis of the underwater radiant energy field is not generally valid in the upper layers of the sea.

The structure of solar radiation in the upper layers of the sea, where an asymptotic radiance distribution does not in general obtain, is at once the most complex, the most interesting and the most difficult to measure accurately. The manner in which the optical properties of the sea distinctively alter the radiance distribution in the upper layers of the sea are of particular interest, and this can best be analyzed by means of the theory of radiative transfer.

III. Equation of Transfer for Radiance

The theoretical study of radiative transfer in the sea has been discussed by many authors (e.g. LeGrand, 1939; Timofeeva, 1957; Jerlov, 1961, 1968; Lenoble, 1961; Preisendorfer, 1961, 1964, 1965, 1968, 1972; Tyler and Preisendorfer, 1962; Tyler, 1968; Feinstein et al., 1970). In the following review, which draws especially from the work of Preisendorfer and Tyler, emphasis is placed on the equation of transfer for radiance and on the optical properties of sea water which are derivable from underwater radiance distributions as a function of depth. This will provide a theoretical framework in which experimental radiometric measurements of solar radiation in the sea can be understood and compared.

The fundamental equation which governs the variation of radiance in a scattering and absorbing medium is the equation of transfer for radiance L:

$$\frac{1}{v}\frac{\delta L(z, \theta, \varphi, t)}{\delta t} + \frac{dL(z, \theta, \varphi, t)}{r} =$$

$$-c(z)L(z, \theta, \varphi, t) + L^*(z, \theta, \varphi, y) + L_\eta(z, \theta, \varphi, t) \qquad (5)$$

where

$$L^*(z, \theta, \varphi, t) = \int_{4\pi} \beta(z, \theta, \varphi; \theta', \varphi')L(z, \theta, \varphi, t)\, d\omega(\theta', \varphi'). \qquad (6)$$

The path function, L^*, is the radiance per unit length in the direction of the line of sight, generated by radiant energy scattered into the line of sight from all directions about the point z. $\beta(z, \theta, \varphi; \theta', \varphi')$ is the volume scattering function at point z for radiant energy incident in the direction (θ', φ') and scattered off in the direction (θ, φ). L_η is the source function, v is the velocity of light in the medium, and c is the total volume attenuation coefficient:

$$c(z) = a(z) + b(z). \qquad (7)$$

Here a is the volume absorption coefficient and the total scattering coefficient, b, is given by

$$b(z) = \int_{4\pi} \beta(z, \theta, \varphi; \theta', \varphi')\, d\omega(\theta', \varphi'). \qquad (8)$$

It is usual to assume that the radiance distribution in the sea is in a steady, or at least quasi-steady, state so that the time dependent term in eq. 5 can be neglected. Before neglecting this term, we note that to do so requires the assumption of a constant input of radiance to maintain the system in a steady state. Thus, we will be dealing with an irreversible dissipative system, analogous to friction, where energy is lost or entropy gained. This is roughly equivalent to saying that the phenomena represented by eq. 5 are inherently statistical events and that we will be viewing the interaction of radiant energy with matter on a macroscopic level.

By assuming that the underwater radiance distribution is in a steady state, supplied by a constant source of radiance on the surface, that the source function is negligible, and that the radiant energy is nearly monochromatic and unpolarized, eq. 5 is reduced to the classical form:

$$\frac{\mathrm{d}L(z, \theta, \varphi)}{\mathrm{d}r} = -c(z)L(z, \theta, \varphi) + L^*(z, \theta, \varphi), \qquad (9)$$

where $-r \cos \theta = z$. The first term on the right gives the space rate of loss of $L(z, \theta, \varphi)$ by attenuation along a direction of travel; the second term gives the space rate of gain of $L(z, \theta, \varphi)$ by rescattering of radiant energy back into the original direction of travel.

Preisendorfer (1968) has discussed in detail the principal methods of formulating and solving the basic equations of radiative transfer theory. Among these methods, which will be briefly outlined below, are iterative procedures and Monte Carlo techniques.

The Monte Carlo method of solving the equation of transfer (Cashwell and Everett, 1959; Meyer, 1954) can be based mathematically on the theory of stochastic processes and, in essence, attempts to follow the probable history of a single photon introduced into a scattering-absorbing medium. This history, by an ergodic argument, can be said to be representative of the instantaneous distribution of an aggregate of particles simultaneously introduced into the system. Plass and Kattawar (1969) and Raschke (1971) have calculated the complete radiation field in the atmosphere-ocean system using a Monte Carlo method. By making a realistic model for each component of scattering and absorption, both in the atmosphere and the ocean, they have calculated radiance values which show many of the essential features of underwater radiance distributions. The full potential of this technique will be realized when experimental measurements of the radiance distribution above and below water, along with the measurement of important optical properties, can be used for comparison with the calculations. For example, if the underwater radiance distribution is sufficiently sensitive to the shape of the volume scattering function, a comparison of trial calculations with experimental radiance distribution data could be used to obtain the optimum shape, including the forward scattering peak, of the volume scattering function.

A first iterative solution of the equation of transfer (eq. 9) has been used by Preisendorfer (1964) to give a model of radiance distribution in natural hydrosols. He develops this model assuming that c and $\beta(\theta)$ are independent of the depth z and that the path function can be approximated, making use of the solution of the two-flow Schuster equations for irradiance by

$$L_*(z, \theta, \varphi) = L_*(0, \theta, \varphi) \exp (-Kz) \qquad (10)$$

where K is independent of depth. The radiance at depth z is then shown to be

$$L_r(z, \theta, \varphi) = L_0(z_t, \theta, \varphi) \exp(-cr) +$$
$$\frac{L^*(z, \theta, \varphi)}{c + K \cos \theta} (1 - \exp(-(c + K \cos \theta)r)). \quad (11)$$

Here $L_r(z, \theta, \varphi)$ is the radiance at depth z from the direction (θ, φ) and $L_0(z, \theta, \varphi)$ the radiance from the same direction evaluated at the target point at depth z_t a distance r away. Equation 11 illustrates the dependence of $L(z, \theta, \varphi)$ on the path function, the volume attenuation coefficient, and the attenuation coefficient for irradiance, K.

The above model, as well as other models developed by Jerlov and Fukuda (1960), Lenoble (1958, 1961, 1963) and Schellenberger (1963), have been reviewed by Jerlov (1968). One basic feature of these models is that they strive to relate the principal optical properties of the sea and the underwater distribution of radiance in a systematic manner.

IV. Optical Properties of the Ocean

Radiance is the most basic radiometric quantity. From radiance distribution data on the natural radiant energy field underwater, many of the important optical properties which relate to radiative transfer processes in the ocean can be calculated. The usefulness of radiance distribution data is outlined in Fig. 9 (Tyler, 1968). In this figure $L(z, \theta, \varphi)$ represents radiance distribution data taken at various depths, z, where θ is the zenith and φ the azimuth angle of a radiance value. From the radiance distribution data, the optical properties shown on the chart in Fig. 9 can be determined.

The irradiance, E, is the radiant flux incident on an infinitesimal element of surface containing the point under consideration, divided by the area of that element. E_d is the downwelling irradiance, i.e. it is the flux per unit area measured by a horizontally oriented cosine collector (Smith, 1969) facing upward. Similarly, E_u is the upwelling irradiance. They are defined in terms of the radiance distribution by the following equations:

$$E_d(z) = \int_{\varphi=0}^{2\pi} \int_{\theta=0}^{\pi/2} L(z, \theta, \varphi) \cos \theta \, d\omega \quad (12)$$

$$E_u(z) = \int_{\varphi=0}^{2\pi} \int_{\theta=\pi/2}^{\pi} L(z, \theta, \varphi) \, |\cos \theta| \, d\omega \quad (13)$$

where $d\omega = \sin \theta \, d\theta \, d\varphi$.

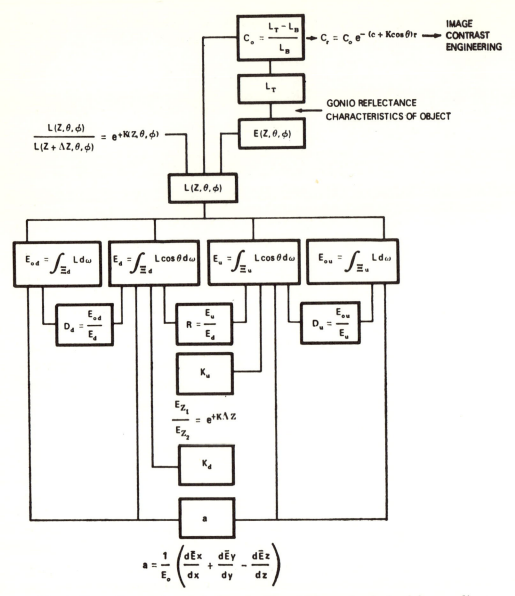

Fig. 9. Chart illustrating optical properties which can be derived from radiance distribution data. These properties are defined and discussed in the text.

The irradiance reflectance, R, at depth z is defined as

$$R(z) = E_u(z)/E_d(z). \tag{14}$$

R may be thought of as the reflectance of a hypothetical plane surface at depth z in the medium. R depends upon and exhibits information about the scattering properties of the entire medium above and below the level z.

The integral of a radiance distribution, at a point at depth z, over all directions about the point gives the scalar irradiance, E_0:

$$E_0(z) = \iint_{4\pi} L(z, \theta, \varphi)\, \mathrm{d}\omega. \tag{15}$$

The scalar irradiances due to radiant energy received separately from the upper and lower hemispheres can be written:

$$E_{0d}(z) = \int_{\theta=0}^{2\pi} \int_{\theta=0}^{\pi/2} L(z, \theta, \varphi)\, \mathrm{d}\omega \tag{16}$$

$$E_{0u}(z) = \int_{\varphi=0}^{2\pi} \int_{\theta=\pi/2}^{\pi} L(z, \theta, \varphi)\, \mathrm{d}\omega \tag{17}$$

where

$$E_0(z) = E_{0d}(z) + E_{0u}(z). \tag{18}$$

The scalar irradiance, when divided by the velocity of light in the medium, yields the total amount of radiant energy per unit volume of space at the given point, i.e. the radiant energy density. Smith and Wilson (1972) have summarized the arguments for considering scalar irradiance, suitably filtered to measure total quanta (350-700 nm) (Jerlov and Nygård, 1969), an optimum measure of the energy available for photosynthesis and they have described a new technique to experimentally measure E_0.

The down- and upwelling distribution functions are defined as:

$$D_d(z) = \frac{E_{0d}(z)}{E_d(z)} \tag{19}$$

$$D_u(z) = \frac{E_{0u}(z)}{E_u(z)} \tag{20}$$

These functions are a simple means of characterizing the depth dependence of the shape of the radiance distribution. In addition, these distribution functions play important roles in the equations of applied radiative transfer theory as developed by Preisendorfer.

The attenuation coefficients for irradiance are defined as the logarithmic depth derivatives of the irradiance functions. Thus,

$$K_d(z) = \frac{-1}{E_d}\left(\frac{\mathrm{d}E_d}{\mathrm{d}z}\right), \tag{21}$$

or alternatively,

$$\frac{E_d(z_2)}{E_d(z_1)} = \exp\left(-K_d(z_2 - z_1)\right), \tag{22}$$

where K_d has units of reciprocal length, and z is the depth at which E_d is measured. Similar K-type functions can be defined for upwelling irradiance and for the scalar irradiances. Physically, the K functions are the quantities that specify the individual depth dependence of irradiance functions. Historically, the K functions were derived from the experimental fact that, in general, radiant energy decreases exponentially with depth. These functions have been used in connection with both theoretical and experimental aspects of radiant energy in natural waters.

Preisendorfer (1958) has shown by a method which makes use of the divergence relation of the radiant energy field (Gershun, 1939), that the volume absorption function, a, can be computed without the requirement of previous knowledge of the volume attenuation function, c, or the volume scattering function, b. This may be done by dividing the rate of change with depth of the net upwelling irradiance by the scalar irradiance, that is,

$$a(z) = \frac{-1}{E_0(z)} \frac{\mathrm{d}}{\mathrm{d}z} (E_d(z) - E_u(z)). \tag{23}$$

This relation holds for an emission-free optical medium that is in the steady state and in which the index of refraction is constant within the medium.

The upper part of Fig. 9 summarizes the application of radiance measurements to the problems of image contrast engineering. For an object to be radiometrically detectable, it must have a different radiance from its background in the direction of view of the sensor. Whatever difference does occur constitutes an optical signal. This optical signal at the object can be described by its inherent contrast (Duntley, 1952, 1962, 1963) defined as:

$$C_0 = \frac{L_T - L_B}{L_B}, \tag{24}$$

where L_T is the inherent radiance of the object and L_B is the inherent radiance of the background against which the object is seen. Thus the first major step in any detection problem is the assessment of the magnitude of the inherent contrast. In computing the inherent contrast of a submerged object, the background radiance, L_B, can be obtained directly from the radiance distribution. The inherent radiance of the object, L_T, depends on the distribution of the flux incident upon the object and the angular reflectance of the object. Thus knowledge of the gonioreflecting properties of the object and the radiance distribution of the light field in which the object is immersed leads directly to the determination of inherent contrast for any path of sight.

Once the inherent contrast has been determined, the equation of contrast reduction (which can be derived from eqs. 24, 11, and 10),

$$C_R = C_0 \exp\left(-(c + K \cos\theta)r\right) \qquad (25)$$

can then be applied to problems of the visibility of submerged objects. Here C_R is the apparent contrast seen by a sensor at a distance, r, along a path, c is the volume attenuation function, and K is the attenuation function for radiance, defined as:

$$K(z, \theta, \varphi) = \frac{-1}{L(z, \theta, \varphi)} \frac{dL(z, \theta, \varphi)}{dz}, \qquad (26)$$

which can be calculated from the radiance distribution data. The θ, as used in eq. 25 and Fig. 9, is the angle between the vertical and the line of sight. Using eq. 25, the apparent contrast can be calculated along any path of sight once the inherent contrast of the object has been determined. Thus, the key to the problem of determining the contrast and detectability of submerged objects is a knowledge of the natural radiance distribution in the water in question.

It should be clear from the summary provided in Fig. 9 that the spatial distribution of radiance at various depths provides one of the most useful descriptions of the underwater radiant energy field. This usefulness was recognized by early workers, concerned with the optics of the ocean, and has inspired a number of them to construct instruments for the measurement of the spatial distribution of the sun's radiant energy underwater.

V. Radiance Instruments

Early studies of the angular distribution of submarine daylight using a screened photometer were reported by Pettersson (1938) and by Jerlov and Liljequist (1938). By use of this instrument they were able to measure the radiance in concentric azimuthal zones of the upper hemisphere and to observe that the submarine light field tended to become more vertical in direction with increasing depth. Jerlov and Liljequist (1938) also made use of a radiance or Gershun tube (1939) photometer which directly limited the solid angle of acceptance of the incident radiant energy. Since this early work, radiance tube photometers have been used with ever increasing sensitivity and sophistication by numerous workers (Whitney, 1941a,b; Timofeeva, 1951, 1957, 1962; Duntley et al., 1955; Tyler, 1960; Jerlov and Fukuda, 1960; Jerlov, 1965; Sasaki et al., 1955, 1958, 1960, 1962; Schellenberger, 1967; Lundgren, 1971). Underwater orientation of the radiance tube instruments have been by means of servo systems and/or ingenious gimbaling principles. Reviews of these instruments and the resulting radiance data have been given by Jerlov (1968) and by Tyler (1968).

With the use of underwater radiance tube photometers these workers have discovered fine details in and the general behaviour of the angular structure of underwater radiant energy. In addition, they have shown that the experimental data is consistent with simple theoretical radiative transfer models and have demonstrated that radiance distributions approach an asymptotic state at great depths. However, in general, most of the reported experimental work to date has been limited to radiance measurements in only a few selected azimuthal planes and the horizontal plane. It was not until 1957 that sufficiently complete sets of data, so that all the optical properties summarized in Fig. 9 could be calculated, were obtained by Tyler (1960). Computations from these data were described by Tyler et al. (1959) and by Tyler and Shaules (1964). In addition, relatively complete radiance data for the Sargasso Sea and comparison of these data with the theory of radiative transfer have been reported by Lundgren and Højerslev (1971). These workers have confirmed the usefulness of radiance distribution data and stimulated further efforts to devise a more simple method for its measurement.

Underwater radiance tube photometers have proven to be accurate, to have high sensitivity, and to be versatile in the sense that they can be constructed so as to make radiance measurements for various spectral bandwidths or so as to measure the polarization of the incident radiation. The principal disadvantages of these instruments have been the problem of orientation and the time required to accumulate a set of data. These disadvantages arise because of the inherent nature of radiance distribution data.

To specify a radiance value requires giving both a magnitude and a direction. It has been found in practice that the maximum radiance for any fixed zenith angle occurs in the azimuth of the sun, which can then be used as an azimuthal reference direction. Securing an accurate zenith orientation, particularly in the oceanographic environment, has proven more difficult. Compounding the difficulty is the quantity of information in a complete radiance distribution.

A complete radiance distribution contains a great quantity of information. For example, to completely record a radiance distribution incident on a single point from the total hemisphere about that point with a radiance tube of $7\cdot2°$ angular resolution would require about one thousand radiance measurements, i.e. taking into account a magnitude and two direction components, about three thousand individual bits of information. Increase the angular resolution and the quantity of information rapidly increases. Radiance tube instruments are slow in obtaining such large quantities of information. Here "slow" is used in its most classical definition, to mean that the sun has moved appreciably during the time interval required to make the desired set of measurements. The full potential of radiance distribution measurements, detailed comparison with radiative transfer theory and completed knowledge of the ocean optical properties, depends upon obtaining accurate radiance distribution as a function of depth in a time small compared to any changes in the input radiance distribution of the sun and sky or to changes in the optical properties of the water itself.

The radiance distribution camera system, discussed briefly at the beginning of this chapter and in detail by Smith *et al.* (1970), was designed to record underwater radiance distributions as a function of depth accurately, rapidly and completely. The information-detecting capacity of photographic emulsions is very high (Zweig *et al.*, 1958; Jones, 1961) and thus the photographic approach is well adapted to the acquisition of radiance data. In addition, a complete depth profile of radiance distribution can be obtained in a time which is short compared to appreciable changes in the sun zenith angle. The accuracy of the radiance distribution camera system is limited by the techniques of photographic photometry (Mees, 1954). Recent preliminary analysis of radiance camera data show that the goal of 25% absolute accuracy and 10% relative precision for the radiance measurements can consistently be attained. The ultimate sensitivity of the instrument depends upon the emulsion speed and latitude of the best available commercial photographic films. Lengthy exposure times, to increase sensitivity of measurement, are of course not possible without reintroducing some of the disadvantages of the radiance tube instruments. To date the camera system has obtained radiance distributions to about seven optical depths ($K_d z \approx 7$) in a wide variety of water types and under a variety of environmental conditions.

The principal advantage of the photographic method of recording radiance, the rapid and complete accumulation of data, dictates that rapid and efficient data processing techniques be developed. The Visibility Laboratory has developed techniques for interfacing a high-speed microdensitometer with a computer for the purpose of processing radiance data stored on photographic film. Computer programs are used for computing radiance distributions and the optical properties outlined in Fig. 9 from the data film. It should be noted that the photographic film is a compact and convenient method of storing raw data. The original data film can be easily rescanned numerous times. Thus, the data can initially be reduced using a relatively coarse resolution and later rescanned with a finer resolution as circumstances dictate. This allows one to maximize the useful information, while minimizing the burden of data quantity.

Radiance distributions as a function of depth are the most basic radiometric quantities for describing the structure of solar radiation in the upper layers of the sea. Radiance, through the equation of transfer and its relation to other optical properties of the ocean, systematically unifies theoretical concepts and experimental results. Through the evolution of radiance instruments we are on the threshold of obtaining even more complete information on the angular structure of underwater radiant energy. This will provide a more complete knowledge of the optical properties of natural waters and should increase our understanding of the processes of radiative transfer in the ocean.

REFERENCES

Arvesen, J. C., Griffin, R. N. and Pearson, B. D. (1969). *Appl. Opt.*, **8**, 2215–2232.

Cashwell, E. D. and Everett, C. J. (1959). "A Practical Manual on the Monte Carlo Method for Random Walk Problems." Pergamon Press, Oxford.

Cox, C. S. and Munk, W. (1954). *J. Opt. Soc. Amer.*, **44**, 838–850.

Cox, C. S. and Munk, W. (1955). *J. Mar. Res.*, **14**, 63–78.

Cox, C. S. (1958). *J. Mar. Res.*, **16**, 199–230.

Duntley, S. Q. (1952). "The Visibility of Submerged Objects." Visibility Lab., Mass. Inst. of Technology, Cambridge, Mass.

Duntley, S. Q. (1954). *J. Opt. Soc. Amer.*, **44**, 574.

Duntley, S. Q., Uhl, R. J., Austin, R. W., Boileau, A. R. and Tyler, J. E. (1955). *J. Opt. Soc. Amer.*, **45**, 904(A).

Duntley, S. Q. (1962). *In* "The Sea" (M. N. Hill, ed.), Vol. 1, pp. 452–455. Wiley-Interscience, New York.

Duntley, S. Q. (1963). *J. Opt. Soc. Amer.*, **53**, 214–233.

Feinstein, P. L., Piech, K. R. and Leonard, A. (1970). *In* "Electromagnetics of the Sea", pp. 38–1–38–10. Agard Conference Proceeding, No. 77, Paris.

Gershun, A. (1939). *J. Math. Phys.*, **18**, 51–151.

Gordon, J. I. and Church, P. V. (1966a). *Appl. Opt.*, **5**, 793–801.

Gordon, J. I. and Boileau, A. R. (1966b). *Appl. Opt.*, **5**, 803–813.

Gordon, J. I. and Church, P. V. (1966c). *Appl. Opt.*, **5**, 919–923.

Gordon, J. I. (1969). *In* "Directional Radiance of the Sea Surface", Ref. 69–20. Scripps Institution of Oceanography.

Ivanoff, A. (1957). *Ann. Geophys.*, **13**, 22–53.

Jerlov (Johnson), N. G. and Liljequist, G. (1938). *Svenska Hydrograf. Biol. Komm. Skrifter, Ny Ser. Hydrog.*, **14**, 1–15.

Jerlov, N. G. and Fukuda, M. (1960). *Tellus*, **12**, 348–355.

Jerlov, N. G. (1961). *Medd. Oceanog. Inst. Göteborg, Ser. B.*, **8**, 1–40.

Jerlov, N. G. (1965). *In* "Progress in Oceanography" (Mary Sears, ed.), Vol. 3, pp. 149–157. Pergamon Press, New York.

Jerlov, N. G. (1968). "Optical Oceanography." Elsevier Publishing Co., Amsterdam.

Jerlov, N. G. and Nygård, K. (1969). Kobenhavns Universitet, *Rep. Inst. Fys. Oceanog.*, **10**, 1–19.

Jones, R. C. (1961). *J. Opt. Soc. Amer.*, **51**, 1159–1171.

LeGrand, Y. (1939). *Ann. Inst. Oceanog.*, **19**, 393–436.

Lenoble, J. (1958). *Ann. Inst. Oceanog.*, **34**, 297–308.

Lenoble, J. (1961). *Compt. Rend.*, **252**, 2087–2089.

Lenoble, J. (1963). *Compt. Rend.*, **256**, 4638–4640.

Lundgren, B. and Højerslev, N. (1971). Københavns Universitet, *Rep. Inst. Fys. Oceanog.*, **14**, 1–33.

Lundgren, B. (1971). Københavns Universitet, *Rep. Inst. Fys. Oceanog.*, **17**, 1–34.

Mees, C. E. K. (1954). "The Theory of Photographic Processes." Macmillan, New York.

Meyer, H. A. (1954). *In* "Monte Carlo Methods", Symposium. Statistical Laboratory, University of Florida.

Minnaert, M. (1954). "The Nature of Light and Color in the Open Air." Dover Publications, New York.

Miyamoto, K. (1964). *J. Opt. Soc. Amer.*, **54**, 1060–1061.

Pettersson, H. (1938). *I.C.E.S. Rapports et Proces-Verbaux des Reunions*, **108**, 9–12.

Plass, G. N. and Kattawar, G. W. (1969). *Appl. Opt.*, **8**, 455–466.

Plass, G. N. and Kattawar, G. W. (1971). *J. Atmos. Sci.*, **28**, 1187–1198.

Preisendorfer, R. W. (1958). *Scripps Inst. Oceanog.*, Ref. Rep., 58–41.

Preisendorfer, R. W. (1959). *J. Mar. Res.*, **18**, 1–9.

Preisendorfer, R. W. (1961). *Union Geod. Geophys. Inst. Mon.* **10**, 11–30.

Preisendorfer, R. W. (1964). *In* "Physical Aspects of Light in the Sea" (J. E. Tyler, ed.), pp. 51–60. University Hawaii Press, Honolulu, Hawaii.

Preisendorfer, R. W. (1965). "Radiative Transfer in Discrete Spaces." Pergamon Press, New York.

Preisendorfer, R. W. (1968). *J. Quant. Spectrosc. Radiat. Transfer*, **8**, 325–338.

Preisendorfer, R. W. (1972). "Hydrologic Optics." Gordon and Breach, New York.

Raschke, E. (1971). *Beitr. Phys. Atmos.*, **45**, 1–19.

Sasaki, T., Okami, N., Watanabe, S. and Oshiba, G. (1955). *J. Sci. Res. Inst.*, **9**, 103–106.

Sasaki, T., Watanabe, S., Oshiba, G. and Okami, N. (1958). *J. Oceanog. Soc. Jap.*, **14**, 1–6.

Sasaki, T., Watanabe, S., Oshiba, G. and Okami, N. (1960). *Rec. Oceanog. Works Jap.*, (Special Number 4), 197–205.

Sasaki, T., Watanabe, S., Oshiba, G. and Kajihara, (1962). *Bull. Jap. Soc. Sci. Fish.*, **28**, 489–496.

Schellenberger, G. (1963). *Gerlands Beitr. Geophys.*, **72**, 315–327.

Schellenberger, G. (1965). *Acta Hydrophys.*, **10**, 79–105.

Schellenberger, G. (1967). *Gerlands Beitr. Geophys.*, **76**, 69–82.

Schellenberger, G. (1967). *Gerlands Beitr. Geophys.*, **76**, 321–333.

Smith, R. C. and Tyler, J. E. (1967). *J. Opt. Soc. Amer.*, **57**, 589–595.

Smith, R. C. (1969). *J. Mar. Res.*, **27**, 341–351.

Smith, R. C., Austin, R. W. and Tyler, J. E. (1970). *Appl. Opt.*, **9**, 2015–2022.

Smith, R. C. and Wilson, W. H. (1972). *Appl. Opt.*, **11**, 934–938.

Thekaehara, M. P., Kruger, R. and Duncan, C. H. (1969). *Appl. Opt.*, **8**, 1713–1732.

Thekaehara, M. P., ed. (1970). *In* "The Solar Constant and the Solar Spectrum Measured from a Research Aircraft", pp. 85. NASA Technical Report NASA TR R-351.

Timofeeva, V. A. (1951). *Dokl. Akad. Nauk SSSR*, **76**, 831–833. (English translation.)

Timofeeva, V. A. (1957). *Dokl. Akad. Nauk SSSR*, **113**, 556–559. (English translation.)

Timofeeva, V. A. (1962). *Izv. Akad. Nauk SSSR, Ser. Geofiz.*, **6**, 1843–1851. (English translation.)

Tyler, J. E. (1958). *J. Mar. Res.*, **16**, 96–99.

Tyler, J. E., Richardson, W. H. and Holmes, R. W. (1959). *J. Geophys. Res.*, **64**, 667–673.

Tyler, J. E. (1960). *Bull. Scripps Inst. Oceanog.*, **7**, 363–412.

Tyler, J. E. and Preisendorfer, R. W. (1962). *In* "The Sea" (M. W. Hill, ed.), Vol. 1, pp. 397–451. Wiley-Interscience, New York.

Tyler, J. E. and Shaules, A. (1964). *Appl. Opt.*, **3**, 105–110.

Tyler, J. E. (1968). *J. Quant. Spectrosc. & Radiat. Transfer*, **8**, 339–354.

Tyler, J. E. and Smith, R. C. (1970). "Measurements of Spectral Irradiance Underwater." Gordon and Breach, New York.

Tyler, J. E., Smith, R. C. and Wilson, W. H. (1972). *J. Opt. Soc. Amer.*, **62**, 83–91.

Ursell, F. (1956). *In* "Surveys of Mechanics" (G. K. Batchelor, ed.). Cambridge University Press, London.

Whitney, L. V. (1941a). *J. Mar. Res.*, **4**, 122–131.

Whitney, L. V. (1941b). *J. Opt. Soc. Amer.*, **31**, 714–722.

Wu, J. (1969). *J. Geophys. Res.*, **74**, 444–455.

Zweig, H. J., Higgins, G. C. and MacAdam, D. L. (1958). *J. Opt. Soc. Amer.*, **48**, 926–933.

19

Originally published by the University of California Press and reprinted by permission of the Regents of the University of California from *Scripps Inst. Oceanogr. Bull.* 7(5):369–386 (1960)

RADIANCE DISTRIBUTION AS A FUNCTION OF DEPTH IN AN UNDERWATER ENVIRONMENT

BY

JOHN E. TYLER

[*Editor's Note:* In the original, material precedes this excerpt.]

DISCUSSION AND TREATMENT OF DATA FOR
CLEAR SUNNY CONDITIONS

Data recording on 28 April was begun at 0850 at a depth of 66.1 m and continued until 1441. The order of depth stations is important because the structure of the light field is a function of depth as well as sun position. The order used is given in table 3.

TABLE 3

ORDER OF DEPTH STATIONS FOR 28 APRIL 1957

(All times are Pacific Standard; sun noon occurred at very nearly 1140 P.S.T.)

Depth (meters)	Time (P.S.T.)		Elapsed time (minutes)	Sun altitude (degrees)		Sun azimuth (degrees)	
	Start	Stop		Start	Stop	Start	Stop
66.1....................	0852	0924	32	41.0	45.5	119.0	127.5
53.7....................	0927	0947	20	46.0	48.5	128.0	134.0
41.3....................	0949	1009	20	48.5	51.0	135.0	142.0
29.0....................	1012	1033	21	51.0	53.0	143.0	150.5
16.6....................	1039	1103	24	53.5	55.0	153.0	162.5
10.4....................	1105	1126	21	55.0	55.5	163.5	172.5
4.2....................	1128	1152	24	55.5	55.5	173.5	183.5
Vertical run..............	1158	1210	12	.2 change		186.0	191.5
66.1....................	1211	1249	38	55.5	53.5	192.0	207.5
53.7....................	1252	1313	21	53.0	51.5	208.5	216.0
41.3....................	1316	1338	22	51.0	48.5	217.0	224.5
29.0....................	1341	1402	21	48.5	45.5	225.5	232.0
16.6....................	1405	1425	20	45.0	42.5	233.0	238.5
Vertical run..............	1432	1441	9	1.2 change		241.0	243.0

The original data clearly show the features of the environment. The image of the sunlit barge wall is obvious at the shallow stations and the shadow of the barge can be seen at all stations although it is not obvious at the deeper ones (see fig. 1). The position of the "bright spot" is always recognizable and at the shallow stations the edge of the "man hole" can be seen, as can the shadow of the instrument itself. In addition, changes in the sun's azimuth position and in its elevation can be detected in the data.

In order to remove the effect of these unwanted parameters from the data the following procedural steps were adopted:

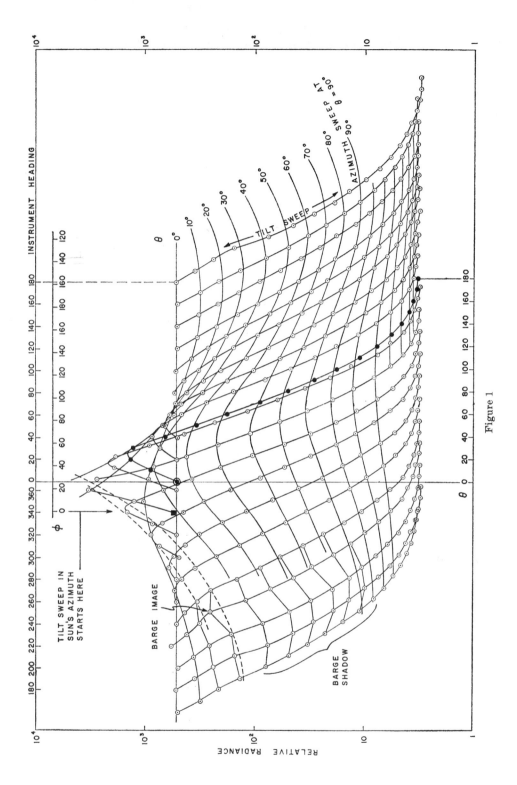

Figure 1

235

1. *Calibration correction.*—The original data, which are very nearly linearly proportional to the log of the radiance, were read at 10-degree intervals of tilt with a special "rule" which converted the data to radiance units and at the same time removed the small departures from linearity that were known to be present. Each information channel, consisting of the multiplier phototube, the chassis and the recorder, has its own calibration "rule." The data for these rules were obtained at the site of the experiment just before and just after the measurements.

2. *Changes in ambient light levels.*—Inspection of the data for the nadir direction indicated the extent of the ambient light-level change owing to changes in the position of the sun. When necessary this has been corrected by normalizing all tilt sweeps for the station to the average nadir reading.

3. *Azimuth motion of the sun.*—The data thus obtained were replotted on semi-log paper, as shown in figure 1. The known azimuth angle between the sun and the instrument heading has been used to locate each tilt sweep on the plot. Thus the azimuth motion of the sun relative to the instrument is not superimposed on the data. (For a complete explanation of figure 1, see step 5.)

4. *Barge image and shadows.*—The data points resulting from step 3 were joined by smooth curves to give azimuth sweeps at constant tilt as well as tilt sweeps at constant azimuth (see fig. 1). The image of the barge, its shadow, etc., can be positively identified in these plots. Sections of the data which were distorted by these spurious signals were not used in the data reduction. The physical location of the instrument was such that a maximum of only about 54° of the horizontal sweep was distorted by the presence of the barge. Data on the left of the sun's plane in figure 1 could be checked by superposing the data on the right of the sun's plane. The position of the sun was located by the maxima of the azimuth sweeps and checked with the known position of the sun in each instance.

5. *Graphical smoothing and interpolation of data.*—In figure 1 the values of the angles at the top, marked "instrument heading," are relative to a fixed compass direction. These angular values are related to the actual angle between the sun and the "instrument heading" through elapsed time. In the depth station illustrated the instrument was rotating "with the sun" and the actual angle was always less than the indicated angle by the amount of angular change in the sun's position during the time interval between the beginning of one tilt sweep and the next. As an example, in the time it took to complete a sequence of tilt sweeps from an indicated instrument heading of 0° to one of 360° the sun's position had changed 8°. Actual rotation of the instrument from the sun was therefore 352°. Thus in the plot the "instrument heading" of 360° is plotted at 352° along the abscissa. The marks on the "instrument heading" scale therefore represent the actual angle between the instrument heading and the sun. Each mark coincides with the first datum point of a tilt sweep. The instrument heading of 0° coincides with the tilt sweep whose first datum point is double circled in figure 1. The data points on this particular tilt sweep (solid circles) are identified with the scale of tilt angles θ at the bottom of figure 1.

Each of the twenty-one tilt sweeps included in figure 1 will have a datum point at $\theta = 90°$ (for example). The procedure that has been used to plot the twenty-one tilt sweeps has placed the $\theta = 90°$ points in proper relation to one another so that a line joining them graphically represents an azimuth sweep at $\theta = 90°$. The other points of the figure have been similarly joined permitting a sort of two-dimensional smoothing of all the data at once.

During the experiment it was not practical to orient the instrument heading with the sun's position. Thus in the working plots the sun's position, which is indicated by the maxima of the azimuth sweeps, bears no relationship with the "instrument heading" scale at the top. The scale marked ϕ (also at the top) has been drawn so that $\phi = 0°$ is in the sun's direction for the azimuth sweep marked $\theta = 0°$ (a straight line). (In figure 1 the sun's direction, for the $\theta = 0°$ curve only, solid square, happens to coincide very nearly with the 340° instrument heading.)

In order to obtain radiance data at equal intervals on each side of the sun's position and for all tilt angles, the ϕ scale is moved right 10° for each successive azimuth sweep. For each such setting the desired data can be interpolated from the azimuth curves.

6. *Duplicate runs.*—Both information channels functioned perfectly during the entire experiment and consequently duplicate runs were available at all depth stations. These were treated independently through step 5, above, and the interpolations were then averaged. A double run was made at the 66.1-m station giving a total of four complete determinations at this depth. All four determinations are averaged together in this instance.

7. *Depth-difference correction.*—The ends of the brightness tubes are about 0.5 m from the center of rotation of the measuring head. As a result the station depth does not remain quite constant but changes continuously with tilt angle according to the equation

$$Z_t = Z \pm r \cos \theta .$$

Where Z_t is the true depth, Z is the reported station depth to the center of rotation of the instrument and r is the distance from the center of rotation to the end of the radiance tube.

The averaged data from step 6 were corrected to give the radiance distribution at a point by determining the slope of the curve of path radiance vs. depth for every pair of values of tilt angle and azimuth angle and making the proper correction along this slope.

8. *Sun altitude changes.*—In addition to changes in ambient light level at the surface of the water, large changes in the altitude of the sun result in a change in the ratio of the zenith to nadir path radiance and a reorientation of the whole radiance distribution solid in the direction of the sun. The depth stations were taken in the order shown in table 3 so that the shallow stations should be at noon and the others clustered around noon in such a way that a complete set of stations could be slected from those nearest noon. Changes in the shape of the distribution solid owing to changes in the sun's altitude were this way minimized.

9. *Normalization to the vertical run.*—Before the seven morning runs could be normalized to the single vertical run at noon it was necessary to demonstrate that the radiance distribution solid for the 66.1-m station was substantially the same in shape at 0908 as it would have been at 1200, that the radiance distribution solid for the 53.7-m station was substantially the same in shape at 0937 as it would have been at 1200, and so on. To do this the ratio of the average zenith to average nadir reading for each depth station was compared with the ratio of the zenith to nadir

reading found from the noon vertical run. For all depth stations, duplicate ratios were found, indicating that no significant error is introduced by adjusting these runs to a single sun altitude. It was also found that the complete data for the 66.1-m and 53.7-m morning stations duplicated the data for the 66.1-m and 53.7-m afternoon stations. Since the former were completed two hours before sun noon and the latter were obtained within about two hours after sun noon, this is further evidence that changes in sun elevation did not significantly affect the shape of the distribution solid at the deep stations. For the remaining morning stations the change in sun altitude between station time and 1200 is very small indeed. After refraction the change in sun angle from 1000 to 1200 is only four-tenths of the resolving power of the radiance tube. It is progressively true throughout the morning data that the large incremental changes in the sun's altitude coincide with the depth stations where such changes have the least effect on the shape of the distribution solid, and the near-zero changes in sun's altitude coincide with the near-surface data where large changes would have had a very large effect.

On the strength of the above evidence, the data for the seven stations given in tables 4, 5, 6, 7, 8, 9, and 10 are presented as data for one sun altitude.

DATA FOR CLEAR SUNNY CONDITIONS

Data representing clear sunny conditions were obtained on 28 April 1957. The voice-recorded notes for the day read as follows:

28 April 1957. It would not be possible to have a more perfect day for the sunny-sky case than today. Between seven in the morning and three-thirty in the afternoon there were no overhead clouds whatsoever. The few small clouds that did appear just over the mountain peaks rapidly evaporated. At no time was it possible to see evidence of a high altitude cirrus layer. In addition to this the lake was practically flat calm all day.

Later computations placed the clouds mentioned at an altitude of 13° in the southern sky. They appeared one at a time and evaporated within 10 minutes. Their angular subtense was never more than 1.5°. The optical state of the lake surface is shown in the photograph, plate 5, taken on 28 April at about 1500.

The radiance distribution data for a clear sunny sky are given for seven depth stations in tables 4 through 10, inclusive.

DISCUSSION OF DATA AND EVALUATION OF SOURCES OF ERROR

In the body of tables 4 through 10 the over-all variation in the value of radiance at any one setting is ±5 per cent of the radiance at that setting. This variation includes instrument errors, reading and plotting errors, errors made in setting and holding azimuth positions during the experiment, and in fact all errors that have entered the measurements before their presentation in the tables.

In the direction of the sun, experimental azimuth steps of 20° move the acceptance cone of the instrument a distance that is almost equal to the base diameter of the acceptance cone. If the air-water boundary were flat the sun's image would therefore be within the cone only once during a complete set of azimuth settings. However, the direction of the sun is a glitter pattern whose size depends on the optical state of the surface (which in turn depends on the wind velocity). During the measurements of 28 April the wind velocity was less than 1 m per second which would indicate a glitter pattern considerably smaller than 10° in angular subtense

TABLE 4

Radiance Distribution under Clear Sunny Sky

(Depth, 4.24 meters; Sun altitude, 56.6°)

Tilt angle (Θ)	Azimuth angle (Φ)									
	0	20	40	60	80	100	120	140	160	180
0	204,000	204,000	204,000	204,000	204,000	204,000	204,000	204,000	204,000	204,000
10	541,000	481,000	374,000	286,000	220,000	174,000	139,000	119,000	108,000	104,000
20	4,300,000	1,320,000	545,000	277,000	168,000	118,000	93,000	79,600	72,100	69,100
30	7,980,000	1,100,000	401,000	198,000	123,000	87,200	69,400	59,700	54,500	52,400
40	573,000	427,000	234,000	135,000	90,500	68,300	56,300	48,700	44,100	42,300
50	207,000	164,000	106,000	69,500	49,300	37,700	31,000	26,500	23,800	23,200
60	114,000	91,800	66,400	47,700	35,300	27,300	22,300	19,000	17,100	16,600
70	61,300	55,800	45,100	34,300	26,700	21,200	17,500	14,800	13,200	12,400
80	41,500	38,200	31,500	25,100	20,100	16,100	12,900	11,300	10,000	9,460
90	26,900	25,300	21,500	17,600	14,200	11,700	9,840	8,560	7,820	7,480
100	17,000	16,000	13,900	11,700	9,940	8,590	7,620	6,860	6,390	6,140
110	11,200	10,700	9,280	8,060	7,180	6,490	5,990	5,600	5,340	5,220
120	7,430	7,170	6,630	6,020	5,590	5,250	5,000	4,800	4,670	4,590
130	5,360	5,220	4,950	4,710	4,520	4,320	4,180	4,040	4,010	3,990
140	4,230	4,170	4,040	3,960	3,850	3,780	3,710	3,650	3,620	3,600
150	3,570	3,560	3,520	3,480	3,440	3,380	3,340	3,300	3,280	3,260
160	3,250	3,250	3,250	3,250	3,250	3,240	3,240	3,240	3,230	3,230
170	3,110	3,110	3,110	3,110	3,110	3,110	3,110	3,110	3,110	3,110
180	3,050	3,050	3,050	3,050	3,050	3,050	3,050	3,050	3,050	3,050

TABLE 5
Radiance Distribution under Clear Sunny Sky
(Depth, 10.4 meters; sun altitude, 56.6°)

Tilt angle (θ)	Azimuth angle (Φ)									
	0	20	40	60	80	100	120	140	160	180
0	127,000	127,000	127,000	127,000	127,000	127,000	127,000	127,000	127,000	127,000
10	274,000	258,000	215,000	166,000	129,000	103,000	86,800	78,300	73,300	71,400
20	1,970,000	610,000	284,000	158,000	101,000	72,000	57,300	49,600	46,500	45,400
30	2,540,000	472,000	208,000	110,000	68,500	49,200	39,000	33,400	30,900	30,300
40	298,000	207,000	118,000	71,000	46,900	34,400	27,200	22,900	20,300	19,200
50	118,000	104,000	69,900	45,900	31,300	23,000	17,700	14,800	13,500	13,000
60	55,900	51,400	41,000	29,400	20,400	14,700	11,600	9,670	8,700	8,360
70	32,500	30,100	24,000	17,900	13,200	10,100	8,150	6,800	6,100	5,860
80	18,100	17,200	14,700	11,400	8,740	6,900	5,660	4,800	4,320	4,150
90	10,600	10,100	8,700	7,020	5,630	4,640	3,960	3,530	3,320	3,270
100	6,500	6,160	5,540	4,800	4,000	3,380	3,000	2,730	2,590	2,540
110	4,320	4,120	3,750	3,320	2,880	2,500	2,260	2,110	2,030	2,010
120	2,940	2,830	2,640	2,440	2,220	1,970	1,840	1,750	1,700	1,690
130	2,020	1,980	1,930	1,860	1,770	1,720	1,630	1,570	1,540	1,530
140	1,630	1,620	1,610	1,570	1,540	1,500	1,450	1,420	1,390	1,390
150	1,410	1,410	1,400	1,370	1,350	1,330	1,310	1,280	1,260	1,260
160	1,240	1,240	1,240	1,240	1,240	1,240	1,240	1,240	1,230	1,230
170	1,180	1,180	1,180	1,180	1,180	1,180	1,180	1,180	1,180	1,180
180	1,180	1,180	1,180	1,180	1,180	1,180	1,180	1,180	1,180	1,180

TABLE 6
RADIANCE DISTRIBUTION UNDER CLEAR SUNNY SKY
(Depth, 16.6 meters; sun altitude, 56.6°)

Tilt angle (θ)	Azimuth angle (Φ)									
	0	20	40	60	80	100	120	140	160	180
0	59,100	59,100	59,100	59,100	59,100	59,100	59,100	59,100	59,100	59,100
10	121,000	111,000	93,100	74,100	59,900	48,700	41,000	36,500	34,000	33,200
20	350,000	183,000	109,000	71,500	49,400	36,900	29,100	24,400	21,900	21,200
30	385,000	169,000	86,200	53,000	36,300	26,700	20,400	16,700	14,600	13,800
40	88,200	75,900	53,500	35,700	24,500	17,900	13,600	10,300	9,380	8,780
50	45,300	41,000	31,700	21,700	14,900	11,100	8,550	6,860	5,880	5,490
60	24,400	22,500	17,400	12,500	9,030	6,950	5,500	4,590	3,960	3,670
70	12,400	11,200	9,100	7,230	5,670	4,520	3,460	3,040	2,670	2,530
80	6,750	6,360	5,490	4,510	3,640	2,990	2,500	2,130	1,870	1,750
90	4,050	3,770	3,280	2,790	2,340	1,970	1,660	1,440	1,310	1,250
100	2,360	2,280	2,070	1,800	1,540	1,330	1,180	1,070	990	964
110	1,520	1,470	1,350	1,210	1,090	982	891	832	796	778
120	1,010	995	946	884	816	758	711	684	666	656
130	740	732	717	685	658	633	606	584	572	564
140	595	587	581	566	550	536	522	512	510	510
150	502	499	497	495	486	480	472	465	462	462
160	451	451	451	450	447	443	439	434	433	433
170	422	422	422	422	422	422	422	422	422	422
180	418	418	418	418	418	418	418	418	418	418

241

TABLE 7
Radiance Distribution under Clear Sunny Sky
(Depth, 29.0 meters; sun altitude, 56.6°)

Tilt angle (θ)	Azimuth angle (Φ)									
	0	20	40	60	80	100	120	140	160	180
0	9,630	9,630	9,630	9,630	9,630	9,630	9,630	9,630	9,630	9,630
10	14,300	13,300	12,000	10,600	9,460	8,410	7,540	6,850	6,330	6,080
20	22,100	16,500	12,100	9,530	7,730	6,440	5,490	4,750	4,220	3,980
30	20,000	13,500	9,280	6,990	5,560	4,620	3,900	3,320	2,880	2,680
40	9,970	8,090	6,080	4,740	3,820	3,110	2,590	2,180	1,820	1,660
50	5,110	4,610	3,750	3,040	2,490	2,030	1,650	1,350	1,100	993
60	2,780	2,490	2,070	1,730	1,440	1,210	1,010	842	708	651
70	1,440	1,330	1,140	986	839	719	618	530	455	425
80	799	739	657	579	505	444	389	343	303	281
90	470	447	412	365	323	287	254	225	202	191
100	293	278	256	231	210	192	175	159	148	143
110	191	185	172	159	150	138	129	121	113	111
120	135	130	123	116	110	105	98.9	94.2	90.0	88.6
130	99.2	97.0	93.6	90.1	87.3	84.1	80.5	78.3	76.4	75.5
140	79.6	78.4	76.4	74.6	73.4	72.1	70.5	69.1	67.8	66.5
150	68.3	67.6	66.3	65.8	64.7	63.7	62.8	61.8	61.3	60.6
160	60.4	59.9	59.6	59.1	58.9	58.5	58.1	58.0	57.4	57.3
170	56.2	56.2	56.2	56.2	56.2	56.2	56.2	56.2	56.2	56.2
180	55.2	55.2	55.2	55.2	55.2	55.2	55.2	55.2	55.2	55.2

TABLE 8
RADIANCE DISTRIBUTION UNDER CLEAR SUNNY SKY
(Depth, 41.3 meters; sun altitude, 56.6°)

Tilt angle (Θ)	Azimuth angle (Φ)									
	0	20	40	60	80	100	120	140	160	180
0	1,380	1,380	1,380	1,380	1,380	1,380	1,380	1,380	1,380	1,380
10	1,650	1,600	1,510	1,410	1,320	1,250	1,180	1,120	1,070	1,050
20	1,680	1,580	1,420	1,270	1,120	1,010	910	822	750	721
30	1,260	1,180	1,049	921	811	716	631	557	496	474
40	859	800	700	614	533	465	408	361	322	306
50	510	478	426	376	333	294	259	229	202	191
60	300	285	254	222	194	170	149	132	117	111
70	165	156	143	126	113	101	90.2	80.6	73.2	70.8
80	88.9	84.9	79.1	72.5	66.6	60.8	55.7	50.9	47.3	45.2
90	51.3	50.4	47.7	44.9	41.7	38.6	36.0	33.4	31.4	30.4
100	33.5	32.5	30.7	28.9	27.2	25.6	24.1	22.8	21.7	20.9
110	21.8	21.5	20.9	19.9	19.0	18.1	17.4	16.7	16.0	15.7
120	15.8	15.8	15.4	14.9	14.3	13.8	13.3	12.9	12.5	12.3
130	12.3	12.2	12.0	11.7	11.5	11.2	11.0	10.8	10.6	10.5
140	10.2	10.1	10.0	9.90	9.78	9.65	9.57	9.45	9.33	9.02
150	8.75	8.73	8.66	8.63	8.57	8.53	8.46	8.40	8.35	8.34
160	7.88	7.88	7.88	7.88	7.87	7.87	7.87	7.87	7.86	7.86
170	7.53	7.53	7.53	7.53	7.53	7.53	7.53	7.53	7.53	7.53
180	7.43	7.43	7.43	7.43	7.43	7.43	7.43	7.43	7.43	7.43

TABLE 9
RADIANCE DISTRIBUTION UNDER CLEAR SUNNY SKY
(Depth, 53.7 meters; sun altitude, 56.6°)

Tilt angle (θ)	Azimuth angle (Φ)									
	0	20	40	60	80	100	120	140	160	180
0	202	202	202	202	202	202	202	202	202	202
10	219	218	212	205	194	184	173	163	157	155
20	194	192	180	168	156	143	132	123	116	113
30	139	137	130	120	110	98.8	89.8	82.7	77.6	75.5
40	88.7	87.2	82.9	77.0	70.7	63.9	58.2	53.3	49.9	48.6
50	52.7	52.3	50.1	47.1	43.7	39.6	36.2	33.4	31.2	30.0
60	32.0	31.2	29.6	27.4	25.1	23.1	21.3	19.9	18.8	18.4
70	17.8	17.6	16.9	16.0	14.9	13.8	12.8	12.0	11.4	11.2
80	10.3	10.2	9.87	9.32	8.80	8.28	7.82	7.48	7.14	7.02
90	6.27	6.17	5.98	5.70	5.50	5.27	5.06	4.87	4.67	4.58
100	4.00	3.98	3.91	3.78	3.63	3.49	3.37	3.26	3.18	3.17
110	2.67	2.65	2.63	2.60	2.54	2.46	2.41	2.34	2.31	2.30
120	1.99	1.98	1.97	1.94	1.91	1.86	1.82	1.79	1.76	1.75
130	1.54	1.53	1.53	1.52	1.51	1.48	1.47	1.46	1.45	1.44
140	1.30	1.30	1.30	1.29	1.28	1.28	1.28	1.27	1.27	1.27
150	1.16	1.16	1.16	1.15	1.15	1.15	1.15	1.15	1.15	1.15
160	1.05	1.05	1.05	1.05	1.00	1.05	1.05	1.05	1.00	1.05
170	1.00	1.00	1.00	1.00	1.00	1.00	1.00	1.00	1.00	1.00
180	0.990	0.990	0.990	0.990	0.990	0.990	0.990	0.990	0.990	0.990

TABLE 10
Radiance Distribution under Clear Sunny Sky
(Depth, 66.1 meters; sun altitude, 56.6°)

Tilt angle (θ)	Azimuth angle (Φ)									
	0	20	40	60	80	100	120	140	160	180
0	28.8	28.8	28.8	28.8	28.8	28.8	28.8	28.8	28.8	28.8
10	29.8	29.6	29.2	28.4	27.7	26.9	26.1	25.3	24.5	24.1
20	26.1	25.7	24.7	23.8	22.8	21.6	20.6	19.6	18.9	18.5
30	19.7	19.3	18.5	17.5	16.5	15.6	14.6	13.8	13.1	12.8
40	12.7	12.4	11.9	11.3	10.6	10.1	9.44	8.87	8.41	8.19
50	7.64	7.54	7.20	6.83	6.46	6.09	5.70	5.40	5.14	5.01
60	4.43	4.34	4.18	3.98	3.78	3.58	3.38	3.20	3.08	2.98
70	2.46	2.44	2.37	2.29	2.20	2.11	2.01	1.92	1.86	1.81
80	1.43	1.42	1.39	1.35	1.29	1.25	1.19	1.14	1.11	1.09
90	0.858	0.855	0.839	0.817	0.788	0.757	0.727	0.700	0.677	0.661
100	0.531	0.525	0.518	0.506	0.496	0.482	0.468	0.453	0.445	0.439
110	0.357	0.355	0.349	0.342	0.334	0.325	0.317	0.311	0.305	0.301
120	0.252	0.250	0.247	0.244	0.241	0.239	0.235	0.233	0.231	0.226
130	0.186	0.184	0.184	0.181	0.180	0.179	0.177	0.175	0.173	0.172
140	0.146	0.145	0.145	0.144	0.144	0.143	0.142	0.141	0.141	0.139
150	0.124	0.124	0.124	0.122	0.122	0.121	0.121	0.120	0.120	0.120
160	0.108	0.108	0.108	0.108	0.108	0.108	0.108	0.108	0.108	0.108
170	0.100	0.100	0.100	0.100	0.100	0.100	0.100	0.100	0.100	0.100
180	0.0975	0.0975	0.0975	0.0975	0.0975	0.0975	0.0975	0.0975	0.0975	0.0975

(Cox and Munk, 1955), but larger than a single sun image. Owing to the optical state of the surface on 28 April this glitter pattern probably consisted of widely separated points of light. Because of the geometry discussed above it is possible that the solid angle of acceptance never covered more than half of the glitter pattern. This would mean that the reading obtained in the sun's direction could be low by a factor of 2.

In certain directions, and especially at the shallow stations, the presence of the glitter generates a noise signal that varies in both frequency and amplitude as a function of depth. Typical high and low values of this noise, together with its frequency, are given in table 11 for the direction of the sun. In tables 4 through 10 average experimental values for radiance are reported for all directions.

TABLE 11

AMPLITUDE AND FREQUENCY OF THE NOISE SIGNAL IN THE
DIRECTION OF THE SUN

Depth (meters)	Maximum variation from mean radiance (%)	Approximate frequency (per min.)
4.2	±93.6	128
10.4	±75.0	74
16.6	±47.4	60
28.9	±10.5	50

DISCUSSION AND TREATMENT OF DATA FOR OVERCAST CONDITIONS

The diffuse nature of the surface lighting for overcast conditions makes it practical to take data during a longer interval around noon than can be used for the clear sunny sky. For the same reason the order in which the stations are run is far less critical. The near-surface data for overcast sky shows the image features of the barge and its shadow at lower contrast than before. The zenith readings at the near-surface stations may exhibit greater variability with an overcast sky because of the time variability in the zenith thickness of the overcast. In addition, the generally lower light level puts the deeper stations experimentally beyond reach. Except for these slight differences the experimental procedure was the same for obtaining the overcast data as it was for obtaining clear sunny data.

The overcast data have been treated by the same procedure used for the clear sunny day data as described in steps 1 through 9, inclusive, with slight modification. Steps 8 and 9 do not, of course, apply as critically to diffuse lighting as they do to the combination of diffuse plus specular lighting provided by a clear sunny day. In step 6, two complete runs of overcast sky data at each depth station were averaged.

DATA FOR OVERCAST CONDITIONS

The voice recorded notes for 16 March, the overcast day, read in part as follows:

The data from about 1030 on are all excellent overcast data. The instrument was working perfectly. . . . Today's wind velocities ranged from 10 knots at about 1100 to about 2 knots at 1215. At 1400 the water on the south side of the barge where we are operating was as nearly calm as I have seen it. . . .

The radiance distribution data for the overcast condition are given for five depth stations in tables 12 through 16, inclusive.

TABLE 12
Radiance Distribution, Overcast Sky
(Depth, 6.1 meters; sun altitude, 40.0°)

Tilt angle (Θ)	Azimuth angle (Φ)									
	0	20	40	60	80	100	120	140	160	180
0	321,000	321,000	321,000	321,000	321,000	321,000	321,000	321,000	321,000	321,000
10	343,000	342,000	338,000	332,000	325,000	319,000	311,000	306,000	302,000	300,000
20	327,000	326,000	321,000	312,000	302,000	290,000	281,000	272,000	267,000	265,000
30	277,000	275,000	268,000	258,000	243,000	229,000	216,000	207,000	202,000	200,000
40	199,000	196,000	189,000	179,000	167,000	152,000	141,000	132,000	127,000	126,000
50	96,800	95,700	92,500	87,300	81,700	74,900	69,500	65,900	63,600	62,900
60	53,500	52,800	51,100	48,800	45,600	42,800	39,700	37,500	36,100	35,400
70	32,500	32,100	31,300	30,000	28,200	26,100	24,100	22,800	21,900	21,600
80	20,000	19,700	19,100	18,300	17,100	15,800	14,700	13,900	13,500	13,400
90	12,600	12,400	12,000	11,500	10,800	9,950	9,170	8,710	8,380	8,310
100	8,060	7,980	7,780	7,490	7,080	6,640	6,190	5,860	5,670	5,620
110	5,620	5,580	5,450	5,250	4,970	4,670	4,390	4,180	4,050	4,010
120	4,130	4,080	3,990	3,850	3,700	3,500	3,330	3,200	3,120	3,090
130	3,200	3,190	3,150	3,070	2,970	2,860	2,740	2,630	2,570	2,550
140	2,650	2,640	2,590	2,540	2,480	2,390	2,300	2,240	2,190	2,180
150	2,300	2,290	2,280	2,250	2,200	2,150	2,090	2,040	2,000	1,980
160	2,070	2,070	2,060	2,030	2,000	1,960	1,910	1,890	1,890	1,880
170	1,860	1,860	1,860	1,860	1,860	1,860	1,860	1,860	1,860	1,860
180	1,830	1,830	1,830	1,830	1,830	1,830	1,830	1,830	1,830	1,830

TABLE 13

RADIANCE DISTRIBUTION, OVERCAST SKY
(Depth, 18.3 meters; sun altitude, 40.0°)

Tilt angle (θ)	Azimuth angle (Φ)									
	0	20	40	60	80	100	120	140	160	180
0	26,800	26,800	26,800	26,800	26,800	26,800	26,800	26,800	26,800	26,800
10	27,500	27,400	27,200	26,900	26,600	26,000	25,200	24,600	24,100	23,900
20	23,600	23,400	23,000	22,500	21,900	21,000	20,100	19,400	18,800	18,600
30	17,500	17,400	17,200	16,800	16,200	15,400	14,400	13,400	12,900	12,600
40	11,900	11,800	11,600	11,100	10,600	9,800	9,080	8,460	8,040	7,860
50	7,520	7,460	7,310	7,090	6,750	6,250	5,710	5,240	4,880	4,720
60	4,600	4,540	4,440	4,280	4,070	3,740	3,440	3,230	3,050	2,970
70	2,780	2,750	2,690	2,590	2,450	2,300	2,130	1,990	1,910	1,860
80	1,650	1,640	1,590	1,550	1,470	1,380	1,280	1,220	1,160	1,150
90	1,040	1,020	995	948	892	830	784	745	722	713
100	645	638	627	603	583	557	533	514	502	500
110	454	449	439	426	409	391	378	366	357	356
120	334	331	324	315	302	289	280	272	268	265
130	258	255	250	243	235	228	218	214	209	209
140	206	205	202	198	193	188	185	182	179	179
150	176	175	174	170	167	164	162	161	161	160
160	156	156	156	155	152	151	150	149	147	147
170	145	145	145	145	145	145	145	145	145	145
180	141	141	141	141	141	141	141	141	141	141

TABLE 14

RADIANCE DISTRIBUTION, OVERCAST SKY

(Depth, 30.5 meters; sun altitude, 40.0°)

Tilt angle (Θ)	Azimuth angle (Φ)									
	0	20	40	60	80	100	120	140	160	180
0	2,430	2,430	2,430	2,430	2,430	2,430	2,430	2,430	2,430	2,430
10	2,500	2,490	2,470	2,430	2,380	2,310	2,210	2,120	2,070	2,030
20	2,190	2,170	2,130	2,080	2,010	1,910	1,790	1,680	1,600	1,570
30	1,690	1,670	1,640	1,600	1,550	1,420	1,300	1,180	1,090	1,050
40	1,150	1,130	1,090	1,050	998	927	840	757	702	676
50	688	683	670	649	629	596	544	498	466	453
60	413	409	400	389	375	353	332	302	284	275
70	254	252	247	240	231	217	203	189	178	173
80	153	151	148	144	136	130	121	114	110	108
90	93.9	93.4	91.3	88.6	85.2	81.1	75.9	72.3	69.7	68.4
100	61.2	60.7	59.7	57.9	56.0	53.4	50.6	48.3	46.4	45.3
110	42.9	42.7	41.8	40.7	39.3	37.8	35.8	34.2	33.3	32.7
120	31.1	30.9	30.5	29.8	29.1	28.1	27.2	26.3	25.8	25.4
130	23.8	23.7	23.5	23.1	22.8	22.2	21.6	21.2	21.0	20.9
140	20.4	20.3	20.0	19.8	19.5	19.0	18.4	18.1	17.6	17.5
150	18.1	18.0	17.8	17.5	17.3	16.9	16.7	16.5	16.3	16.3
160	16.2	16.1	16.1	16.1	16.0	15.8	15.5	15.3	15.2	15.2
170	14.9	14.9	14.9	14.9	14.9	14.9	14.9	14.9	14.9	14.9
180	14.8	14.8	14.8	14.8	14.8	14.8	14.8	14.8	14.8	14.8

TABLE 15
RADIANCE DISTRIBUTION, OVERCAST SKY
(Depth, 42.8 meters; sun altitude, 40.0°)

Tilt angle (Θ)	Azimuth angle (Φ)									
	0	20	40	60	80	100	120	140	160	180
0	250	250	250	250	250	250	250	250	250	250
10	254	252	250	247	242	239	233	227	223	221
20	229	226	223	216	209	201	190	182	176	174
30	175	173	167	161	155	146	138	131	125	123
40	119	117	113	109	104	99.1	92.5	87.4	83.9	82.6
50	76.0	74.9	73.4	70.4	67.0	63.6	59.8	56.9	54.7	53.8
60	45.7	45.3	44.0	42.6	40.8	39.1	38.2	36.6	35.4	34.9
70	27.6	27.2	26.6	25.9	24.9	24.0	22.9	22.0	21.2	20.9
80	16.8	16.6	16.2	15.8	15.4	14.7	14.1	13.6	13.2	13.1
90	10.5	10.4	10.1	9.87	9.55	9.21	8.87	8.58	8.38	8.20
100	6.82	6.78	6.61	6.48	6.28	6.06	5.82	5.61	5.40	5.35
110	4.66	4.63	4.55	4.46	4.36	4.24	4.08	3.95	3.86	3.82
120	3.46	3.42	3.39	3.33	3.27	3.16	3.08	3.02	2.96	2.94
130	2.67	2.66	2.65	2.61	2.58	2.52	2.45	2.39	2.34	2.32
140	2.22	2.22	2.20	2.16	2.13	2.10	2.06	2.01	1.99	1.97
150	1.93	1.93	1.93	1.92	1.91	1.89	1.88	1.86	1.83	1.82
160	1.76	1.76	1.76	1.76	1.76	1.76	1.76	1.76	1.76	1.76
170	1.67	1.67	1.67	1.67	1.67	1.67	1.67	1.67	1.67	1.67
180	1.65	1.65	1.65	1.65	1.65	1.65	1.65	1.65	1.65	1.65

TABLE 16

RADIANCE DISTRIBUTION, OVERCAST SKY
(Depth, 55.0 meters; sun altitude, 40.0°)

Tilt angle (Θ)	Azimuth angle (Φ)									
	0	20	40	60	80	100	120	140	160	180
0	29.6	29.6	29.6	29.6	29.6	29.6	29.6	29.6	29.6	29.6
10	29.9	29.7	29.3	28.8	28.2	27.6	27.0	26.5	25.9	25.7
20	24.7	24.6	24.0	23.4	22.8	22.2	21.5	20.9	20.5	20.4
30	18.8	18.5	18.1	17.7	17.3	16.6	16.0	15.4	14.9	14.6
40	13.0	12.8	12.6	12.3	11.9	11.6	11.2	10.8	10.5	10.4
50	8.11	8.00	7.86	7.72	7.54	7.32	7.07	6.86	6.69	6.61
60	4.99	4.96	4.85	4.76	4.64	4.55	4.42	4.28	4.19	4.17
70	3.00	2.98	2.93	2.87	2.81	2.75	2.69	2.63	2.59	2.56
80	1.90	1.88	1.84	1.82	1.78	1.75	1.71	1.67	1.65	1.63
90	1.19	1.18	1.16	1.15	1.12	1.09	1.07	1.05	1.04	1.03
100	0.765	0.761	0.749	0.736	0.723	0.708	0.689	0.678	0.665	0.661
110	0.512	0.509	0.503	0.495	0.488	0.481	0.469	0.460	0.452	0.449
120	0.363	0.360	0.358	0.355	0.352	0.346	0.341	0.337	0.333	0.331
130	0.273	0.272	0.271	0.268	0.266	0.264	0.259	0.256	0.254	0.253
140	0.215	0.215	0.214	0.213	0.210	0.208	0.206	0.204	0.203	0.203
150	0.181	0.181	0.181	0.181	0.181	0.181	0.181	0.181	0.181	0.181
160	0.158	0.158	0.158	0.158	0.158	0.158	0.158	0.158	0.158	0.158
170	0.146	0.146	0.146	0.146	0.146	0.146	0.146	0.146	0.146	0.146
180	0.143	0.143	0.143	0.143	0.143	0.143	0.143	0.143	0.143	0.143

[*Editor's Note:* Material has been omitted at this point.]

REFERENCE

Cox, C., and W. Munk. 1955. Some problems in optical oceanography. *J. Mar. Res.* **14:**63–78.

20

Reprinted from *Sears Found. J. Mar. Res.* **18**(1):1–9 (1959)

THEORETICAL PROOF OF THE EXISTENCE OF CHARACTERISTIC DIFFUSE LIGHT IN NATURAL WATERS[1]

By

RUDOLPH W. PREISENDORFER

Scripps Institution of Oceanography, University of California
La Jolla, California

ABSTRACT

This paper develops a mathematical model for the radiance distribution of light penetrating a homogeneous hydrosol on the basis of the general theory of radiative transfer. It is proved that the radiance distribution approaches an asymptotic pattern at great depths. This is in accord with previous field measurements of the directional patterns in underwater light and with L. V. Whitney's conjecture that there is at some depth in natural waters a characteristic diffuse light symmetrically distributed around the vertical. The angular form of this equilibrium light pattern is derived in terms of the mathematical model presented.

INTRODUCTION

Recent experimental evidence (Tyler, 1958) forms the basis for fresh support of the long-standing conjecture that the radiance distribution about a point in an optically deep natural hydrosol approaches, with increasing depth, a characteristic form which is independent of the external lighting conditions and of the optical state of the medium's surface and which depends only on the inherent optical

---

[1] This paper represents results of research which has been supported by the Bureau of Ships, U. S. Navy.

253

properties of the medium. This conjecture was apparently given its first definitive formulation by Whitney (1941a, 1941b), who referred to the asymptotic radiance distribution as *characteristic diffuse light*. (Some early experimental evidence for this conjecture is cited in Whitney's papers.) In this note we complement experimental evidence in favor of this conjecture by supplying a simple mathematical proof of the existence of characteristic diffuse light in all homogeneous optically deep natural waters. The paper concludes with a derivation of the integral equation governing the angular structure of characteristic diffuse light as well as a brief discussion of an interesting and tractable example for the case of isotropic scattering.

We note in passing that the applicability of Whitney's hypothesis has been widened considerably since its formulation. The problem of a limiting angular distribution has since been encountered in modern neutron transport theory, basically as an abstract mathematical problem rather than as an experimental phenomenon. A similar type of problem has long been extant in astrophysical radiative transfer. A general proof of the existence of an asymptotic radiance distribution which covers all of these contexts recently has been devised (Preisendorfer, 1958a).

However, the hypothesis still retains its greatest usefulness in the context of geophysical optics. For in this field, unlike the others mentioned above, the trend to a characteristic limiting form is a directly observable phenomenon. Furthermore, the existence of such a form is of inestimable importance to all experimental research work dealing with the determination of optical properties of natural waters. In many important instances, knowledge that an asymptotic radiance distribution exists will obviate the necessity of experimental probings to extremely large depths; for such knowledge will allow, by means of relatively simple formulae, accurate prediction of the geometrical structure of the light field in great-depth ranges. Some of these practical consequences of the asymptotic radiance hypothesis have been formulated recently (Preisendorfer 1958b).

PHYSICAL BACKGROUND OF THE METHOD OF PROOF

The argument used by Whitney in establishing experimental evidence for the asymptotic radiance hypothesis went basically as follows: he showed that when experimentally obtained plots of radiance distributions at various large depths were blown up to the same size (more precisely, the zenith readings were normalized to a common value), they formed a set of nearly congruent figures. Now, an interesting feature of such distributions is that they assume the same *shape* and decrease in size with increasing depth at nearly the same exponential rate. This fact can be stated precisely as follows:[2]

$$N(z, \theta, \phi) = g(\theta, \phi)e^{-kz}. \tag{1}$$

From this we see that the asymptotic radiance hypothesis is equivalent to the statement that *the directional and depth dependence of radiance distributions multiplicatively uncouple at great depths.* That is, the radiance function N may be represented as the product of two functions: g which gives the shape or directional structure common to all distributions and an exponential function which gives the depth dependence of the distributions.

Each factor on the right-hand side of (1) has special physical significance. The function g evidentally defines the angular form of characteristic diffuse light. The exponent k of the exponential function has the following interesting interpretation.

Define *scalar irradiance* $h(z)$ at depth z as follows:

$$h(z) = \int_{\phi=0}^{2\pi} \int_{\theta=0}^{\pi} N(z, \theta, \phi) \sin \theta \, d\theta \, d\phi. \tag{2}$$

The quantity $h(z)$ is then a measure of the volume density of radiant energy at depth z. Measurements of $h(z)$ over the years in many hydrosols have shown that $h(z)$ varies essentially in an exponential manner with depth. That is, semilog plots of $h(z)$ *vs* depth show an unmistakable trend toward linearity as depth increases. In any event, $h(z)$ may be accurately represented by a general formula of the type

$$h(z) = h(o) \exp \left\{ -\int_0^z k(z') \, dz' \right\}, \tag{3}$$

where $k(z)$ is the negative logarithmic derivative of $h(z)$. (Here and below, a primed symbol refers to a dummy variable of integration.) As depth increases, experimental evidence gathered in the field indicates that $k(z)$ approaches a constant value. Denote this limit value by k_∞. Now, assuming that an asymptotic radiance distribution is approached by the radiance distributions in a particular body of water, we see from (1), (2) and (3) that

$$h(z) = h(z_0)e^{-k_\infty(z-z_0)} = e^{-kz} \int_{\phi=0}^{2\pi} \int_{\theta=0}^{\pi} g(\theta, \phi) \sin \theta \, d\theta \, d\phi, \tag{4}$$

[2] It is implicit in the definition of radiance that it applies to an arbitrary but fixed wavelength of radiant flux. This is also true of all the other radiometric quantities used in this note.

where z_0 is the depth below which we may assume that $k(z) = k_\infty$. From this we conclude that

$$k = k_\infty. \tag{5}$$

Hence, under the above assumption (1) we see that, at great depths, the size of a radiance distribution plot decreases exponentially with increasing depth and that the rate of this decrease is precisely that of scalar irradiance (or energy density).

The close connection between the depth dependence of scalar irradiance and that of the radiance distributions, as summarized in (5), suggests the following mode of representation of the radiance distributions for any depth: Define, for each direction (θ, ϕ),

$$K(z, \theta, \phi) = \frac{-1}{N(z, \theta, \phi)} \frac{dN(z, \theta, \phi)}{dz}. \tag{6}$$

Then, in analogy to (3), $N(z, \theta, \phi)$ at any depth z may be represented exactly by

$$N(z, \theta, \phi) = N(0, \theta, \phi) \exp\left\{- \int_0^z K(z', \theta, \phi)\, dz'\right\}. \tag{7}$$

Now suppose there is some depth z_0 below which we have $K(z, \theta, \phi) = k_\infty$ for all directions (θ, ϕ). Then (7) may be written

$$N(z, \theta, \phi) = N(0, \theta, \phi) \exp\left\{- \int_0^{z_0} K(z', \theta, \phi)\, dz' - \int_{z_0}^z K(z', \theta, \phi)\, dz'\right\}$$

$$= N(z_0, \theta, \phi) \exp\left\{- k_\infty(z - z_0)\right\}.$$

If we set

$$g(\theta, \phi) = N(z_0, \theta, \phi) \exp\left\{k_\infty z_0\right\},$$

then we may write

$$N(z, \theta, \phi) = g(\theta, \phi)\, e^{-k_\infty z}, \tag{8}$$

for all depths z below z_0.

The similarity between (1) and (8) is unmistakable and it points out a method of attack we may follow in order to prove the asymptotic radiance hypothesis: we must show that the quantities $K(z, \theta, \phi)$ approach a limit as depth is increased and that this limit is independent of the directions (θ, ϕ). Furthermore, this limit, in accordance with the preceding discussion, should be none other than the limit k_∞ of $k(z)$, as defined in (3).

THE PROOF

We make use of the equation of transfer for radiance:

$$\frac{dN(z, \theta, \phi)}{dr} = -\alpha N(z, \theta, \phi) + N_*(z, \theta, \phi), \qquad (9)$$

where

$$N_*(z, \theta, \phi) = \int_{\phi'=0}^{2\pi} \int_{\theta'=0}^{\pi} \sigma(\theta, \phi; \theta', \phi') \, N(z, \theta', \phi') \sin \theta' \, d\theta' \, d\phi' \qquad (10)$$

defines the path function N_*; σ is the volume scattering function (which governs the law of scattering in the water) and α is the volume attenuation coefficient.   The formal solution of (9) is readily obtained:

$$N(z, \theta, \phi) = N^0(z, \theta, \phi) + \int_0^r N_*(z', \theta, \phi) \, e^{-\alpha(r-r')} \, dr'. \qquad (11)$$

The first term represents the component of N consisting of unscattered light.   The second represents the space light over the path of length r

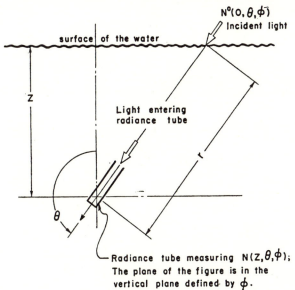

Figure 1

(Fig. 1) which has been generated by light scattered into the path of sight all along its extent.   The formal solution (11) has been written for a general downward direction of flow of light (see Fig. 1) so that $N^0(z, \theta, \phi)$ is interpreted as the directly transmitted light from the upper boundary of the medium and is of the form

$$N^0(z, \theta, \phi) = N^0(0, \theta, \phi)\, e^{-ar},$$

where

$$-r \cos \theta = z.$$

We now turn eq. (11) into a useful inequality by means of the following three steps:

First, since $N(z, \theta, \phi)$ clearly exceeds its spacelight component at all depths, we write

$$N(z, \theta, \phi) > \int_0^r N_*(z', \theta, \phi)\, e^{-a(r-r')}\, dr'.$$

Second, using the definition of N_*, we strengthen the inequality when we write

$$N(z, \theta, \phi) > \sigma_{\min} \int_0^r h(z')\, e^{-a(r-r')}\, dr',$$

where $\sigma_{\min}$ is the minimum value of the volume scattering function; that is, we have used (10) to deduce that

$$N_*(z, \theta, \phi) > \sigma_{\min} \int_{\phi'=0}^{2\pi} \int_{\theta'=0}^{\pi} N(z, \theta', \phi')\, \sin \theta'\, d\theta'\, d\phi' = \sigma_{\min} h(z).$$

Finally, since $h(z)$ decreases with increasing depth, we certainly strengthen the inequality by writing

$$N(z, \theta, \phi) > \sigma_{\min} h(z) \int_0^r e^{-a(r-r')}\, dr'.$$

That is, we have

$$N(z, \theta, \phi) > \frac{\sigma_{\min}}{\alpha} h(z)(1 - e^{-ar}) \tag{12}$$

for all depths z. From this we see that, as depth increases indefinitely, the exponential rate of decrease $K(z, \theta, \phi)$ of the radiance cannot eventually exceed the $k(z)$ of scalar irradiance and remain larger by any finite amount; if it did, the plot of N would eventually fall and remain below that of h. This observation is stated symbolically as follows:

$$\lim_{z \to \infty} K(z, \theta, \phi) \leqslant \lim_{z \to \infty} k(z) = k_\infty \tag{13}$$

for all downward directions (θ, ϕ). We now show that strict equality must hold in (13). We achieve this by initially assuming the contrary; that is, we assume that there is a nonzero solid angle Ω_0 of directions over which

$$\lim_{z \to \infty} K(z, \theta, \phi) \leqslant k_\infty - \epsilon,$$

where ϵ is *any* small positive number. Then it is clear that the radiances in this set of directions decrease at a definitely smaller rate than scalar irradiance; so much smaller in fact that, by our assumption, for some depth z_1 we must have

$$\int_{\Omega_0} N(z_1, \theta, \phi) \sin \theta \, d\theta \, d\phi > h(z_1).$$

However, this conclusion clearly contradicts (2) since a part cannot exceed the whole. We have reached a contradiction which leaves only one other possibility:

$$\lim_{z \to \infty} K(z, \theta, \phi) = k_{\infty} \tag{14}$$

for all downward directions (θ, ϕ). In the light of the preceding discussion [*cf* (8)], this means that the shape of the radiance distributions impinging on the upper boundaries of deep layers of water eventually assume a fixed form. But it is known that the shape of the *reflected* radiance distribution at the upper boundary of a scattering layer is determined by the shape of the incident radiance distribution at that boundary. Hence if the incident radiance distribution approaches a fixed shape, so does that of the reflected distribution. This completes the proof.

We observe that the present proof can also be applied in all natural waters which eventually become homogeneous. That is, the preceding arguments are basically unchanged if the medium is inhomogeneous over any initial finite depth range below the surface. Even more general media exist which give rise to asymptotic radiance distributions, namely media in which the ratio σ/α eventually becomes independent of depth (Preisendorfer, 1958a).

THE EQUATION FOR CHARACTERISTIC DIFFUSE LIGHT

Using the equation of transfer, definition (6), and the relation between z and r, we write the equation of transfer in the following general form:

$$N(z, \theta, \phi) = \frac{N_*(z, \phi, \theta)}{\alpha + K(z, \theta, \phi) \cos \theta}. \tag{15}$$

From (14) and (8) we see that the limiting form of (15) (as depth increases indefinitely) is

$$g(\theta, \phi) = \frac{\int_{\phi'=0}^{2\pi} \int_{\theta'=0}^{\pi} \sigma(\theta, \phi; \theta', \phi') \, g(\theta', \phi') \sin \theta' \, d\theta' \, d\phi'}{\alpha + k_{\infty} \cos \theta}; \tag{16}$$

this is the equation governing the angular form of characteristic diffuse light. It is a property of equations of the type shown in (16) that the function g is independent of ϕ for all real physical situations. Thus characteristic diffuse light is always represented by a surface of revolution whose axis of symmetry is vertical.

The theory of the solution of such equations as (16) is fairly well understood (*e.g.*, see Davison, 1957). The present note, therefore, will not discuss (16) in any detail. However, there is one simple special case which is immediately solved and which can shed much light on some of the salient details of the structure of the asymptotic radiance distributions. This is the case of isotropic scattering, where the volume scattering function σ is independent of direction and has the form

$$\sigma(\theta, \phi; \theta', \phi') = \frac{s}{4\pi},\tag{17}$$

where s is the total scattering coefficient.

To observe the resulting structure of the asymptotic radiance distribution it is convenient in the present case to turn to (15). With assumption (17) and definitions (2) and (10), we have

$$N(z, \theta, \phi) = \frac{s}{4\pi}\frac{h(z)}{\alpha + K(z, \theta, \phi)\cos\theta},$$

which at great depths approaches the form

$$N(z, \theta, \phi) = \frac{1}{4\pi}\left(\frac{s}{\alpha}\right)\frac{h(z_0)\,e^{-k_\infty(z-z_0)}}{1 + \left(\dfrac{k_\infty}{\alpha}\right)\cos\theta}.\tag{18}$$

Here z_0 is the depth below which $h(z)$ decreases exponentially with increasing depth. Comparing (18) with (8), we see that for the present case

$$g(\theta, \phi) = \frac{1}{4\pi}\left(\frac{s}{\alpha}\right)\frac{h(z_0)\,e^{k_\infty z_0}}{1 + \left(\dfrac{k_\infty}{\alpha}\right)\cos\theta}.\tag{19}$$

We have written (19) in the indicated form to point up the following geometric fact: A polar plot of $g(\theta, \phi)$ is generally a prolate ellipsoid of revolution with vertical axis and of eccentricity k_∞/α. When there is no absorption in the medium, it is easy to deduce then that $k_\infty = 0$ and that characteristic diffuse light is represented by a sphere. On the other hand, if there is little scattering as compared to absorption, the figure assumes a narrow pencil-like shape. In the limit of no scattering, k_∞ approaches α, and the figure degenerates into a vertical line.

The structure of the expression in (18) is related to the limiting form of a simple model for the radiance distribution discussed elsewhere (Preisendorfer, 1958c) and to a formula derived by Poole (1945). We conclude with the observation that (19) predicts a different limiting ratio of horizontal to upward radiance than that derived by Whitney (1941a) under the same circumstances (*i.e.*, isotropic scattering). Instead of the ratio 2:1, as suggested by Whitney, the present formula yields

$$\frac{g(\pi/2, \phi)}{g(0, \phi)} = 1 + \left(\frac{k_\infty}{\alpha}\right) \leqslant 2. \tag{20}$$

In other words, the ratio in (20) is not a fixed magnitude but depends on the optical properties of the medium in the manner shown.

The distribution (19) can serve as a convenient standard reference distribution against which experimentally determined radiance distributions can be compared. The amount of departure of the experimental distributions from (19) would then serve as a measure of the anisotropy of scattering in the real medium.

REFERENCES

DAVISON, B.
  1957. Neutron transport theory. Clarenden Press, Oxford. 450 pp.
POOLE, H. H.
  1945. The angular distribution of submarine daylight in deep water. Sci. Proc. R. Dublin Soc., *24*: 29–42.
PREISENDORFER, R. W.
  1958a. A proof of the asymptotic radiance hypothesis. Visibility Laboratory Report, Scripps Institution of Oceanography, University of California. SIO Ref. 58–57.
  1958b. Some practical consequences of the asymptotic radiance hypothesis. Visibility Laboratory Report, Scripps Institution of Oceanography, University of California. SIO Ref. 58–60.
  1958c. Model for radiance distributions in natural hydrosols. Visibility Laboratory Report, Scripps Institution of Oceanography, University of California. SIO Ref. 58–42.
TYLER, J. E.
  1958. Radiance distribution as a function of depth in the submarine environment. Visibility Laboratory Report, Scripps Institution of Oceanography, University of California. SIO Ref. 58–25.
WHITNEY, L. V.
  1941a. The angular distribution of characteristic diffuse light in natural waters. J. Mar. Res., *4*: 122–131.
  1941b. A general law of diminuation of light intensity in natural waters and the percent of diffuse light at different depths. J. opt. Soc. Amer., *31*: 714–722.

21

Reprinted from pp. 283, 284–307 of *Sears Found. J. Mar. Res.* **16**(3):283–307 (1958)

FACTORS, MAINLY DEPTH AND WAVELENGTH, AFFECTING THE DEGREE OF UNDERWATER LIGHT POLARIZATION[1]

By

ALEXANDRE IVANOFF

École Supérieure de Physique et de Chimie Industrielles, Paris

AND

TALBOT H. WATERMAN

Department of Zoology, Yale University, New Haven, Conn.

[*Editor's Note:* The abstract and introduction have been omitted.]

INSTRUMENTATION

The Photoelectric Polarimeter. Since the three methods we have used to measure underwater light polarization have already been described (Waterman, 1955; Ivanoff, 1956b, 1957b; Waterman, 1958) they need not be considered in detail. However, the principle of the photoelectric polarimeter must be understood because it provided most of the data presented here; consequently it is diagrammed (Fig. 1) and briefly described. Essentially the photocell, with the aid of the revolving mirror (M), scans the horizontal lines of sight underwater through 360° at a rate determined by the mirror's rota-

---

[1] Contribution No. 245 from the Bermuda Biological Station.

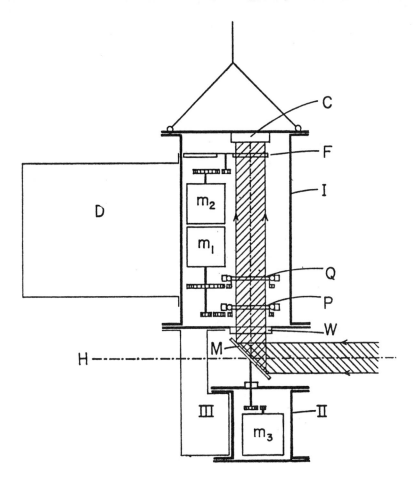

Figure 1. Photoelectric polarimeter. I, II, water-tight compartments suspended by cable and solidly connected by stiff metal plate, III; C, Westaphot photocell; D, azimuth stabilizing fin; F, changeable interference filters; G, weight to maintain vertical; H, horizontal plane; M, 45° front surface mirror to reflect horizontal light rays into window, W, of chamber I; m_1, motor to rotate Polaroid P filter (P) at about 2 RPM and a half-wave mosaic (Q) at about 500 RPM; m_2, motor to change or remove filters in light path; m_3, motor to rotate mirror at 0.22 RPM.

[*Editor's Note:* The author has eliminated G, the weight to maintain vertical, from figure 1.]

tions. The revolving Polaroid filter (P) acts as an analyzer for any linear polarization present in the incoming light, and the rapidly rotating half wave (for λ = ca. 550 mμ) mosaic (Q) depolarizes the light from the Polaroid to eliminate the possibility of an artifact arising from variations of photocell sensitivity with plane of polarization. The fact that the Polaroid filter is not a perfect analyzer will result in measured degrees of polarization somewhat smaller than the actual ones, an effect which will be increased to some extent by the finite angular field of the instrument.

The further possibility of artifacts due to pressure birefringence originating in the window (W) of chamber I is minimized or excluded by using a special type of plexiglass. This shows but little birefringence under pressure, and even this small amount would be appreciable only for rays deviating considerably from the optical axis. The photocell is connected by an insulated cable (not shown in Fig. 1) to a shock mounted galvanometer on the ship's deck. Because of the rotation rate of the Polaroid, a fast indicating instrument is required with a time constant shorter than 2 sec.

The dimensions of the optical system are such that the beam being measured illuminates nearly the whole photocell surface (about 61 mm in diameter). The sensitivity of the latter, which is maximal for a horizontal pencil of parallel light, decreases to 75% for a ray inclined to the horizontal by 7°. It is well known that the current produced by a barrier photocell is proportional to the irradiance so long as the latter is low and the cell feeds into a low resistance. In the case of the Westaphot cell and galvanometer (internal resistance = 680Ω) used here, the response is linear within 1% so long as the illuminance does not exceed 12 lux and the loading current is less than 2μA. This was generally the case in our measurements, made either with colored filters or without filters at considerable depths. On the other hand, without filters at shallow depths the conditions necessary for a linear response were not present. However, the error so introduced for I_{max} and I_{min} is always reduced in the calculation for p (see p. 291).

Between the depolarizer and the photocell a series of interference filters (Table I) can be inserted as desired by remote control. One of these is a narrow band blue-green element (Balzers Filtraflex B-40); the other five are broad band filters (Balzers Filtraflex Series K). To study the effects of wavelength accurately, the transmission curves of the various colored filters used should be corrected for spectral distribution of the submarine light, which itself varies with h and θ. However, since the influence of λ on underwater polarization is weak, the maximum transmission of each filter for white light is used here. The slight displacement of the observed points relative to λ which

results from the corrections concerned would scarcely affect our conclusions.   On the other hand, the effect of λ on the polarizing properties of the Polaroid filter in the instrument is a significant one and is
discussed below.

Comparison with Earlier Methods.   While nearly all of the present
data were obtained with the photoelectric polarimeter, a series of
comparative measurements were made both in the field and in the
laboratory to contrast this instrument with the two others previously
used.   The questions of present interest are: first, whether the photographic and visual techniques provide essentially the same information on submarine polarization as does the photoelectric method;
second, whether all of these various instruments are equally versatile
and accurate in obtaining the desired data.

To the first question the evidence of five parallel series of field
measurements made in inshore waters down to 18 m provides essentially an affirmative answer.   In general the p determined by each
of the three methods under as closely identical conditions as possible
showed satisfactory agreement when average values are considered
(these varied from 7–22% at the different stations and under the prevailing meteorological conditions).   However, the relatively low
degree of polarization of the waters studied, and particularly the variable cloudiness of the sky (often inevitable), had an adverse effect on
the precision of all of these measurements.   Nevertheless, in the only
set of data where p was as much as 20%, the differences between the
three methods just exceed 5% in absolute value.

These comparative measurements also showed clearly the relative
advantages and disadvantages of the three methods.   In accuracy
the visual and photoelectric techniques were superior to the photographic; in either of the first two techniques the absolute error of a
single measurement does not exceed 5% whereas in the photographic
technique it may exceed 5%.   The photoelectric instrument is more
convenient for collecting much data since, unlike the other two, it
does not require a SCUBA diver to operate it; this also removes any
restriction to relatively shallow depths.   On the other hand, the
present photoelectric instrument is less versatile than the others because it operates only in a horizontal plane; furthermore, it does not
permit measurement of the polarization plane.   However, since data
on p are particularly needed and since many measurements are required to determine the influence of h and λ on p, this is the instrument
of choice for present purposes.

Almost perfect agreement between the visual and photoelectric
methods of determining p was found in the laboratory (Fig. 2).   Here

a light source whose degree of polarization could be varied by changing the angle of a pile of glass plates in the beam was measured. This suggests that measurements made in the field by these methods are also closely comparable although the difficulties of field work usually increase considerably the variance of such data.

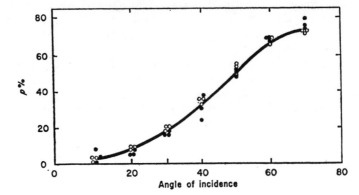

Figure 2. Laboratory comparison of visual and photoelectric methods of measuring degree of polarization (p%). Abscissa shows angle of inclination of a pile of glass plates used to produce partially polarized light. Photoelectric measurements, open circles; visual measurement, solid circles (see text).

Effects of Incomplete Polarization. While these comparative studies were being made, the same laboratory set-up was used with the photoelectric polarimeter to test the influence of the various filters in the instrument on the polarizing properties of its Polaroid P analyzer. Average results of five measurements at one inclination angle of the glass plates are shown in the last column of Table I. The results are between 51.7 and 52.1% without filter and with filters K_2, K_3 and K_5,

TABLE I

Filter	Approx. max. transmission λ (mμ)	Band width for 40% transmission (mμ)	% polarization* by Polaroid P analyzer
K_1 (violet)	415	380–430	48.0
K_2 (blue)	445	430–465	51.8
K_3 (blue green)	490	460–500	51.9
K_4 (yellow green)	550	515–570	50.4
K_5 (orange)	600	580–610	51.7
B–40 (blue green)	500	494–506	50.4
Without filter	—	—	52.1

* See text for explanation of this partially polarized test source.

but at 500 mμ (B-40) and 550 mμ (K_4) the Polaroid's polarizing properties apparently diminish slightly; at 450 mμ (K_1) they definitely
diminish.[2] If the Polaroid actually transmits at "extinction" a fraction kI_p of the incident linearly polarized light and if k is zero without
filter as well as with filters K_2, K_3 and K_5, then the above figures give
the following coefficients: $k = 0.019$ for K_4 and B-40, $K^. = 0.052$ for
K_1. These permit corresponding corrections to be made in the measured p;[3] this has been done in all data given below.

In the next section the new information obtained with the photoelectric polarimeter serves not only to illustrate the capabilities of the
instrument but to provide quantitative data on several aspects of
submarine light polarization.

RESULTS

1. GENERAL FACTORS

Two examples of the type of records obtained with the photoelectric
polarimeter are shown in Figs. 3 and 4. While the mirror (M) sweeps
through the horizontal plane (one revolution taking about 4.5 min)
and if the incoming light is polarized linearly, the photocell output
to the galvanometer undergoes a series of oscillations from maximum
to minimum that arise from rotations of the Polaroid (one turn taking
about 32 sec). If these maxima and minima are plotted, the two
curves so obtained in turn oscillate between maxima and minima
which occur respectively in the bearings of the sun and the anti-sun.
Provided the azimuthal position of the polarimeter does not change,
the time intervals between these points of zero slope are equal; when
they were not, boat drift was usually responsible; this was minimized, when necessary and possible, by fore and aft anchors. When
the mirror faces the metal plate (Fig. 1, III) connecting the lower
pressure case of the instrument to the upper one, there is a sudden
decrease in the apparent light level,[4] but this artifact is usually easy
to eliminate, as shown in Figs. 3 and 4.

[2] The decreased polarization of violet light by the filter used is in accord with
the average data published by its manufacturer although the slight dip at 500 and
550mμ is not. The latter, however, is not crucial for our argument.

[3] The increased submarine polarization at the extreme violet end of the spectrum
was not noticed in the first measurements with the photoelectric polarimeter (Ivanoff,
1957b) because the decreased polarizing properties of the Polaroid filter at these
wavelengths was not considered.

[4] In the earlier version of the polarimeter (Ivanoff, 1957b) three brackets connected
the two main parts of the polarimeter. The resulting three minima in the records
made the analysis of the results more difficult and less certain. The replacement of
the three brackets with a single supporting plate greatly increases the accuracy of
the present measurements.

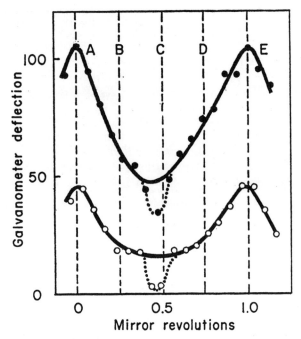

Figure 3. Photoelectric measurements obtained in very clear water with solar zenith angle about 33°. Data taken at 1345, 27 August 1957, St. 1 (Fig. 5); bottom, 2500 m; Secchi disc reading, 50 m. Polarimeter at 16 m using the 500 mμ narrow band filter. Maximal galvanometer readings during rotation of Polaroid, solid circles; minimal ones, open circles. Dotted line indicates shadowing effect of metal plate which connects two chambers of polarimeter. Broken vertical lines A–E indicate five positions of mirror for which the degree of polarization was calculated from equation $p = I_{max} - I_{min}/I_{max} + I_{min}$. The angular difference (θ) between azimuth of instrument's line of sight and sun's bearing is 0° for A and E, ±90° for B and D, and 180° for C. Computations for p are as follows:

$$\text{A. } p = (106 - 45)/(106 + 45) = 40.4\%$$
$$\text{B. } p = (61 - 20)/(61 + 20) = 50.6\%$$
$$\text{C. } p = (48 - 16)/(48 + 16) = 50.0\%$$
$$\text{D. } p = (74 - 24)/(74 + 24) = 51.0\%$$
$$\text{E. } p = (105 - 46)/(105 + 46) = 39.1\%$$

Within errors of measurement, p is the same at this station for $\theta = \pm90°$ and 180° (av. 50.5%), but it is definitely smaller at $\theta = 0°$ (av. 39.7%). If the effect of wavelength on the polarizing properties of the Polaroid filter is adjusted, the averages become 40.8% for $\theta = 0°$ and 52.0% for $\theta = \pm90°$, 180°. The ratio of maximal to minimal horizontal radiances which occur at $\theta = 0°$ and $\theta = 180°$ respectively, is $(106 + 46)/(48 + 16) = 2.4$.

For a given θ the ordinates for the two curves connecting the maxima and minima are proportional respectively to $I_{max} = I_{n/2} + I_p$ and $I_{min} = I_{n/2}$, where I_n and I_p are respectively the unpolarized and linearly polarized fractions of the incident light. Hence the polarization factor is

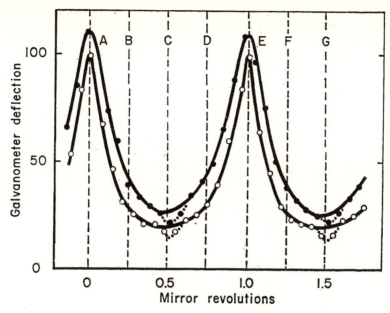

Figure 4. Photoelectric measurements obtained in less clear water (Fig. 5, St. 3) than those in Fig. 3 and with a greater solar zenith distance (about 67°). Data taken at 1630, 26 August 1957. Polarimeter at 5.4 m, using the 500mμ narrow band filter; bottom, 15 m; Secchi disc reading, 10 m. Symbols and method of plotting as in Fig. 3. The corresponding degrees of polarization in various azimuths are: A. 4.7%, B. 18.3%, C. 12.5%, D. 18.4%, E. 4.3%, F. 18.2%, and G. 13.1%. Obviously, although the percent polarization is less over-all, it varies considerably more with azimuth here than in Fig. 3. On an average, $p = 4.5\%$ when $\theta = 0°$, 12.8% when $\theta = 180°$, and 18.3% for $\theta = \pm 90°$. With the λ correction for the Polaroid, these become 4.6%, 13.1% and 18.8% respectively. The ratio of maximal to minimal radiance is $(111 + 101)/(27 + 21) = 4.4$.

$$p = \frac{I_p}{I_n + I_p} = \frac{I_{max} - I_{min}}{I_{max} + I_{min}}$$

Theoretically then the two curves allow p to be computed in any horizontal line of sight so long as variation in this parameter is greater than errors of measurement. These curves also indicate the angular distribution of radiance in the horizontal plane, since this is proportional to $I_n + I_p = I_{max} + I_{min}$.

Effects of Azimuth. Results have been analyzed mainly for $\theta = 0$, ± 90 and 180°, although for measurements around sunset (Fig. 9) $\theta = \pm 45°$ is also included. Three different patterns are clear in the various series studied. 1) In some cases p is maximal when $\theta = \pm 90°$, intermediate when $\theta = 180°$, and minimal when $\theta = 0°$ (Figs. 8, 11, 13, 15). 2) In other cases $\theta = 0°$ is also minimal, but p is essentially equal for $\theta = \pm 90°$ and $\theta = 180°$. Consequently these latter two have been averaged (Figs. 7, 10, 12, 14, 16). 3) Still others, mainly in turbid waters, show the same p, within errors of measurement, at all azimuths.

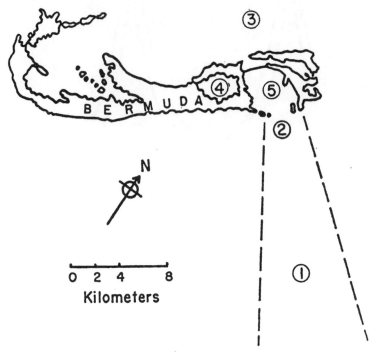

Figure 5. Five stations where measurements of submarine polarization were made. (1) South of Bermuda, with bottom depths greater than 400 m. Secchi disc visibility more than 18 m and as much as 50 m at appreciable distances from the islands. (2) South shore, with bottom depths between 18 and 40 m. Secchi disc readings from 14–18 m. (3) North shore, with bottom depths varying from 15 to 20 m. Secchi disc visibility down to 10–17 m. (4) Harrington Sound; maximum depth about 21 m. Secchi disc visibility 10–13 m. (5) Castle Harbour; maximum depth about 13 m. Secchi disc depths around 6–8 m.

Effect of Transparency. The measurements here reported can be used incidentally to illustrate the influence of the water's clarity on p. The Bermuda area is of considerable interest in this respect since a wide range of transparencies from very clear Sargasso Sea water to highly turbid inshore regions can be conveniently studied. The stations occupied in the present measurements are numbered from 1 to 5 on a scale of increasing turbidity (see Fig. 5). Since transparency measurements were not a primary part of our 1957 summer program, Secchi disc readings provide the only information available on water

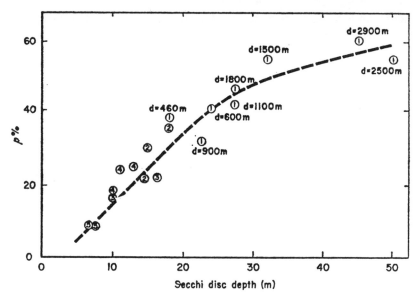

Figure 6.   Maximum *p* measured at 9 m depth using the 500mµ narrow band filter as a function of Secchi disc depth (*S*).   Circled numbers indicate water types in Fig. 5.   Bottom depths (*d*) are shown for stations in area 1.   Clarity of water increases, on an average, with bottom depth *d*.

transparency at the time of polarization measurements (Fig. 6).   As is well known, such data are limited among other things by the observer, the state of the water surface, and the relative solar bearing (*e.g.*, the sunny side of the ship in one instance gave a reading of 19 m, the shady side 25 m; in another case the readings were 45 and 58 m respectively).   Nevertheless, the correlation between Secchi disc readings and degree of polarization is striking.

Note that, to provide reasonably comparable data, the results plotted are limited to $h = 9$ m and $\lambda = 500$ mµ.   At other depths somewhat different results would be expected.   In deeper more turbid water the decrease in p with h is probably greater so that the average curve would bend still more.   If h is less than 9 m, a straighter curve would likely be found to fit the data.   In addition, measurements in Fig. 6 include only the case when $\theta = \pm 90°$.   Usually this is maximal, but as already noted, it may be equalled under certain conditions at $\theta = 180$ or even 0°.

Influence of Cloud Cover.   Whenever possible, measurements were made while the sun was shining (Figs. 3–21).   However, there usually are some scattered clouds in the sky in Bermuda so that the measurements were frequently disturbed by the passage of a cloud in front of the sun. This always decreases p, but the actual effects, both on p and on the plane of polarization, depend on the resulting change in directionality of underwater illumination, which varies with different optical conditions of the sky.

2. Effect of Depth

In addition to the direct effect of h on p_1, a secondary influence arises where the bottom is close enough to affect significantly the light that is present. As earlier observations showed (Waterman, 1958), p decreases near the bottom, and more so if the latter is light rather than dark. For this reason the present data fall into two categories: a) where there is no bottom influence; b) where there is bottom influence.

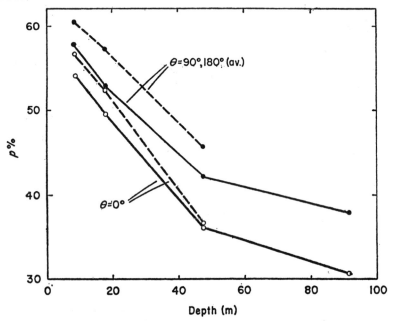

Figure 7. Influence of h on p in clear water at St. 1 (Fig. 5); bottom, 2900 m; no Secchi disc reading; solar zenith distance (i) = 60–78°. Solid lines connect points determined without colored filter; broken lines connect points measured with the 500mμ narrow band filter. The points indicated for θ = ±90 and 180° are averages of the two. Data of 22 July 1957.

Far from Bottom. Two series of measurements were made in 2500–3000 m of water where the optical effect of the bottom is certainly negligible (Figs. 7 and 8). Some of these data were taken without colored filters to increase the instrument's sensitivity to the low in-

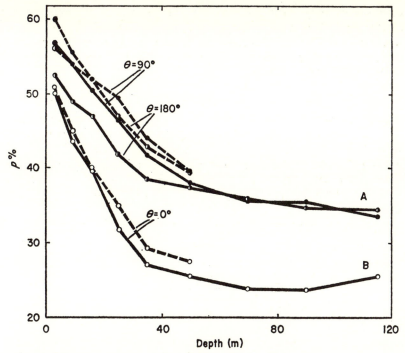

Figure 8. Influence of *h* on *p* in clear water at St. 1 (Fig. 5); bottom, 2500 *m*; *S* = 50 *m*; *i* between 29 and 33°.   Broken and solid lines as in Fig. 7.   Data of 27 August 1957.

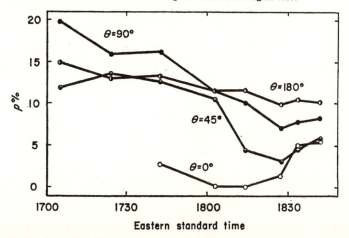

Figure 9.   Degree of polarization around sunset at St. 2 (Fig. 5); measured at 5.4 m without filter in moderately turbid water; bottom about 12 m.   Sun's altitude at 1700 was 18°50'; sunset occurred at 1852.   At 1815 the solar disc was dim enough to look at with the naked eye; after 1845, the underwater light was too weak for measurement with the polarimeter.   Data of 26 August 1957.

tensities present; the rest were taken with the 500 mμ filter (narrow band) which reduces the influence of depth on the band width error. In Fig. 7, where the sun's zenith distance was less than 20° throughout the measurements, the percent polarization is the same within the errors of measurement at both 90 and 180° to the sun's bearing, but it is clearly less in the solar bearing. The data in Fig. 8, which extend down to 115 m, show in addition that at this depth maximal and minimal radiance in the horizontal plane differ by a factor of two toward and away from the sun, with i between 40 and 45°. The degree of polarization follows the same general trend as in the previous case, but in addition it illustrates two new points.

1) While the differences in p when $\theta = $ 90 and 180° do not exceed the experimental errors at depths greater than 60 m, polarization at 180° becomes weaker than at ±90° in shallower depths. This point is discussed later in relation to the effects of azimuth on polarization.

2) Down to 90 m p decreases more and more slowly when $\theta = $ 0°, but at greater depths it appears to increase again. This matter will also be discussed later where it is shown to be similar to some phenomena observed around sunset (Fig. 9).

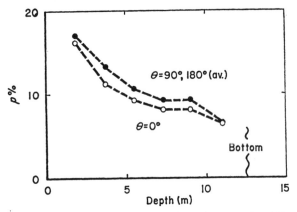

Figure 10.   Effect of h on p in turbid water at St. 5 (Fig. 5); bottom, 12.5 m; 500mμ narrow band filter used; $S = 6.5$ m; i between 17 and 32°.   Note that decrease in p with h accelerates slightly beyond 9 m, probably as a result of increased turbidity.   Data of 19 August 1957.

Near Bottom. Several series of measurements (Figs. 10–13) indicate that in shallow water, when p is determined in a narrow spectral band centered at 500mμ, it decreases sharply close to bottom. The extent and degree of this polarization decrease probably depend on the clarity of the water, the albedo of the bottom, and the fact that turbidity often increases near the latter. The effect of transparency

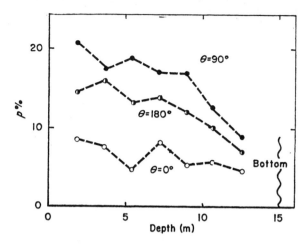

Figure 11.   Effect of *h* on *p* in moderately turbid water at St. 3 (Fig. 5); bottom, 15 m; measured with 500mμ narrow band filter;   *S* = 10 m; *i* between 55 and 57°.   Note that decrease in polarization with depth is accelerated at depths greater than 9 m, at least when *θ* = ±90 or 180°.   Data of 26 August 1957.

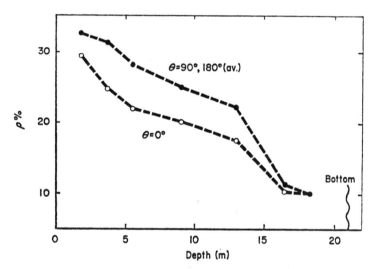

Figure 12.   Effect of *h* on *p* in moderately turbid water at St. 4 (Fig. 5); bottom, 21 m; measured with 500mμ narrow band filter; *S* = 11 m; *i* between 16 and 22°.   Decrease in *p* with depth clearly accelerates with depths in excess of 13 m.   At about 18 m, submarine illumination appears to reach an equilibrium condition where *p* becomes nearly independent of azimuth.   Data of 17 August 1957.

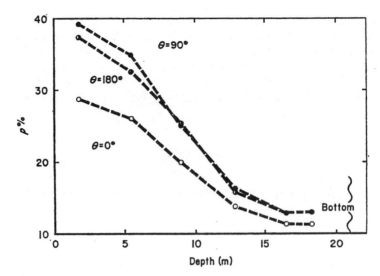

Figure 13. Effect of *h* on *p* in moderately turbid water at St. 4 (Fig. 5); bottom, 21 m; measured with 500mμ narrow band filter; *S* = 13 m; *i* between 27 and 32°. Decrease in *p* with *h* accelerates below 6 m. At about 16 m, submarine illumination appears to reach an equilibrium and *p* stops decreasing. Data of 30 August 1957.

may be seen by comparing the above figures where the extent of the bottom influence ranges from 3.5 to 15 m. Close to the bottom *p* seems to approach a constant value (Figs. 12 and 13).

3. Effect of Wavelength

In general, measurements of *p* as a function of λ (Figs. 14–16; Table II) show that *p* is greater at the two ends of the visible spectrum, with our highest readings usually at 600mμ. The minimum occurs between 445 and 500mμ, with the actual p_{min} apparently being correlated with the color of the water mass concerned. Thus the blue waters of the Sargasso Sea (Fig. 14, St. 1) and of Castle Harbour (Table II, St. 5) have p_{min} near 450mμ, while the waters of Harrington Sound (Figs. 15, 16, St. 4), which appear greener, have p_{min} close to 500mμ. The data also show that the change in *p* due to λ, (Δ*p*), is apparently greatest where *p* is large, that is, in clear water and near the surface. As would be expected from these facts, the variation of Δ*p*/*p* as a function of λ, which averages 0.2, seems to be practically independent of turbidity and depth.

In addition, the ratios of maximal (*θ* = 0°) to minimal (*θ* = 180°) radiance measured in various azimuths decrease toward the short wavelengths.   Although the variations of i in the different series of data prevent a quantitative analysis, the ratio appears to decrease toward shorter wavelengths down to the lower spectral limit of our filters.   This should be compared with the fact that the ratio of submarine irradiance coming from above to that coming from below is known to pass through a minimum at a λ shorter than that of maximum transparency (Jerlov, 1951).

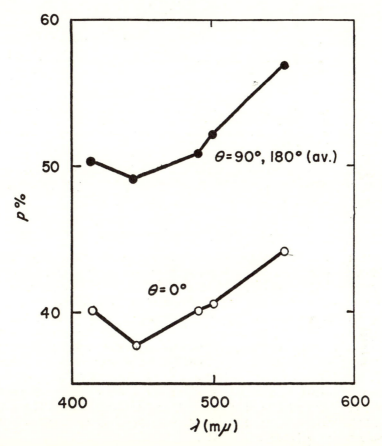

Figure 14.   Effect of λ on p when h is 16 m in clear water at St. 1 (Fig. 5); bottom, 2500 m; $S =$ 50 m; i between 29 and 33°.   Without colored filters, p at $\theta = 0°$ was 39.8%, and at $\theta = \pm 90$ and 180° p was 49.4% (av.).   Data of 27 August 1957.

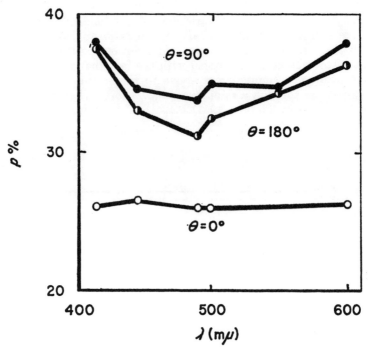

Figure 15. Effect of λ on *p* when *h* = 5.5 m in moderately turbid water at St. 4 (Fig. 5); bottom, 21 m; *S* = 13 m; *i* about 24°. Without colored filters, *p* at θ = 0° was 25.7%, and at θ = ±90 and 180° *p* was 34.3% (av.). Data of 30 August 1957. *Cf.* Fig. 16, where conditions were essentially the same except for *h*.

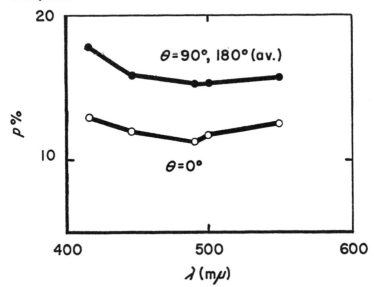

Figure 16. Effect of λ on *p* when *h* = 14.5 m in moderately turbid water at St. 4 (Fig. 5); bottom, 21 m; *S* = 13 m; *i* about 25°. Without colored filters, *p* at θ = 0° was 12.4%, and at θ = ±90 and 180° it was 15.8% (av.). Data of 30 August 1957.

TABLE II. Summary of the effects of λ on p

Station*	Bottom depth (m)	h (m)	Clarity	S** (m)	i (°)	p (no filter) $\theta = 0°$ (%)	$\theta = 90°,180°$ (Av. %)	p_{min} (%)	$p_{415m\mu}$ (%)	$p_{546m\mu}$ (%)	Δp† (%)	$\Delta p/p_{min}$
1 (Fig. 14)	2500	16	clear	50	29–33	39.8	49.4	48.9 (at 445mμ)	50.2	56.8	9.2	0.19
1	2900	18	clear	45	21–23	—	—	52.2 (at 445mμ)	54.0	60.7	10.3	0.20
4 (Fig. 15)	21	5.5	moderately turbid	13	24	25.7	34.3	33.7 (at 490mμ)	37.8	34.6	5.0	0.15
4 (Fig. 16)	21	14.5	moderately turbid	13	25	12.4	15.8	15.3 (between 490 and 500mμ)	17.8	15.6	2.8	0.18
4	21	5.5	moderately turbid	11	20	25.8	29.1	28.1 (at 500mμ)	33.2	30.4	7.4	0.26
4	20	9	moderately turbid	10	44–54	—	—	18.0 (at 500mμ)	19.5	20.2	3.7	0.21
4	16	14.5	moderately turbid	—	23–34	9.1	10.1	10.2 (between 445 and 500mμ)	10.8	11.1	1.5	0.15
5	12.5	9	turbid	6.5	28	—	—	7.6 (at 445mμ)	8.2	9.3	2.3	0.30

* See Fig. 5.
** Secchi disc reading in meters.
† $\Delta p = (p_{415m\mu} - p_{min}) + (p_{546m\mu} - p_{min})$.

DISCUSSION

The preceding data may be discussed from several points of view; probably the most important points relate to 1) the angular distribution of p underwater, and 2) the angular distribution of submarine light itself.

1. *Effects Dependent on the Angular Distribution of* p.   The angular distribution of percent polarization in the sea is still poorly known since the photographic (Ivanoff, 1956b, 1957a) and visual (Waterman and Westell, 1956; Waterman, 1958) measurements which have been made are few and quite incomplete.   However, the function may roughly approximate Rayleigh's equation, at least beyond the critical angle.   If so, $p = \sin^2 \alpha / 1 + \cos^2 \alpha$, where α is the angle between the parallel rays of underwater sunlight and the direction of scattering (*i.e.*, line of sight).   Provided this equation holds and if only horizontal lines of sight are considered (as in the above measurements), p should be maximal for all solar zenith distances when $\theta = \pm 90°$ and it should be minimal for $\theta = 0$ and 180°, except when $i = 0°$ and when p is the same for all azimuths.   At both minima, percent polarization should be equal and directly related to the sun's altitude above the horizon. The difference between maximal and minimal p is least when three parameters have particular values: 1) The zenith distance must be minimal so that the axis of best symmetry[5] of submarine illumination makes a maximal angle with the horizontal plane.   2) The depth should be great because it is well known that the axis of best symmetry of submarine light's angular distribution approaches the vertical at increasing depths (see Jerlov, 1951).   3) The water must be turbid because this produces vertically symmetrical light distribution at shallower depths.

These various points have largely been verified by measurements in the field.   Thus the degree of polarization observed in horizontal directions normal to the sun's bearing is independent of i (Waterman and Westell, 1956: fig. 6), but in the sun's bearing p decreases as i increases (Waterman and Westell, 1956: fig. 7).   Similarly, the present results demonstrate that variation in p with azimuth increases as the sun approaches the horizon (for example, compare Figs. 3 and 4), decreases as depth increases (see Fig. 8 at ± 90 and 180° to the sun's bearing), and may become practically zero in turbid water.

[5] The underwater direction of maximum radiance (apparent direction of the sun) may differ from the axis of best symmetry for the angular distribution of submarine illumination, but it is not known whether the direction or the axis is the more significant for underwater polarization. Here the two terms will not be distinguished even though they may not be strictly equivalent.

On the other hand, a relation not predicted by Rayleigh's equation appears again in the present results, namely that lower degrees of polarization occur more consistently in the sun's bearing than they do in that of the anti-sun. This agrees with results obtained previously with the photographic technique (Ivanoff, 1956b: 54; Ivanoff, 1957a: figs. 9 and 10) but not with those found visually (Waterman and Westell, 1956: fig. 8).[6]

In summary then, Rayleigh's equation appears to hold for submarine polarization only as a first order approximation; this we have assumed also in our analysis of underwater elliptical polarization (Ivanoff and Waterman, 1958). It is not surprising that the equation provides only a rough approximation; in the first place, most of the light scattering by sea water is not Rayleigh scattering; in the second place, the combination of sun, sky and ocean have a plane of symmetry which passes through the sun's bearing rather than an axis of symmetry. This plane of symmetry may disappear when the sun's contribution becomes small relative to that of the sky, as in cloudy weather or at sunrise or sunset. The distribution of radiance in the sky must then be important. Thus the percent polarization may be greater at 180° than at ±90° (Fig. 9) when the solar zenith distance is greater than 80°. This could be the effect of some clouds which increased the radiance of the sky in an azimuth different from the sun's bearing.

Much of the problem of angular distribution of submarine polarized light remains to be studied from both experimental and theoretical points of view.

2. *Effects Related to the Angular Distribution of Submarine Light.* As a result of field observations on underwater polarization, a theory was proposed which accounts simultaneously for the effects of tur-

---

[6] In the latter data the polarization for $\theta = 0°$ did in fact average less than for $\theta = 180°$, but the variation hardly seems significant there, since maxima at $+90$ and $-90°$ differ by greater amounts.

[7] Recently measurements of the degree of polarization have been made in sea water samples illuminated by a parallel beam of artificial light (Ivanoff, 1958a, 1958b). In these waters, which were collected in the Mediterranean Sea between Nice and Corsica, polarization varied from 45 to 67% for surface waters but was as much as 75, or even 77%, in deep waters. Optically pure water under the same conditions would produce 80% polarization. Furthermore, some samples which were rather turbid and had the milky appearance of a colloidal suspension gave readings of 88% polarization. In certain of these samples, spheroidal particles $1-3\mu$ in diameter were visible under the microscope. Thus the shape of the particles may have an important influence since the degree of polarization is known to be greater for spheroidal particles.

bidity, sky overcast, depth, nearness to bottom, and wavelength (Ivanoff, 1957a). The theory may be summarized as follows: 1) With parallel light, the degree of polarization in light scattered by sea water depends mainly on the optical properties and dimensions of the suspended particles; its actual value increases as the dimensions of the particles decrease.[7] 2) In daylight, submarine polarization decreases for a given water type with the diffuseness of the light.

To support this theory an attempt has been made to derive the variation in the amount of polarization from changes in the angular distribution of submarine illumination. These rough calculations were intended to determine merely whether or not the order of magnitude of the theoretical effect is comparable to the measurements. Such analysis is misleading if it suggests that polarization is dependent on the angular distribution of light. In fact, both phenomena are dependent on the same parameters, such as surface irradiance and optical properties of the water, rather than on each other. However, their variations usually occur together in the same direction. Thus, when underwater illumination becomes more diffuse, the degree of polarization diminishes. This should be the case for more turbid waters, overcast skies, increased depth, greater bottom reflection and the most penetrating wavelength. The data here reported support these predictions of the theory.

Previous work had shown that the degree of submarine polarization may reach 60% in quite clear water at shallow depths (Ivanoff 1957a), decreasing proportionately at greater turbidities (Waterman, 1954; Waterman and Westell, 1956; Ivanoff, 1957a). This decrease is illustrated from the present data by Fig. 6. These data also seem to show that in the clearest water the percent polarization varies least as a function of the Secchi disc reading. Previous evidence had suggested the contrary, namely that p changes most rapidly in the clearest waters (Ivanoff, 1957a: 50–52). However, to evaluate these discrepancies, one must consider precisely what is meant by clarity and how it is determined. Both the amount of underwater polarization and the Secchi disc reading are complex functions of the optical properties of sea water, and the data in Fig. 6 permit no conclusions to be drawn relative to the way in which these functions are interrelated. All that may be said from present evidence is that, for clear waters, the depth to which the Secchi disc is visible changes more rapidly than does the degree of polarization. Contrary to a previous assumption, p does not seem to provide a particularly sensitive means of distinguishing various clear waters one from another. On the other hand, the measurement of polarization is more precise with greater p, which in fact occurs in the clearest waters.

As mentioned above, a decrease in the degree of polarization was regularly observed whenever the sun was covered by clouds, which agrees with previous findings (Waterman, 1954; Ivanoff, 1957a). Similar effects occur at sunset (Fig. 9) when the relative importance of sky light increases and when irradiance becomes more diffuse. However, it is important to emphasize that, although a completely overcast sky and a blue sky with no sun (at sunrise and sunset, or with a dense cloud covering merely the sun) produce less polarization underwater than does the direct sunlight itself, their influences are not identical. Thus, at angles less than the critical angle, sky polarization is not visible in the former but is visible in the latter case; also, the plane of polarization is more or less horizontal in overcast weather but it may behave in quite a different manner at sunset with a clear sky, possibly as a result of the latter's polarization (Ivanoff and Waterman, 1958).

Our data show that p decreases with depth (irregularities in these measurements being due to either errors in measurement or optical heterogeneity of the water) and approaches a limit of about 30% in very clear sea water. Some previous observations (Waterman, 1955) suggested that polarization in deep water is as great as that near the surface, but the strong interference pattern still recorded in the crystal analyzer at 200 m may have a more likely interpretation in the much narrower deep water light spectrum (Ivanoff, 1957a: 41). Other measurements made by the visual method (Waterman, 1958) indicated a slight increase in the percent polarization at certain depths. On the basis of present data this could be due to an error in measurement, to a change in the sky overcast during measurements, or possibly to the effect of a local increase in the water's clarity. But the last possibility implies that in a clearer layer of water the illumination becomes less diffuse again, which is unlikely.

Wavelength has a definite but small influence on the degree of submarine polarization. Present results agree with the sole previous measurement (made by the photographic method; see Ivanoff, 1957a) and with the fact that it is for the most penetrating wavelengths that underwater light is most diffuse (Johnson and Liljequist, 1938). Since diffuseness of illumination and polarization by scattering are inversely related, one would thus expect that minimum polarization in very clear sea water would occur near 475mμ. The apparent correlation observed between the λ for p_{min} and the color of the water in the present results is mentioned above.

Two circumstances have already been cited which influence the degree of polarization in the sun's azimuth in a special way involving both the angular distribution of p and that of submarine radiance. In deep water, p for the sun's bearing passes through a minimum and then increases again (Fig. 8).   The same thing happens at sunrise or sunset (Fig. 9), when this has also been observed at $\theta = \pm 45°$,   Such phenomena could result from opposing effects of an increase in the diffuseness of underwater illumination, which decreases polarization, and a reduced obliquity in the light, which augments polarization in the sun's bearing.

Actually the diffuseness of the light increases with depth and with the sun below the horizon while the axis of best symmetry of the angular distribution of radiance tends toward the vertical with depth and with i near 0°.   Hence the degree of polarization would decrease while the first of these effects predominated and would then increase again when the second factor became predominant.   In Fig. 8, curves A and B may come together horizontally where the angular distribution of submarine illumination becomes a symmetrical figure of rotation around the vertical.   Degree of polarization would then be independent of azimuth regardless of the sun's zenith distance.   Unfortunately this equilibrium depth is too great in the Sargasso Sea to be reached with our present instruments.

ACKNOWLEDGMENTS

The authors are grateful to Mr. Eric Kindwall and the crew of the PANULIRUS, especially Mr. Brunell Spurling, for their great assistance in effecting these measurements.   Thanks are also due Miss Jeanie Caldwell and Dr. W. H. Sutcliffe, JR., Director of the Bermuda Biological Station, for their constant cooperation.

REFERENCES

IVANOFF, ALEXANDRE

    1955.   Au sujet du facteur de polarisation de la lumière solaire dans la mer. C. R. Acad. Sci. Paris, *241*: 1809–1811.

    1956a. Degree of polarization of submarine illumination.   J. opt. Soc. Amer., *46*: 362.

    1956b. Facteur de polarisation du résidu sous-marin de lumière du jour.   Ann. Geophys., *12*: 45–56.

    1957a. Contribution à l'étude des propriétés optiques de l'eau de mer en Bretagne et en Corse, et à la théorie de la polarisation sous-marine.   Ann. Géophys., *13*: 22–53.

    1957b. Un polarimètre sous-marin à cellule photoélectrique.   Premiers résultats obtenus.   Bull. d'Infor. C. O. E. C., *9*: 491–499.

    1958a. Au sujet de l'utilisation d'un diagramme p-β pour caractériser les masses d'eau océanique.   C. R. Acad. Sci. Paris, *246*: 2636–2639.

    1958b. Essai d'hydrologie optique entre Nice et la Corse.   C. R. Acad. Sci. Paris, *246*: 3492–3496.

IVANOFF, ALEXANDRE AND T. H. WATERMAN
   1958.   Elliptical polarization of submarine illumination.   J. mar. Res., *16* (3):
           255–282.
JERLOV, N. G.
   1951.   Optical studies of ocean waters.   Rep. Swedish Deep Sea Exped. 1947–8,
           3(1): 1–59.
JOHNSON (JERLOV), N. G. AND G. LILJEQUIST
   1938.   On the angular distribution of submarine daylight and the total submarine
           illumination.   Svenska hydrograf.-biol. Komm. Skr., N. S. *14:* 1–15.
WATERMAN, T. H.
   1954.   Polarization patterns in submarine illumination.   Science, *120:* 927–932.
   1955.   Polarization of scattered sunlight in deep water.   Deep-Sea Research,
           Suppl. *3:* 426–434.   Bigelow Festschrift.
   1958.   Polarized light and plankton navigation.   *In:* "Perspectives in Marine
           Biology".   (Ed.: A. A. Buzzati-Traverso)   University of California
           Press, Berkeley.   pp. 429–450.
WATERMAN, T. H. AND W. E. WESTELL
   1956.   Quantitative effects of the sun's position on submarine light polarization.
           J. mar. Res., *15:* 149–169.

Part VI

APPARENT OPTICAL PROPERTIES

Editor's Comments
on Papers 22 Through 25

22   **TYLER and SMITH**
Submersible Spectroradiometer

23   **BAUER and IVANOFF**
Spectro-Irradiance-Metre

24   **TYLER and SMITH**
Spectroradiometric Characteristics of Natural Light Under Water

25   **SMITH and WILSON**
Photon Scalar Irradiance

Figure 1 in Part II offers a method for calculating the apparent optical properties and the absorption coefficient from radiance distribution by integration to obtain vector irradiances $H+$ and $H-$, and scalar irradiances $h+$ and $h-$. These apparent optical properties can then be used to calculate the absorption coefficient, a, as indicated by the equation at the bottom of that same figure. Vector and scalar irradiances can also be obtained by direct experimental measurement using optical integrators that perform the intergrations indicated for $H(\pm)$ and $h(\pm)$ in figure 1 in Part II.

Photosensitive surfaces are not intrinsically cosine collectors, nor do flush-mounted flat diffusing glass or plastic surfaces collect naturally, according to a cosine function. At angles of incidence greater than about 50°, a flush-mounted collector will exhibit increasing error in cosine collection as the angle of incidence of the impinging radiation increases. The effect of such errors on the value of the reading will depend upon the geometrical distribution of radiant flux surrounding the collector. A well-designed vector irradiance collector for both underwater and above water use is described by R. C. Smith (1969).

Instruments for measuring irradiance in the ocean can be designed to have any spectral response within the sensitivity limits of the detector employed. Quantum irradiance meters are designed to be equally sensitive to quanta at all wavelengths. Since the fixation of carbon by phytoplankton is basically a quantum process, the quanta meter has been recommended by the Scientific Committee on Oceanic Research (1974) for radiant energy measurements that are to be related to photosynthesis. Quantum irradiance meters are described by L. Prieur (1969) and N. G. Jerlov and K. Nygård (1969). For problems involving human vision underwater, photometric irradiance meters are used. These have spectral response characteristics that duplicate the response of the average normal human eye; the readings obtained are thus directly comparable to human eye response.

For research purposes, the monochromatic irradiance meter is by far the most useful since it permits quantitative studies involving any chosen spectral band width. Research-type spectral irradiance meters for underwater measurements have been described by J. E. Tyler and R. C. Smith, Paper 22, and by D. Bauer and A. Ivanoff, Paper 23. These two instruments have been used extensively to obtain quantitative spectral irradiance data (watts/meter^2) at various depths and locations in the world's oceans, and also in lake water, (J. E. Tyler and R. E. Smith, 1970; A. Morel and L. Caloumenos, 1971; J. E. Tyler, editor, 1973). Typical data are given in Paper 24.

A. Morel and L. Caloumenos (1974) have critically compared the spectral irradiance attenuation coefficients of clean lake water (Crater Lake, Oregon, USA) with clean Sargasso sea water and have found them to be the same within the precision of measurement for all wavelengths between 400 and 650 nm. Their graphical comparison (figure 1), together with the analytical data in table 1, demonstrates the correlation between spectral irradiance and the concentration of plant pigments related to chlorophyll-a.

Spherical irradiance $(h_{4\pi})$ and scalar irradiance (h) are related to each other by the equation

$$h_{4\pi} = \frac{1}{4}h$$

Scalar irradiance is the total flux arriving at a point from all directions around the point and is essentially a geometrical concept. The common method used for determining scalar irradiance depends on the direct measurement of spherical irradiance. In order to conform to the formal definition given above, it is ideally necessary to employ a sphere of diffusing material that exhibits cosine collection for radiant energy at every point of its surface. If such a sphere were tested by means of a parallel beam of light having a cross-sectional area larger than the sphere, the sphere would exhibit equal response for all orientations within the beam. In practice, a sphere having identical, but not cosine, collecting properties at every point of its surface will be equally useable since the losses due to non-cosine collection will be constant for all directions and will be accounted for by calibration. Spherical irradiance collectors have been constructed in a two-hemisphere form by R. I.

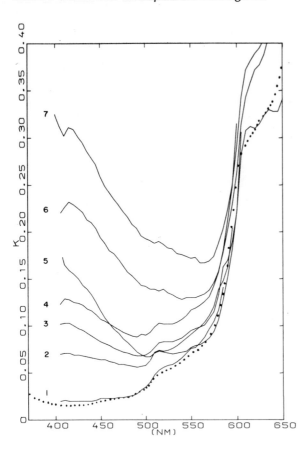

Fig. 1. Values of the attenuation coefficient as a function of wavelength, at the seven stations, between the surface and the depths indicated for each curve in Table 1. The dotted curve is the one obtained by Tyler and Smith (1967–1970) in the Waters of Crater Lake (Oregon). Table 1 lists the characteristics of the different stations examined. *Reprinted from A. Morel and L. Caloumenos, Variabilité de la répartition spectral de l'énergie photosynthétique, pp. 97 and 101 in* Tethys *6(1–2): 93–104 (1974), by permission of the Director of La Station Marine d'Endoume et du Centre d'Oceanographie.*
[*Editor's Note:* The senior author of this paper has made minor corrections in *Tethys*.]

Table 1. Characteristics of the different stations examined.

N°	Stations	(1)	z prof. (m)	E_z / E_o en % (2) Q	w	λ_z (nm) (3)	Chl a mg m^{-3}	(4)
1	DISCOVERER 21 1-6-70	25°44 N 65°38 W (mer des Sargasses)	47	11,2	13,7	443 (et 460)	0,02	(0–65)
2	DISCOVERER 10 14-5-70	5°39 S 89°06 W (Pacifique Sud)	29	9,0	9,8	481 (et 492)	0,22	(0–33)
3	DISCOVERER 9 13-5-70	3°55 S 84°44 W (courant des Galapagos)	20	10,6	12,1	493 (et 481)	0,38	(0–18)
4	DISCOVERER 16 21-5-70	6°27 N 79°57 W (golfe de Panama)	15	11,3	12,9	491 (et 481)	0,43	(0–15)
5	CINECA 72 17-4-71	18°49 N 16°25 W (Mauritanie)	20	11,3	12,5	501	1,10	(0–21)
6	CINECA 87 20-4-71	18°45 N 16°36 W (Mauritanie)	10	11,5	12,6	546	1,52	(0–10)
7	HARMATTAN 1 4-5-71	12°03 N 17°26 W (Sénégal)	8	12,5	13,0	550–560		

Currie (1961) and T. Sasaki et al. (1966). Most recently, R. C. Smith and W. H. Wilson (Paper 25) have designed and constructed a scalar irradiance meter employing refracting optics and C. R. Booth (1976) has built a complement of three compact spherical irradiance collector-type instruments, having different configurations, for use in natural waters, on deck as a radiation monitor, and in laboratory or *in situ* culturing experiments.

REFERENCES

Booth, C. R. 1976. The design and evaluation of a measurement system for photosynthetically active quantum scalar irradiance. *Limnol. Oceanogr.* **21**(2):326–336.

Currie, R. I. 1961. Scalar irradiance as a parameter in phytoplankton photosynthesis and a proposed method for its measurement. *Int. Union Géod. Géophys.* N. G. Jerlov, ed., Symposium on Radiant Energy in the Sea, Monogr. 10. 107–112.

Jerlov, N. G., and K. Nygård. 1969. A quanta and energy meter for photosynthetic studies. Københavns Universitet, Institut for Fysisk Oceanografi. 10:1–19.

Morel, A. and L. Caloumenos. 1971. Mesures d'eclairements sous marins flux de photons et analyse spectrale. *Report No. 11,* Laboratoire d'Oceanographie Physique, Centre de Recherches Oceanographie de Villefranche-sur-Mer, France, 160 pp.

———. 1974. Variabilité de la répartition spectral de l'énergie photosynthétique. *Tethys* **6** (1–2):93–104.

Prieur, L. 1970. Photometre marin mesurant un flux de photons. *Cah. Oceanogr.* **22**:493–501.

Sasaki, T., G. Oshiba, and M. Kishino. 1966. A 4π-underwater irradiance meter. *Oceanogr. Soc. Jpn. J.* **22**(4):1–6.

Smith, R. C. 1969. An underwater spectral irradiance collector. *J. Mar. Res.* **27**(3):341–351.

Tyler, J. E., and R. C. Smith. 1967. Spectroradiometric Characteristics of Natural Light Underwater. *Opt. Soc. Am. J.* **57**(5):595–601.

———. 1970. Measurements of spectral irradiance underwater. Gordon and Breach Science Publishers, New York, 103 pp.

Tyler, J. E. (ed.). 1973. Data report—SCOR discoverer expedition, SIO ref. 73–16: (Volume 1): pp F1–F341; (Volume 2): pp G1–G160.

22

Reprinted from *Opt. Soc. Am. J.* **56**(10):1390–1396 (1966)

Submersible Spectroradiometer*

John E. Tyler and Raymond C. Smith

*Visibility Laboratory, Scripps Institution of Oceanography,
University of California, San Diego, California 92152*

(Received 15 April 1966)

This paper describes a submersible spectroradiometer for measuring the spectral distribution of the multiply scattered natural light in ocean or lake water. Because of the anticipated large range of flux levels in the spectrum at the exit slit, strong measures have been taken to reduce stray light within the instrument. These methods are described in some detail. Procedures for spectral alignment and calibration, determination of bandwidth, absolute calibration, and over-all testing of the instrument are described in detail and the probable errors introduced by various components are estimated. It is estimated that the precision of measurements, limited by random errors of data taking, will be within ±2.5% and that the absolute accuracy is between 5% and 12%. The major limitations on the accuracy are the uncertainties of the standard of spectral emittance and the measurement of the bandwidth of the instrument.

The instrument makes possible a mode II determination of the optical properties of natural water as a function of wavelength and can furnish data on the spectral distribution of the flux available for photosynthesis and animal stimulation.

INDEX HEADINGS: Spectroradiometry; Oceanography; Scattering.

THE instrument described herein was designed to measure the spectroradiometric properties of the spacelight generated by multiple scattering of natural light underwater. The basic optical system of the instrument is shown in Fig. 1. As can be seen, it is a double Ebert[1] monochromator with a double objective system and rotating sector mirror to irradiate the entrance slit. The design features and general performance of the Ebert-type monochromator have been discussed by Fastie[2] and apply to this system. However, the unconventional application to *in situ* measurements in ocean and lake water has placed special requirements on the instrument's design.

Within the visible region of the spectrum, the range of monochromatic radiant energy levels that occur naturally at a specified depth in the ocean can be very great. By way of illustration, assume that Johnson's[3] tabulation of solar spectral irradiance (his Table I) represents the irradiance at the surface of the ocean (H_0) and that Hulburt's[4] values for the total attenuation coefficient of distilled water can be used to repre-

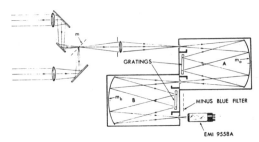

Fig. 1. Diagram of the optical system of the Scripps Spectroradiometer. The rotating sector mirror is indicated at m.

* The development of this instrument has been supported by the U. S. Office of Naval Research, Physics Branch and the N. S. F. Earth Sciences Division.

[1] H. Ebert, Wied. Ann **38**, 489 (1889).
[2] W. G. Fastie, J. Opt. Soc. Am. **42**, 641 (1952).
[3] F. S. Johnson, J. Meteorol. **11**, 431 (1954).
[4] E. O. Hulburt, J. Opt. Soc. Am. **35**, 698 (1945).

sent the diffuse attenuation coefficient (K) of clear ocean water; then using the equation $H_z = H_0 e^{-Kz}$ we calculate the monochromatic irradiance (H_z) at a depth of 50 m to be approximately 0.03 W cm^{-2} μ^{-1} at 470 nm and 3.0×10^{-10} W cm^{-2} μ^{-1} at 670 nm. Thus the monochromatic radiant energy level at the exit slit of a submerged spectroradiometer can be expected to vary over eight log cycles, while the radiant energy passing the entrance slit remains constant. This situation places a stringent requirement on the level of scattered light that can be tolerated within the instrument. If for example, observations are to be made at a wavelength of 670 nm with an EMI 9558A multiplier phototube (S20 surface), at a depth of 50 m, and the internal scattering in the monochromator is 10^{-7} of the input flux and constant with wavelength, then the maximum flux available at 670 nm would be 3.0×10^{-10} W cm^{-2} μ^{-1} and the scattered flux from wavelength 470 nm would be 0.03×10^{-7} W cm^{-2} μ^{-1}. The ratio of wanted flux to unwanted flux from 470 nm would therefore be $\frac{1}{10}$. But the ratio of the spectral sensitivity of the S20 surface at these two wavelengths is about 4 and the total flux at the entrance slit is perhaps 10 times the flux from 470 nm alone. The wanted to unwanted signal ratio would therefore be more like 0.0025.

REDUCTION OF SPURIOUS FLUX

The present instrument follows Ebert's original design in that a single spherical mirror is used to collimate the flux from the entrance slit onto the grating and also to focus the dispersed flux onto the exit slit. It is, of course, necessary to restrict the angular spread of the entering flux to about half the aperture of the mirror in order to prevent direct irradiation of the exit slit. Ebert accomplished this by using longitudinal baffles beside the grating mount. In the present instrument, use has been made of a transverse baffle located about 2 cm in front of the grating.

A disadvantage of Ebert's original design is that, at some angular positions of the grating, the collimated zero order leaving the grating irradiates the mirror; this flux is refocused on the grating where it is scattered again to the mirror and can then be directed to the exit slit as collimated flux. In this form it passes unhindered through the second monochromator with very little further dispersion. The amount of spurious flux from this source is generally supposed to be small, but it is not necessarily small compared to the amount of flux available underwater in the red region of the spectrum.

In order to reduce this spurious flux, the transverse baffle, mentioned above, has an opaque horizontal strip which covers the central third of the grating. This strip and the surrounding surfaces of the baffle are covered on both sides with black Norzon cloth to absorb the zero-order image.

Much of the dispersed flux in an Ebert monochromator strikes the walls of the tubular housing at a large

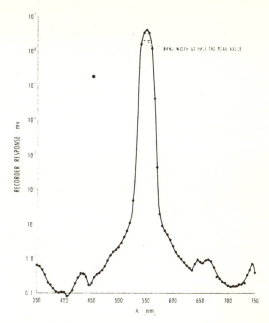

FIG. 2. Recorder response as a function of wavelength when the input flux to the Scripps spectroradiometer was centered at 550 nm and had a bandwidth at half the peak value of about 18 nm.

angle of incidence and can find its way to the edges of slits or mirrors and can appear as spurious flux at the exit slit. To control this, an aperture stop or baffle is installed in front of each concave mirror and the housing wall between this baffle and the grating baffle is lined with Norzon.

Final tests for stray light were performed as follows: The Scripps spectroradiometer was placed at the exit slit of a Hardy[5] monochromator, of the type used in the G. E. spectrophotometer (see Fig. 2). The Hardy[5] monochromator was adjusted to irradiate the entrance slit of the Scripps spectroradiometer with flux having a bandwidth of approximately 18 nm, with a peak value at 550 nm. While the Scripps spectroradiometer was operated at its lowest gain, calibrated neutral density filters were placed in front of the entrance slit and the magnitude of the input signal to the Scripps spectroradiometer was recorded. The neutral density filters were then removed and the Scripps spectroradiometer was operated at maximum gain to record the spurious signal falling on the phototube at wavelengths outside the input bandwidth. The results of this test are shown in Fig. 2. These data were obtained without using an auxiliary filter to reduce the effect of overlapping orders; Fig. 2 therefore displays any overlapping that may have occurred. From Fig. 2 it can be seen that the magnitude of the phototube signal in the blue and red regions of the spectrum varies by a factor of about 10 as a function

[5] A. C. Hardy, J. Opt. Soc. Am. 25, 305 (1935).

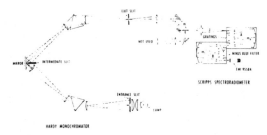

FIG. 3. Arrangement of Hardy monochromator and Scripps spectroradiometer for obtaining wavelength and bandwidth calibration.

of the grating angle, but that on the average its value is about 10^{-7} of the peak signal recorded at 550 nm.

Since an absolute spectral calibration for the instrument is available, the ordinates of Fig. 2 can be converted from recorder response to W cm^{-2} nm^{-1}; from those data the ratio of scattered flux to total input flux can be calculated for the particular test conditions used. Thus the integrated input flux would have been roughly the peak value (101 μW cm^{-2} nm^{-1}) times the bandwidth of the input band (18.3 nm), or about 1.8×10^3 μW cm^{-2}, and similarly the scattered flux would have been 3×10^{-5} μW cm^{-2} nm^{-1} times 5.5 nm or about 16.5×10^{-5} μW cm^{-2} in the wavelength region of interest (650 to 700 nm). The fraction scattered was therefore 9×10^{-8}. This is not final, however, because in actual use the input bandwidth is established by the body of water being measured and changes, becoming narrower, with depth; of course, the signal level decreases with depth as well.

From absolute spectral data and data for the diffuse attenuation coefficient obtained near the surface, it is possible to estimate both the signal level and the scattered-light level to be expected at greater depths and to determine the limiting depth beyond which meaningful results cannot be obtained.

For measurements in deep water a Wratten No. 24 filter can be introduced, in front of the phototube, by remote control. This filter would be expected to reduce the spurious flux in the 650 to 700 nm region by an additional factor of 10^{-3}.

SPECTRAL ALIGNMENT, CALIBRATION, AND BANDWIDTH

Because of the presence of absorption bands in sun and skylight, because of a requirement to take data along the steep slope of the absorption band of water, and because the radiometric environment beneath the surface of water is likely to exhibit rapid changes in magnitude, it was decided at the outset to utilize a step-type wavelength drive by means of which the wavelength setting could be either held fixed or changed on command in 5-nm steps. To accomplish this, twin step cams were cut simultaneously. Each cam has 90

steps and each step has a fixed radius. The two cams are mounted on the opposite ends of a single sturdy shaft and when in use, are rotated simultaneously by means of a stepping motor through a worm gear. The cam riders are held against the cams by means of gentle springs. Foam-rubber damping prevents excessive bumping as the cam riders travel from step to step. For ease of adjustment the cam riders are ring-clamped to the shaft of the grating mount. Each rider is also slotted and provided with a set-screw arrangement for making fine adjustments of the angular position of the grating.

The Scripps spectroradiometer was calibrated for wavelength and bandwidth by again making use of the monochromator section of the Hardy spectrophotometer. The essential optical arrangement is shown in Fig. 3. (Both instruments contain double monochromators and the following discussion must make specific reference to one or the other. In order to avoid confusion the names shown in Fig. 3 are adhered to in the text).

The slits of the Hardy monochromator were nearly closed, to pass a narrow band of wavelengths. The Scripps spectroradiometer was then placed to receive the output of the Hardy monochromator. With the Scripps spectroradiometer set at a fixed wavelength, the Hardy monochromator was slowly run through the range of wavelengths from 400 to 700 nm. The resulting recording was a haystack-shaped curve which exhibited the combined bandwidth of the two instruments. By making the slit-width setting of the Hardy monochromator narrower and narrower it was possible to obtain a recording that was very nearly triangular in shape. Curves of this latter type were obtained for all wavelength positions of the Scripps spectroradiometer. In processing the data, we extrapolated the straight slopes of these curves to form a triangle. The peak values of the triangles were taken to indicate the true wavelength for each setting and the width of each recording at half the extrapolated peak value was taken as the bandwidth. Figure 4 shows the character of these recordings.

For this series of tests the wavelength drive of the Hardy monochromator was operated slowly with a synchronous motor and a special cam was mounted on the linear wavelength shaft to operate a microswitch every 10 nm and produce an event mark on the record-

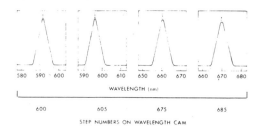

FIG. 4. Typical recordings of bandwidth showing extrapolation lines. This figure has been traced from the data for purposes of reproduction.

ing. Since the recorder paper drive was also synchronous, a linear wavelength scale could be used between event marks to interpolate the wavelength position of the recorded peak.

In addition to these tests, the Hardy monochromator and the Scripps spectroradiometer were both checked for wavelength calibration against a didymium filter.

The bandwidth passed by the Scripps spectroradiometer is a function of the mechanical width of the entrance slit, the aberrations of the optical system, and the alignment of the two monochromators (A and B in Fig. 3).

The gratings have 600 lines/mm. In monochromator A, the dispersion at the exit slit is about 6.7 nm/mm at 550 nm. The exit slit of monochromator A was set at about 1.6 mm, to pass an approximately 10-nm bandwidth. Since the instrument uses straight slits, there is a wavelength error from the center to the ends of the slit, which amounts to about 1 nm. In monochromator B the formation of a curved image on a straight slit operates to reduce the bandwidth at the extremities of the image. The bandwidth of the combined monochromators (A+B) is, of course, further reduced by any misalignment since monochromator B cannot pass a greater range of wavelengths than is available at its entrance slit.

The experimentally measured bandwidth varies more or less systematically over the spectral range, apparently because of minor but fixed irregularities between specific radii of the two cams (Fig. 5). At some wavelength settings the bandwidth has a minimum measured value of 5.3 nm, and at other wavelength settings a maximum value of 5.9 nm was obtained. In the absolute calibration of the Scripps spectroradiometer the average value at each wavelength was used rather than an average value over the whole range of wavelengths.

OBJECTIVE SYSTEM AND ACCESSORY OPTICS

The design of the objective system of the Scripps spectroradiometer incorporates three basic features; a wide range of spectral transmission, solid-angle restriction, and double-path operation. An image of the entrance slit of monochromator A is formed in the plane of a rotating sector mirror which alternately permits this image to be re-imaged in space by one or the other of two quartz objective lenses. The divergence of the slit image beam is about 1.5° measured from the axis of the objective lens when the image is placed 6 m in front of the instrument. At this position the oscillation of the slit image introduced by the rotating mirror is less than 1 min of arc.

The objective system is arranged so that the same number of mirror reflections occur along each path. In spite of this precaution, the 100% line is not straight and must be taken into consideration when processing data.

By means of accessory optical equipment attached in front of one of the objectives, it is possible to make relative spectral observation of two objects, one of which can be a standard. Other accessory equipment permits the measurement of specific spectroradiometric quantities such as irradiance and spherical irradiance.

When measurements are made without resorting to a reference standard, it is usually necessary to keep track of the ambient natural-light level. For this purpose a spherical irradiance collector with separate photodetector is used. This consists of a diffusing plastic sphere mounted over the aperture of a phototube and optically coupled to it by a filter which has approximately the same bandwidth and peak value as those of the underwater flux.

Ambient flux above the water surface and spectroradiometric data below the surface are recorded simultaneously on the two channels of a Sanborn model 320 Dual Channel Recorder.

OPERATING FEATURES

The amount of flux available underwater varies enormously as a function of depth and also as a function of wavelength; it is therefore necessary to provide some kind of stable gain control in order to record data over the ranges of depth and wavelength which are generally of interest. The Scripps spectroradiometer is equipped with an iris diaphragm in a portion of the optical path that is common to both objectives; this can be used to control the flux entering the instrument. Remote control is accomplished by means of a series of precision resistors in the dc supply, which control the high voltage to the multiplier phototube. These resistors are used to vary the gain of the system to accommodate the flux level to the circuit requirement. In addition to this, the recorder provides a wide range of gain for the signal level.

Also on the control panel are a switch for selecting one of three speeds for the sector mirror, a push-button switch to introduce the optical filter which reduces the stray light and overlapping orders, a pilot light to indicate that the filter is in place, a zero-set potentiometer which is used to equate circuit zero with recorder zero, and wavelength controls which permit manual or automatic scanning of the wavelength range in either direction. In order to keep track of the wavelength setting an auxiliary cam is attached to the wavelength cam shaft. The auxiliary cam, by means of a microswitch, activates the recorder's event-marking pen, which marks the record every even 50 nm and at 540 and 560 nm.

ABSOLUTE CALIBRATION

For absolute calibration, the Scripps spectroradiometer was mounted on a photometer bench in front of a standard of spectral output and the response of the instrument was determined as a function of wavelength.

From these data, the specified spectral output of the standard lamp and the bandwidth data for the Scripps spectroradiometer, the absolute response of the Scripps spectroradiometer as a function of wavelength was calculated. The absolute calibration is independent of circuit gain factors, linearity, and other circuit characteristics; however, these characteristics have been measured.

These tests are discussed more fully in the next section.

DISCUSSION OF ERRORS

(A) Electrical

The electrical system of the Scripps spectroradiometer consists of an EMI 9558A photomultiplier tube and its high-voltage power supply, a linear amplifier circuit, and a Sanborn Dual Channel DC Amplifier Recorder. This system must accurately receive and record flux levels that can vary by many orders of magnitude. Four or five orders of magnitude variation in flux are accommodated by varying the voltage on the photomultiplier in reproducible calibrated steps and by using the various sensitivity scales provided on the Sanborn recorder.

The sensitivity scales used on the Sanborn recorder are 0.5, 1, 2, 5, 10, 20, 50, 100, and 200 mV/mm, corresponding to 25, 50, 100, 250, 500, 1000, 2500, and 5000 mV full scale on the 50-mm recording chart. These sensitivity scales have been externally calibrated twice, six months apart, using a Princeton Applied Research voltage/current reference source (model TC-100 AR) which has a rated accuracy of ±0.1 mV. The two calibrations agreed to better than 0.5%. The internal recorder calibration is rated at ±1%. The recorder noise level is 0.25-mm peak-to-peak maximum at 0.5-mV/mm gain setting, and the recorder drift is rated at 0.25 mm maximum as a function of 10°C temperature change in the region of 0° to 40°C and 0.1-mm maximum drift as a function of line-voltage changes from 103 to 127 V. The gain stability is rated at better than 1% under the above conditions. With care, the recording chart can be read to ±0.1 mm, which corresponds to from 0.4 to 0.8% reading precision for data recorded in the upper half of the recording chart. Over-all, under normal operating conditions, the recorder can be expected to have a limit of error less than ±1%.

The stability of the complete electronic system, powered by a Hydro Products ac motor generator, as under field operating conditions, was tested by measuring the output of an "absolute memory" lamp[6] over an extended period of time and under various adverse conditions. The absolute-memory lamp was regulated so as to have a flux output stable to better than ±0.1%. While observing this lamp, we varied the temperature

of the electronics over the range from about 10° to 40°C, the voltage from the generator from 115 to 125 V and the generator frequency from 58 to 62 cps. The stability of the electronics under these adverse conditions was better than ±1%. When temperature changes are modest and the generator is maintained at 120 V and 60 cps the stability is better than ±0.5%.

(B) Mechanical and Optical

To obtain accurate and consistent results it is necessary that the cams which control the grating orientation return to the same position at each setting. This mechanical–optical reproducibility was measured by recording the output of the absolute-memory lamp as a function of wavelength several times. The recorded values at each wavelength were reproduced to better than ±0.75%.

The wavelength assignments for the instrument were made by direct comparison with the Hardy monochromator, as previously described. In addition, the transmission of a didymium filter, calibrated by the U. S. National Bureau of Standards, was measured; the instrument scale was assigned wavelengths from its absorption bands. These two methods were in close agreement and from this test the wavelength assignment is estimated to be accurate to better than ±1.5 nm.

(C) Absolute Calibration of Instrument

An absolute calibration of the instrument was made by use of a 1000-W quartz iodine standard of spectral irradiance supplied by the Eppley Laboratory. The spectral irradiance of these lamps is based upon the spectral radiance of a blackbody as defined by Planck's equation and has been determined[7] by comparison of a group of quartz iodine lamps with: (1) the NBS standards of spectral irradiance, (2) the NBS standards of luminous intensity, and (3) the NBS standards of total irradiance. The maximum uncertainty of the spectral irradiance is estimated[7] to range from about 8% at the shortest wavelengths in the ultraviolet to about 3% in the visible and infrared.

A working standard was prepared at this laboratory by calibrating a 1000-W quartz iodine lamp against the laboratory standard supplied by Eppley. The photometric techniques used in preparing the working standard were such that there was no significant increase in the maximum uncertainty of the spectral irradiance of the working standard over that of the laboratory standard. In addition, the photometric techniques were such that the maximum uncertainty of the instrument calibration is essentially that owing to the maximum uncertainty of the standard lamp.

The instrument was calibrated against the working standard by use of various reproducible photomultiplier voltages. At each photomultiplier-voltage setting, the

---

[6] Developed by George Tate and Richard Johnson at this laboratory for field checking and as a reference level for photometric instruments.

[7] R. Stair, W. E. Schneider, and J. K. Jackson, Appl. Opt. 2, 1151 (1963).

output of the working standard was recorded as a function of wavelength. From these, the relationship between lamp output and instrument response in μW cm^{-2} nm^{-1} mV^{-1} was obtained for each setting. Reproducibility of voltage setting was checked by repeating the calibration several times. The final calibration values were taken as the average of these repetitions; the results were found to be reproducible to $\pm 1\%$.

(D) Bandwidth

The spectral irradiance of the standard lamp is given in μW cm^{-2} nm^{-1}. It is thus necessary to know the bandwidth of the Scripps spectroradiometer in order to obtain the actual spectral irradiance input to the instrument. As mentioned earlier, the bandwidth at each wavelength setting was measured using the Hardy monochromator; the results are shown in Fig. 5. If the explanation for the oscillatory variation of bandwidth with wavelength, given above is correct, then the precision assigned the bandwidth values at each wavelength is about $\pm 2\%$. On the other hand, if the oscillatory variation is due to unknown systematic errors, then the accuracy assigned to the bandwidth values would be about $\pm 5\%$. In this connection it should be noted that data taken to a precision of 1% to 2% and reduced using bandwidths assigned from the oscillatory curve in Fig. 5 do not show these oscillations in the final result. This is a further indication that the oscillatory nature of the bandwidth vs wavelength curve is real and reproducible.

(E) Sun Normalization

In the course of taking data, the ambient light level varies owing to sun elevation or to changes in cloudiness. To correct for these changes the recording from the spherical irradiance meter is used to normalize all data to the maximum sun level for the day. The spherical irradiance meter is constructed so as to be sensitive to roughly the same spectral region as the light which penetrates the ocean. This meter has been found to be linear to $\pm 1\%$ over the light levels of interest; hence the precision of sun normalization is estimated to be

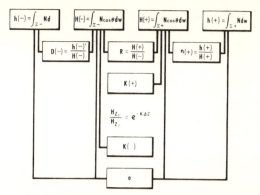

FIG. 6. Diagram of mode II specification of the optical properties of ocean water. h and H are, respectively, the scalar and horizontal irradiance and are shown for the upwelling $(+)$ and downwelling $(-)$ flux fields. The property, a, is the absorption coefficient. (See Preisendorfer[8] for further details).

$\pm 1\%$. If the ambient light level changes frequently over a wide range, as it does on days of intermittent cloudiness, some sort of averages must be made for the readings from the spherical irradiance meter, above the surface, and the instrument readings underwater. This averaging degrades to some extent the precision of sun normalization. In the happy event that the sky is clear and sunny during the period of data taking, the precision of sun normalization can be improved to about $\pm 0.5\%$.

SUMMARY

The above considerations lead to the estimate that precision of measurement owing to the random errors of data taking is less than $\pm 2.5\%$. The largest error arises from the measurement of the instrument bandwidth, followed by the errors introduced by the electrical system. The absolute accuracy of the instrument is estimated to be between 5% and 12% based on the uncertainty assigned to the standard lamp and the possibility of systematic errors in the bandwidth measurements. The lower limit would apply to measurements in the visible and infrared if the bandwidth measurements are correct. The upper limit would apply to the ultraviolet if the bandwidth measurements were improperly interpreted. It can be added that for some of the quantities derived from the data, irradiance K values for example, the contribution to the limit of error due to systematic errors is reduced because only the ratios of irradiances are used.

DISCUSSION

The Scripps spectroradiometer makes possible what Preisendorfer[8] refers to as a mode II determination

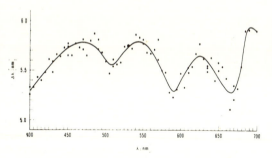

FIG. 5. Bandwidth of the Scripps spectroradiometer as a function of wavelength. Curve is best visual fit to the data points.

[8] R. W. Preisendorfer, Union Géodésique et Géophysique Internationale, International Association of Physical Oceanography; Symposium on Radiant Energy in the Sea, Monographie No. 10, p. 11 (1961).

of the optical properties of ocean water as a function of wavelength. Mode II is illustrated diagrammatically in Fig. 6. The essential measurements for this mode are up- and down-welling horizontal and scalar irradiances. From these it is possible to compute the diffuse attenuation functions $[K(+)$ and $K(-)]$, the reflectance functions $[R(+)$ and $R(-)]$, the distribution functions $[D(+)$ and $D(-)]$ and the absorption functions $[a(+)$ and $a(-)]$.

The Scripps spectroradiometer measures directly the absolute spectral distribution and bandwidth of the natural flux underwater. This is, of course, the flux available for photosynthesis and for the visual activity of marine animals during the daytime hours. Because the spectral distribution of the natural flux exhibits the absorption bands of chlorophyll, the instrument can also be used for detecting and estimating chlorophyll in ocean water.

Spectroradiometric measurements in the ocean are subject to greater operational restrictions than most other oceanographic measurements. The natural light flux at a given point in the ocean varies as a function of the distance to the surface. If the point remains stationary, the change in water path owing to the passing of waves may affect the flux reaching the point. Likewise if the surface is calm but the point moves vertically, the flux reaching the point varies. This means that this particular source of "noise" in the data cannot be entirely avoided either by suspending the instrument from the surface or fastening it to the bottom. Direction of observation is also an important consideration. The change in radiance and in irradiance as a function of the angle of observation have both been quantitatively demonstrated.[9,10]

---

[9] J. E. Tyler, Bull. Scripps Inst. Oceanog. 7, 363 (1960).
[10] J. E. Tyler, Appl. Opt. 3, 105 (1964).

Near the surface, the interaction of sunlight with surface waves can concentrate or diffuse the light flux entering the water and cause large fluctuations in light level at the instrument. To a lesser extent this also happens under overcast lighting conditions.

The effect of clouds has already been mentioned and is obvious. The effect of sun altitude is more subtle. When the angle of incidence of the sun's rays on the water is 45° or more, Fresnel reflection becomes increasingly larger. This effect is not detected by the spherical irradiance collector above the surface and consequently the ratio of ambient flux to underwater flux changes and introduces an error into the results.

Spectroradiometric measurements are most successfully obtained during the four hours around sun noon under clear sunny, or uniformly overcast sky, and with a calm water surface. We have, however, taken data in rough seas with variable lighting and although it is much more troublesome to process the data, we have been pleasantly surprised by the results, which are far more consistent than would be expected.

As a final comment, it should be noted that data taken over a period of perhaps several hours are commonly processed as though they had been taken at a single time when the sun was at its maximum. This is acceptable if the water sample does not change, but if the sample does change, large systematic errors could be introduced. This error could probably be detected and perhaps rectified by long-term temporal averaging.

ACKNOWLEDGMENT

The basic design and construction of the submersible spectroradiometer was done under the direction of W. G. Fastie at the Johns Hopkins University. His long experience with monochromators of the Ebert type and the fine workmanship of his group have contributed greatly to the success of the instrument.

23

Reprinted from Cah. Océanogr. **22**(5):477–482 (1970)

SPECTRO-IRRADIANCE-METRE

par

Daniel BAUER et ALEXANDRE IVANOFF[1]
Laboratoire d'Oceanographie Physique
de la Faculte des Sciences de Paris

R é s u m é

Appareil à monochromateur à réseau fournissant sur une bande de papier photographique la courbe de variation de la densité spectrale de l'éclairement en fonction de la longueur d'onde. Un photomultiplicateur de pilotage permet d'amener la valeur de la haute tension du photomultiplicateur de mesure à une valeur telle que toute la largeur de la bande de papier reste utilisée lorsque l'immersion augmente. La valeur de cette dernière est énregistree egalement, ainsi que celle de la haute tension, et des tops en longueur d'onde. Le spectre est parcouru en moins de trois secondes. Une trentaine d'enregistrements est obtenue en une demi-heure, entre la surface et une centaine de mètres de profondeur.

A b s t r a c t

The instrument makes use of a monochromator, with a grating. The distribution of the spectral concentration of the irradiance is recorded on a roll of photographic paper. A second photomultiplier acting as a pilote regulates the high voltage of the photomultiplier making the measurements in such a way that a full scale reading is still obtained when the depth increases. Values of the depth and of the high voltage are recorded together with the spectral distribution curve. The grating sweeps in less than three seconds. About thirty records are obtained in half an hour, between the surface and a depth of one hundred meters.

L'étude de la répartition spectrale du résidu de lumière du jour à diverses immersions étant utile tant en Optique océanographique qu'en Biologie marine, nous avons voulu réaliser dans ce but un appareil à **enregistrement** incorporé, manoeuvrable à l'aide du classique câble d'hydrologie, d'utilisation rapide (fournissant une trentaine de spectres en une demi-heure environ), et d'une sensibilité suffisante pour étudier la répartition spectrale de l'éclairement "descendant" jusqu'à 150 m d'immersion au moins (dans les eaux limpides).

----

[1]Manuscrit reçu le 1er août 1969.

La figure 1 fournit un schéma synoptique de l'instrument. Le collecteur de lumière C peut être orienté soit vers la surface de la mer en vue de l'étude de l'éclairement "descendant", soit vers le fond, en vue de l'étude de l'éclairement "ascendant". Une image de ce collecteur est formée sur la fente d'entrée d'un monochromateur à réseau M[1]. La fente de sortie de celui-ci est balayée par le spectre en 2,5 secondes seulement à l'aide d'un moteur Mo dont l'axe entraîne en plus du réseau une roue dentée agissant sur un relais électrique qui fournit ainsi des "tops" correspondant à des longueurs d'onde λi connues. Une image de la fente de sortie du monchromateur dont la dispersion est de 6,4 nm/mm et la largeur de la fente de sortie de 0,75 nm est formée sur la photocathode du photomultiplicateur de mesure PMme[2]. Le courant anodique débité par ce dernier traverse deux des oscillographes d'un enregistreur multivoie E[3] à bande de papier photographique, disposés en série, et dont les sensibilités se trouvent dans un rapport égal à deux. La voie la plus sensible se trouve surchargée pour les longueurs d'onde les plus pénétrantes, auxquelles correspond un courant anodique important, mais permet une étude plus précise des deux extrémités du spectre.

Une troisième voie enregistre les "tops" de longueur d'onde λi dont nous avons parlé ci-dessus, rendus indispensables par l'absence de synchronisation entre la rotation du réseau du monochromateur et le déroulement de la bande de papier photographique de l'enregistreur. Une quatrième voie réservée au courant délivré par une jauge de pression P[4] fournit l'immersion correspondant à la mesure. Nous reviendrons ci-dessous sur l'utilisation de la cinquième et dernière voie de l'enregistreur.

Le résidu de lumiere du jour décroissant rapidement avec la profondeur et les sensibilités des voies d'enregistrement étant fixes, il convient de compenser l'effet de la profondeur par une augmentation progressive de la haute tension d'alimentation du photomultiplicateur de mesure. Dans l'appareil présenté, ceci est réalisé d'une manière automatique grâce à l'emploi d'un second photomultiplicateur, identique au premier, PMpi dit de "pilotage" qui ajuste la valeur de la haute tension d'alimentation à une valeur telle que le courant anodique maximal debite par le photomultiplicateur de mesure (obtenu pour la longueur d'onde la plus pénétrante) reste sensiblement constant quelle que soit l'immersion, (jusqu'à une certaine profondeur limite bien entendu), et en conséquence permette à l'enregistreur de fournir une courbe de répartition spectrale couvrant toujours toute la largeur de la bande de papier photographique.

----

[1]Monochromateur Baush et Lomb - Résolution de 5 nm dans l'intervalle de mesure utilisé à savoir entre 375 et 800 nm. Lumière diffusée limitée à 0,05 %

[2]Photomultiplicateur RCA type 7265, à 14 étages, courbe de sensibilité S.20.

[3]Enregistreur SFIM type A.22

[4]Jauge de pression SFIM type H 2930 de 25 bars (soit 250 m de profondeur).

Photographie 1. - Vue d'ensemble de l'appareil.

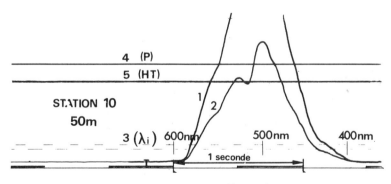

Photographie 2. - Exemple d'enregistrement.

A cet effet, une fraction de l'ordre de 10 % de l'énergie arrivant
sur la fente d'entrée du monochromateur est soustraite au moyen d'une lame de
verre L inclinée à 45° et arrive sur la photocathode du photomultiplicateur
de pilotage. Le courant anodique Ipi débité par ce dernier est maintenu constant et égal a 100 μA grace à un asservissement A qui réagit sur la valeur de
la haute tension d'alimentation HT, qui elle-même est commune aux deux photomultiplicateurs. La haute tension d'alimentation du photomultiplicateur de mesure se trouve ainsi "pilotée" par l'énergie lumineuse atteignant la photocathode du photomultiplicateur de pilotage. Il importe donc que cette énergie
soit filtrée et ajustée en sorte que le courant anodique débité par le photomultiplicateur de mesure pour la longueur d'onde la plus pénétrante permette
de couvrir la largeur de la bande de papier photographique de l'enregistreur.
Il suffit à cet effet de disposer entre la lame de verre séparatrice L et la
photocathode du photomultiplicateur de pilotage un filtre interférentiel F
correspondant à la longueur d'onde la plus pénétrante (de l'ordre de 470 nm
pour les eaux limpides) et un diaphragme D qu'il convient de régler, pour chaque filtre F utilisé, en fonction du rapport des éclairements obtenus sur la
photocathode du photomultiplicateur de pilotage et celle du photomultiplicateur de mesure (pour la longueur d'onde de transmission maximale du filtre F).
Si le filtre interférentiel choisi ne correspond pas tout à fait a la longueur
d'onde la plus pénétrante pour les eaux étudiées, ou si la densité spectrale
de l'éclairement à la surface de la mer passe par un maximum pour une longueur
d'onde trop différente, on risque d'obtenir des courbes d'enregistrement sor-
tant des limites de la bande de papier photographique. Ceci ne peut arriver
toutefois que très exceptionnellement les répartitions spectrales tant du coefficient de pénétration des eaux que de la lumière du jour étant assez "plates"
dans le domaine de longueurs d'onde intéressé (470 - 510 nm).

La cinquième et dernière voie de l'enregistreur est consacrée à la
mesure de la haute tension d'alimentation des photomultiplicateurs. La connais-
sance de celle-ci permet de "situer" les uns par rapport aux autres les spectres obtenus aux diverses immersions, autrement dit, d'effectuer des mesures en
valeur relative non seulement en fonction de la longueur d'onde mais également
en fonction de la profondeur. Pour passer aux valeurs absolues, il suffit par
exemple de connaître pour une longueur d'onde et en valeur absolue la densité
spectrale de l'éclairement à la surface de la mer.

La photographie 1 fournit l'aspect extérieur de l'appareil, cylindre
de 1,10 m de long et de 35 cm de diamètre, pesant une quarantaine de kilos dans
l'air et environ une vingtaine dans l'eau. Il peut être suspendu au classique
câble d'hydrologie. La profondeur désirée étant atteinte, le déclenchement s'effectue à l'aide d'un câble non porteur à deux conducteurs, qui permet de court-
circuiter du pont du navire océanographique une minuterie M (voir figure 1) mettant en marche simultanément l'enregistreur E et le réseau du monochromateur M[1].

----

[1]Par contre l'alimentation de la haute tension des photomultiplicateurs ainsi
que celle de la jauge de pression fonctionnent en permanence.

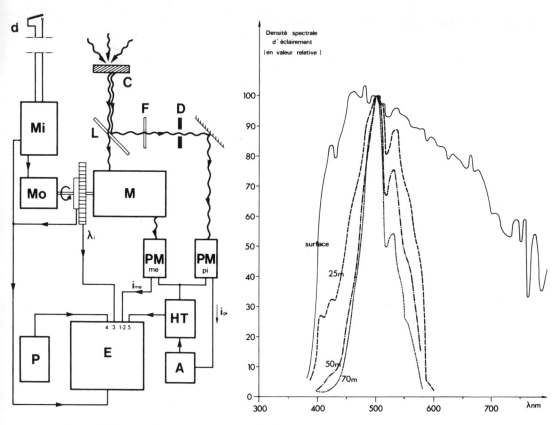

Figure 1. *Schéma synoptique*

Figure 2. *Répartition spectrale de l'éclairement descendant par 0,25,50 et 70 m de profondeur (Golfe de Californie, le 13 mai 1968 à 13 h). Les diverses courbes sont normalisées à 495 nm.*

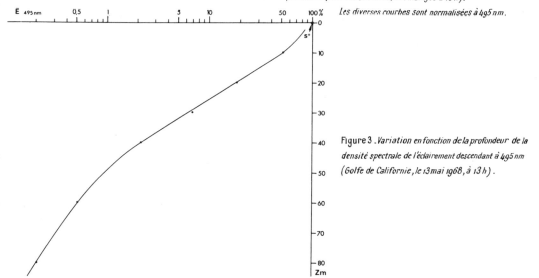

Figure 3. *Variation en fonction de la profondeur de la densité spectrale de l'éclairement descendant à 495 nm (Golfe de Californie, le 13 mai 1968, à 13 h).*

Ainsi que nous l'avons déjà signalé, l'enregistrement d'un spectre ne nécessite que 2,5 s, ce qui permet de travailler même par ensoleillement fluctuant (du moins en valeur relative et à une immersion donnée). En fin de rotation du réseau, l'alimentation de l'enregistreur se trouve coupée, tandis que la minuterie fait effectuer les opérations nécessaires à l'accomplissement d'un nouveau cycle de mesures (retour en arrière du réseau, réarmement, puis arrêt de la minuterie). Une demi-heure environ suffit pour effectuer une étude complète en un point donné (20 spectres d'éclairement descendant et 10 d'éclairement ascendant). Les accumulateurs équipant l'appareil[1] lui donnent une autonomie de 2 heures environ. L'immersion maximale est limitée par le bruit de fond des photomultiplicateurs. La haute tension peut varier entre 1500 et 3000 V, ce qui a permis d'étudier la répartition spectrale des éclairements descendant et ascendant respectivement jusqu'à 160 et 80 m de profondeur dans des eaux du large limpides, par temps ensoleillé, et pour une hauteur du soleil de l'ordre de 60°.

La photographie 2 fournit à titre d'exemple un enregistrement de répartition spectrale d'éclairement descendant obtenu dans le golfe de Californie en mai 1968 par 50 m de profondeur. Les courbes (1) et (2) correspondent aux deux voies de mesures. En (3), les tops de longueur d'onde λi, en (4) le signal de la jauge de pression P, en (5) la valeur de la haute tension. La répartition spectrale de la sensibilité de l'ensemble et en particulier du photomultiplicateur de mesure intervient bien entendu, et le dépouillement des résultats nécessite un étalonnage préliminaire effectué au moyen d'une lampe étalon de spectre d'émission connu. Par ailleurs, pour comparer entre eux les résultats obtenus à diverses immersions, un second étalonnage préliminaire est nécessaire, celui de la valeur de la haute tension en fonction du niveau lumineux, étalonnage effectué au moyen de filtres neutres de densités connues. Ces étalonneges doivent être contrôlés périodiquement.

Les figures 2 et 3 fournissent à titre d'exemple des résultats obtenus dans le golfe de Californie le 13 mai 1968 à 13 h : d'une part la répartition spectrale de l'éclairement "descendant" par 0, 25, 50 et 70 m d'immersion (remarquer la bande d'absorption à 520 nm, déjà observée par J. TYLER en particulier[2], et d'autre part, la variation en fonction de la profondeur (et en valeur relative) de la densité spectrale de l'éclairement descendant a 495 nm.

551.46.083
551.463.5

----

[1]Accumulateurs SAFT type VB 200, 24 V, 3, 2 A.H.

[2]Communication personnelle.

Reprinted from *Opt. Soc. Am. J.* **57**(5):595–601 (1967)

Spectroradiometric Characteristics of Natural Light Under Water*

J. E. TYLER AND R. C. SMITH

Scripps Institution of Oceanography, University of California, San Diego, La Jolla, California 92037

(Received 12 January 1967)

This paper presents spectroradiometric data on the natural radiant flux occurring underwater at three locations in the Gulf of California, in Pacific coastal water near San Diego, and in the plankton-rich water of San Vicente Reservoir (San Diego County). Spectral radiance and irradiance have been measured, and it is shown that under certain circumstances spectral data for these two radiometric quantities are directly proportional. The data have been used to calculate spectral values of the attenuation coefficient and of the reflection function. The various spectra correlate qualitatively with apparent chlorophyll concentration and water color.

INDEX HEADINGS: Oceanography; Spectroradiometry; Color; Reflections; Transmittance; Scattering.

THE color of natural water is a source of important and useful information about its composition, as well as a specification of its ability to absorb radiant energy. The specific type of information revealed by the color of water and its quantitative recovery have been problems of considerable interest to the Scripps Institution of Oceanography for the past several years.

The problem of accurately measuring the color of water *in situ* is an extraordinary problem in spectroradiometry since the spectroradiometer must be submerged in the medium during the measurement. Also, the problem is extraordinary because within the wavelength range 350–750 nm the transmittance of water varies over a very great range, and since measurements in the region of its major absorption band are of considerable interest, it is necessary to use an exceptionally good monochromator for the measurements.

A few workers (see references in Ref. 1) have made limited spectroradiometric measurements of the radiant flux at different depths in lake or ocean water with spectrographic equipment.[1] Detailed *in situ* spectroradiometric measurements for the purpose of determining the presence of absorption spectra were made at S.I.O. as early as 1961,[2] and demonstrated the advantage gained by using multiply scattered light for detecting low-pigment concentrations. In the present work, which is reported here and in the preceding paper,[3] the Scripps Spectroradiometer[4] has been used to measure the spectral-radiometric quantities, radiance and irradiance, for the up- and down-welling streams of radiant energy underwater. The measurements have been made over the visible region of the spectrum from 350 nm to 750 nm at various depths within the euphotic zone, and in a variety of natural water types.

* This research has been supported by the National Science Foundation, Earth Sciences Division.

[1] J. E. Tyler, J. Opt. Soc. Am. **55**, 800 (1965).
[2] J. E. Tyler, Proc. Natl. Acad. Sci. U. S. **47**, 1726 (1961).
[3] R. C. Smith and J. E. Tyler, J. Opt. Soc. Am. **57**, 589 (1966).
[4] J. E. Tyler and R. C. Smith, J. Opt. Soc. Am. **56**, 1390 (1967).

INSTRUMENT

The Scripps Spectroradiometer has been described in detail in Ref. 4. Additional features, particularly applicable to the present work, are discussed in Ref. 3.

As discussed in these references, the instrument can be used to measure various aspects of a radiant-energy field simply by changing the method of collecting radiant flux. Some of the early measurements were made directly through a quartz window covering the objective lens of the instrument. In this configuration the instrument was used to measure horizontal radiance at Crater Lake[1] and radiance in the nadir and zenith directions at other localities.

In order to obtain spectral values of the diffuse-attenuation coefficient and other optical constants, and also to simplify the absolute calibration and assure its accuracy, we have equipped the instrument with a carefully designed irradiance collector[3] which permits the measurement of up- and down-welling irradiance.

Because of the strong optical-filtering action of water, measurements obtained at wavelength settings between 650 and 750 nm in relatively deep water have been examined carefully for evidence of error due to light scattered within the instrument. As pointed out in our previous papers[3,4] this is the spectral region that is most vulnerable to error due to spurious light.

Several new tests, in addition to those described in Refs. 3 and 4, were devised to try to detect stray light in the instrument The most stringent of the new tests was to expose the irradiance collector to direct sunlight through two Wratten green filters (No. 64) and while it was thus exposed, to measure the transmittance of the instrument's internal minus-blue filter (a Wratten No. 24) in the region from 600 to 750 nm. No evidence of stray light was detected.

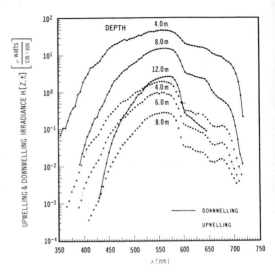

FIG. 2. San Vicente Reservoir (32°55′ N, 116°55′ W), 22 July 1966: spectral up- and down-welling irradiance. The Chlorophyll-a concentration at the surface was 2.6 mg/m³.

We also made a laboratory test which established that the dark current of the phototube was not a source of error in any of the field measurements. In making this test, however, we succeeded once again in demonstrating the linearity of the circuit over the whole range encompassed by our measurements.

We have also further examined the errors due to the finite bandwidth of the instrument. As discussed in Refs. 3 and 4, this type of error is believed to be present in the measurements between 600 and 750 nm and is estimated to be 10–15%. This bandwidth error raises the red end of the data curves above their correct positions but does not greatly change the shape of the radiance or irradiance data. This error does, however, effect the shape (and values) of the calculated curves for the diffuse-attenuation functions, K, and the reflectance functions, R. Sections of these calculated curves which are in doubt owing to possible bandwidth error have been omitted.

RESULTS

Southern California Coastal Water

The relative up-welling (nadir) radiance for San Diego coastal water is shown in Fig. 1. These data were taken approximately 8 km west of the mouth of San Diego harbor in water 70–100 m deep. During the four hours required to complete the data, the boat drifted six km to the south. The data were taken under clear sunny skies in moderate seas with swells of one to two m. It required roughly 45 min at each depth to obtain the spectral distribution of the radiance. This is more than double the time normally required in calm seas and is due to instrument-handling problems

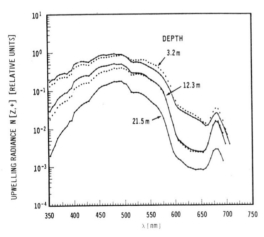

FIG. 1. San Diego coastal water (32°35′ N, 117°25′ W), 6 Jan. 1966: spectral up-welling (nadir) radiance at various depths. The overlapping, broken, curves are data repeated at the same depth two to three hours after the solid-curve data.

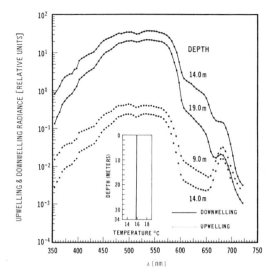

FIG. 3. Isla Coronado (north east cove), Republica Mexicana (29° N, 113.5° W), 24 April 1966: spectral radiance of the nadir (up-welling) and zenith (down-welling).

and the necessity to average the signal over time. These data exhibit strong absorption in the blue region of the spectrum below 435 nm and in the red region at about 665 nm, attributed in both instances to the presence of chlorophyll-a and also, in the blue region, to the presence of other pigments and possibly the dissolved materials known as gelbstoff.[5] The shapes of the curves in Fig. 1 can be compared with the shapes of the up-welling curves in Fig. 2 where the chlorophyll-a concentration has been measured.

The data in Fig. 1 were strongly and adversely affected by environmental conditions. It will be appreciated that in order to measure the nadir radiance at 3.2 m in seas with swells of 1 to 2 m, it is necessary to average the data over time. The greatest loss of precision under these circumstances is in the depth assignment, which was no more than a best estimate of the position of a mark on the instrument cable relative to the "average level" of the water surface. On the other hand, the time averaging of the radiant-energy level is somewhat more precise. Even near the surface, where the light level varies over a wide range, a time recording of the varying up-welling radiance allows a good estimate of the average radiance to be made. At greater depths, where the light-level fluctuations are not as great, the precision is improved. The measurement of near-surface down-welling (zenith) radiance under adverse conditions has such large fluctuations that rational averaging was virtually impossible.

The data of Fig. 1 were taken over a period of several

[5] K. Kalle, Union Géodésique et Géophysique Internationale, Int. Assoc. of Phys. Oceanog., Monograph No. 10, 59 (1961).

hours but were processed[4] as though they had been taken at a single time when the sun was at its maximum. The dashed curves in Fig. 1 represent data from measurements repeated for the same depth two to three hours after measurement of the data represented by the solid curves. The differences between the dashed and solid curves can in part be due to the difficulty of returning the instrument to the same depth. However, it is perhaps more likely that the differences are due to changes in the water during the time the data were being taken.

San Vicente

Figure 2 shows the up- and down-welling irradiance measured in San Vicente reservoir. These data were taken under clear sunny skies, with a calm water surface, and a very light breeze. In this case a surface sample of the water was analyzed; it had a chlorophyll-a concentration of 2.6 mg/m³.

Bahia de los Angeles

Figures 3–5 show up- and down-welling radiance data taken in the Gulf of California near Bahia de los Angeles, Baja California, Republica Mexicana. The figures present several water types in the order of decreasing clarity. All the data were taken under clear sunny skies, with the water surfaces calm to rippled, and the wind zero to a light breeze.

DISCUSSION OF DATA

A feature of this type of measurement is that the results contain information about all of the dyes,

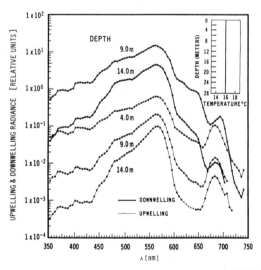

FIG. 4. Isla Coronado (west side), Republica Mexicana (29° N, 113.5° W), 22 April 1966: spectral radiance of the nadir (up-welling) and zenith (down-welling).

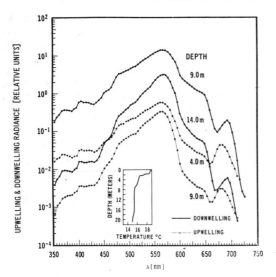

Fig. 5. Puerto Don Juan, Republica Mexicana (29° N, 113.5° W), 24 April 1966: spectral radiance of the nadir (up-welling) and zenith (down-welling).

pigments, and absorbing materials present, including the water. Since water is the base material in which various concentrations of other absorbing materials are to be found, it is of considerable importance to obtain measurements on the clearest and purest water available. To this end, two different types of measurements have been made in Crater Lake and are reported in previous publications.[2,3]

By comparing the measurements reported in this paper with those obtained in Crater Lake, we see that all of the measurements exhibit the characteristic absorption bands of water. This is particularly noticeable in the region of the spectrum around 585–600 nm. However, the measurements reported herein indicate additional absorption superimposed on the water spectrum, which is attributed to the presence of chlorophyll-a and other pigments or dyes associated with the plankton in the water.

It is typical of the irradiance data that absorption bands tend to be more distinct for the up-welling irradiance than for the down-welling. This is because, to reach the instrument, the radiant flux must travel a much longer path to join the up-welling stream, and therefore suffers greater absorption than is the case for the down-welling stream.

It is also typical that as measurements are made at a greater depth the chlorophyll band at 665 tends to be overwhelmed by the water absorption that also occurs in this region.

We have explored the possibility that there is an optimum depth for revealing a specific absorption band, such as the chlorophyll band at 665 nm. Let λ_a be the wavelength of the center of an absorption band and

λ_b the background wavelength, outside of the absorption band, and let K_a and K_b be the diffuse attenuation coefficients for irradiance at λ_a and λ_b. Then the irradiances at these wavelengths at depth z can be written

$$H(\lambda_a) = H_0(\lambda_a)e^{-K_a Z}, \tag{1}$$

$$H(\lambda_b) = H_0(\lambda_b)e^{-K_b Z}, \tag{2}$$

where $H_0(\lambda)$ is the surface irradiance at the wavelengths indicated. As shown in Refs. 3 and 6, the surface irradiance is relatively constant over broad regions of the spectrum so that, as a first approximation, we may assume

$$H_0(\lambda_a) = H_0(\lambda_b) = H_0. \tag{3}$$

It follows that the difference of irradiances at λ_a and λ_b can be written

$$H(\lambda_b) - H(\lambda_a) = H_0(e^{-K_b Z} - e^{-K_a Z}). \tag{4}$$

To optimize this difference, Eq. (4) is differentiated with respect to the depth Z and set equal to zero. This gives

$$K_b e^{-K_b Z_m} = K_a e^{-K_a Z_m}, \tag{5}$$

where Z_m is the depth which displays the maximum irradiance difference for the absorption band at λ_a. Solving Eq. (5) for this optimum depth gives

$$Z_m = (K_a - K_b)^{-1} \ln(K_a/K_b). \tag{6}$$

Thus, once the spectral irradiance-K values have been measured for the water in question, the optimum depth at which measurements should be made to display the desired absorption band can be calculated.

ACCURACY AND PRECISION

The inherent accuracy and precision of the Scripps Spectroradiometer have been discussed in detail in a

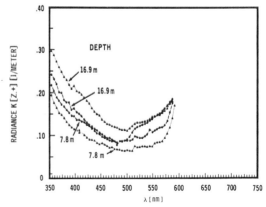

Fig. 6. San Diego coastal water, 6 January 1966. Variability in the spectral-diffuse-attenuation functions for radiance calculated from the radiance of the nadir data of Fig. 1 using Eq. (7).

[6] F. S. Johnson, J. Meteorol. 11, 431 (1954).

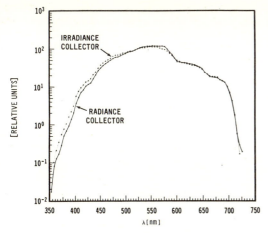

FIG. 7. San Vicente Reservoir, 1 July 1966: spectral down-welling irradiance and the radiance of the zenith measured simultaneously at 4.0 m. The curves have been arbitrarily normalized at 550 nm.

previous paper.[4] The additional tests described earlier in this paper have confirmed our original estimates in every way.

However, the inherent accuracy of the instrument says nothing about the accuracy of the relationship between measurements at any two wavelengths, a relationship which is controlled by the temporal and spatial homogeneity of the hydrosol. Some feeling for the magnitude of the variations that can be encountered in this respect can be obtained from Figs. 1 and 6.

Gross changes in homogeneity can sometimes be detected in the measurements themselves as they are being taken, for example in the San Diego coastal-water data. Other indications of spatial or temporal inhomogeneity can be obtained from measurements of temperature vs depth which locate the thermocline where a sharp change of the optical properties of the water can be expected owing to sharp changes of the particulate or plankton concentration. Beam-trans-mission measurements can also be used to locate inhomogeneities.

APPLICATION OF THE DATA

The spectral-irradiance data directly specify the energy available underwater for life and living require-ments. It is a measure of the flux available to phyto-plankton for purposes of photosynthesis and to those zooplankton and other animals that require light in their search for food. These data can be converted to available quanta as a function of wavelength by dividing by the energy per quantum $E = hc/\lambda$. The total energy (or total quanta) can, of course, be ob-tained by integration of the appropriate functions, i.e., energy (or quanta) vs wavelength.

Spectral irradiance data obtained at two or more depths permit computation of average values of the diffuse-attenuation coefficient as a function of wave-length and these can then be used to predict the spectral irradiance at other depths.

Spectral irradiance data for the up- and down-welling streams of radiant energy obtained at the same depth permit the calculation of the reflection function which is a property of the hydrosol nearly independent of the properties of the natural lighting. This function is useful for checking theoretical equa-tions, is indicative of the particle concentration, and can be used (as can the total energy data) to compute a color specification of the water.

If spectral data are obtained for two or more depths, the spectral energy absorbed per unit volume can be determined and, for a homogeneous hydrosol, can then be calculated for other depths.

An interesting and useful experimental result was discovered when some data were taken using both objectives of the Scripps Spectroradiometer. One ob-jective was operated as a radiance collector, the other as an irradiance collector. Within the precision of measurement of the instrument it was found that the spectral values of the radiance of the nadir or zenith were directly proportional to the up- or down-welling irradiance, respectively. This result is illustrated in Fig. 7 where the data for the radiance of the zenith direction and the down-welling irradiance measured at 4 m have arbitrarily been normalized at 550 nm. The agreement in relative spectral values (i.e., the spectral curve shape) is even closer for deeper measurements of down-welling flux and for measurements of up-welling flux at all depths.

It is possible to imagine radiance distributions for which the above results would not have been obtained, namely, collimated light at large angles of incidence to the collectors (e.g., near-surface down-welling data in clear water). The experimental result is not too sur-prising, however, when we realize that, at depths beyond a few attenuation lengths, the radiance distribution is a monotonically decreasing function with respect to the angle from the sun. When this is true, the up- or down-welling irradiance will be closely proportional to the radiance of the nadir or zenith, respectively.

Theory predicts that, at asymptotic depth, spectral-radiance measurements in a fixed direction will be directly proportional to spectral-irradiance measure-ments. Our experimental result implies that there is a broad near-asymptotic region within which spectral measurements of radiance and irradiance will be very nearly directly proportional to each other. On the other hand, the absolute values of the up- and down-welling irradiances will be correct only for a carefully con-structed collector that has a cosine response.

The color of the water is directly indicated by the spectral radiance or irradiance curves (and by the up/down ratio). The water on the east side of Isla Coronado was observed to be relatively clear with a

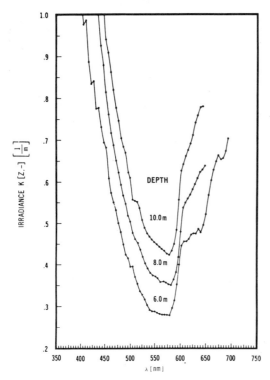

FIG. 8. San Vicente Reservoir, 22 July 1966: spectral diffuse-attenuation function for irradiance calculated from the irradiance data of Fig. 2 using Eq. (7).

blue-green appearance, whereas the water of Puerto Don Juan, a small lagoon on the south entrance to Bahia de los Angeles, was turbid and had a distinct yellow-green appearance. These color judgments are indicated in Figs. 3 and 5.

The density of the standing crop of phytoplankton is also indicated by radiance data. By comparing the 9-m nadir radiance of Figs. 3 and 5 it can be seen that the nadir radiance for Puerto Don Juan is one to two orders of magnitude lower in the blue and nearly one order of magnitude lower in the red than for Isla Coronado (east), whereas in the green-yellow region of the spectrum the radiance values are almost the same for the two waters. This difference is presumably due to a much higher chlorophyll-a concentration in the water of Puerto Don Juan than in the Isla Coronado (east) water. Similar conclusions can be obtained by comparing the radiance-K curves.

CALCULATIONS

In addition to biologically oriented applications, the above data can be used to obtain certain optical properties of the water. Figs. 6 and 9 show the average diffuse attenuation function for radiance derived from the above radiance data using Eq. (7), where $N(z, \pm)$

are the nadir $(+)$ and zenith $(-)$ radiance at depth z,

$$K_N[(z_1+z_2)/2, \pm]$$
$$= (z_2-z_1)^{-1} \ln[N(z_1, \pm)/N(z_2, \pm)]. \quad (7)$$

In order for K_N in Eq. (7) to represent the average diffuse-attenuation function between depths z_1 and z_2, it is more meaningful if no sharp change of radiance occurs between z_1 and z_2. Included with the radiance data are plots of temperature vs depth which were used to locate possible sharp inhomogeneities.

The radiance-K values for San Diego coastal waters are four or more times larger than for pure water[3] in the blue region of the spectrum but are still relatively small compared to the K values shown for San Vicente and the Bahia de los Angeles waters. The coastal-water radiance-K data emphasize the change of water with time as discussed above.

Figure 8 shows the irradiance-K values for various depths in San Vicente. These were calculated by means of Eq. (7) using H in place of N. The strong chlorophyll absorption in the blue and red regions of the spectrum is evident. The data show that the irradiance-K values are larger at 10 m than at 6 m and indicate that the plankton population is changing with depth.

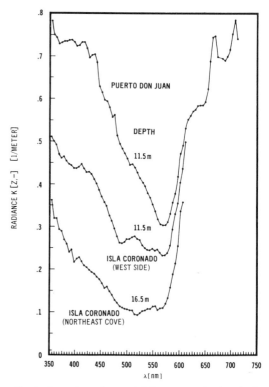

FIG. 9. Comparison of spectral-diffuse-attenuation function for radiance at three locations in the Gulf of California calculated from the radiance of the zenith data of Figs. 3, 4, and 5.

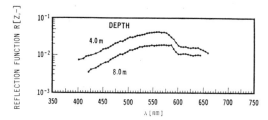

Fig. 10. Spectral-reflection function $R(z, -) = H(z, +) / H(z, -)$ of San Vicente at 4 and 8 m, calculated using the data of Fig. 2.

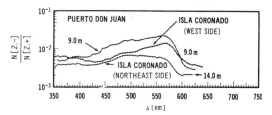

Fig. 11. [1]Ratio of $N(Z, +)$ to $N(Z, -)$ calculated from the data of Figs. 3, 4, and 5.

The radiance-K values for the Bahia de los Angeles waters are shown in Fig. 9. These curves give an objective measure of the subjective appearance mentioned above. The data show that the waters of Isla Coronado (east) have smaller radiance-K values, and hence are clearest, while the Puerto Don Juan waters have the highest radiance-K values, and hence are the most turbid.

In the above discussion of the attenuation-coefficient data and throughout the text we have for didactic reasons distinguished radiance-K from irradiance-K in accordance with the radiometric quantity measured at each site. At asymptotic depth, this is an unnecessary distinction. As discussed above, Fig. 7 indicates that for these experimental results the distinction is also unnecessary at depths less than the asymptotic depth.

Figures 10 and 11 give the calculated values of the reflection function, $R(z, -)$, for San Vicente and the ratio of $N(z, +)$ to $N(z-)$ for the Bahia de los Angeles waters. For the up–down ratios it is necessary at all depths to distinguish between results obtained from irradiance and radiance measurements.

ACKNOWLEDGMENTS

The authors would like to acknowledge the assistance of Dr. R. W. Holmes for the determination of chlorophyll concentration at San Vicente Reservoir.

25

Reprinted from *Appl. Opt.* **11**(4):934–938 (1972)

Photon Scalar Irradiance

Raymond C. Smith and Wayne H. Wilson, Jr.

Photon scalar irradiance h is defined as the total number of photons per unit time and area arriving at a point from all directions about the point when all directions are weighted evenly. Reasons for considering photon scalar irradiance an optimum measure of radiant energy available for photosynthesis are briefly reviewed. A new technique for the measurement of h and considerations for calibrating an underwater photon scalar irradiance collector on an absolute basis are outlined.

Introduction

What aspects of radiant energy are most meaningful with respect to photosynthesis in natural waters? Specifically, is there a parameter that optimally expresses the radiant energy available for photosynthesis? Ideally, such a parameter would satisfy the following requirements. (1) The optimum parameter should be a well-defined and standardized physical quantity, amenable to theoretical description, and possessing standardized units. Underwater energy measurements, in which the geometrical and spectral properties of the collector are unspecified, fail to satisfy this requirement. Also, it can be argued that the *Secchi disk depth* meets this requirement in a circuitous and unsatisfactory manner.[1] (2) It should be a true measure of the energy available for photosynthesis and not be inconsistent with present knowledge of photosynthetic processes. As has been repeatedly pointed out, photometric (*light*) measurements fail this requirement. (3) It should be measureable simply, precisely and accurately, both in the laboratory and at sea. Present methods involving a complete spectral[2] or geometrical[3] analysis of the energy available for photosynthesis are complicated and not practical as routine instruments.

Atkins and Poole[4,5] suggested in their early work that the rate of photosynthesis in seawater probably depends on the total radiation at the given point, as would be measured by an instrument whose response was independent of the direction of incidence of the incoming radiation. There is a great diversity in the shape of phytoplankton organisms, and to date there is little further evidence concerning the orientation and angular response to radiation of these organisms.

The authors are with the Visibility Laboratory, Scripps Institution of Oceanography, La Jolla, California 92037.

Received 29 September 1971.

Add to this the general turbulence in the natural environment and the continuous reorientation of individual organisms and the above suggestion seems to be realistic. Acting on the assumption that it does not matter from which direction radiation reaches the organism, a number of workers[6–9] have constructed 4π or scalar irradiance meters.

The choice of a scalar parameter has the simplicity and reasonableness suggested by Atkins and Poole[5] and the added advantage that the directional geometry of a particular experimental situation is eliminated. Thus, all laboratory and field measurements using photon scalar irradiance could be directly intercompared on an absolute basis. This is not in general true for a cosine collector, for example, since the collection properties of this collector imply a particular orientation with respect to the incident radiation. Thus, measurements with a cosine collector are meaningful only with respect to the particular geometry of the experimental situation under study.

The choice of a correct spectral response for an instrument measuring photosynthetically useful radiant energy in the sea has been of particular concern to Working Group 15, which was established by SCOR for recommending standard techniques for measuring the energy available for photosynthesis in the sea. This group, in its first meeting,[10] considered the ideal concept of the *equivalent detector*, i.e., a detector that, for the same radiant flux input, yields a response equivalent to that of the phytoplankton. However, as suggested by Steemann Nielsen,[11] shown by Pickett and Myers,[12] and discussed by Smith,[13] the response spectrum of phytoplankton is a function of the magnitude of the irradiance, being more or less independent of wavelength at high irradiance levels and approaching an action spectrum type response at low irradiance levels.

Recognizing that it might be impractical to construct a single instrument with ideal spectral response characteristics, Working Group 15[14] recommended that a practical detector should measure total energy or, preferably, total quanta available for photosynthesis and specified the range 350–700 nm. This concept of a *quantum meter* has since been developed into a practical instrument by Jerlov and Nygard.[15]

It is now widely acknowledged that a measure of the total number of photons in the spectral region from approximately 350 nm to 700 nm, with a geometrical response independent of direction, approaches a practical optimum for assessing photosynthetically useful radiant energy. That this wide acknowledgment is not attended by equally wide usage is in large part due to difficulties in making a practical instrument that incorporates the three requirements outlined above. In the following, photon scalar irradiance is accepted as the desired optimum parameter to measure, a new technique for its measurement is described, and the principal considerations for absolute calibration of such instruments are outlined.

Photon Scalar Irradiance

Tyler and Preisendorfer[16] rank scalar irradiance as a major radiometric concept and define it as follows. If $N(p,\theta,\phi)$ is the field radiance at point p, arriving from the direction (θ,ϕ), then the scalar irradiance $h(p)$ at point p is

$$h(p) = \int_{\theta=0}^{\pi} \int_{\phi=0}^{2\pi} N(p,\theta,\phi)d\Omega, \qquad (1)$$

where

$$d\Omega = \sin\theta d\theta d\phi. \qquad (2)$$

The scalar irradiance (watts·m^{-2}) is a quantitative measure of the total energy per unit time and area arriving at a point from all directions about the point where all directions are weighted evenly. The scalar irradiance divided by the velocity of light in the medium is the total amount of radiant energy per unit volume of space at the given point, i.e., the energy density (joule · m^{-3}).

Since in photosynthetic mechanisms it is the number of photons interacting that determines the rates of the process, it has been deemed preferable to use total quanta rather than total energy.[14] It is only very recently[17] that standard nomenclature to include quantum or photon flux in the International System of Units (S.I. units) for Radiometry and Photometry has been proposed. In keeping with the recommended radiometric nomenclature for photon quantities, scalar irradiance, when one is working in quantum units, becomes photon scalar irradiance ($h\nu$·sec^{-1}·m^{-2}); i.e., the total number of quanta per unit time and area (see also Ref. 18).

In passing, it should be noted that an instrument designed to measure photon scalar irradiance in quantum units (a spectral energy sensitivity proportional to wavelength) can in general be converted to measure scalar irradiance in radiometric units (a spectral energy

sensitivity constant for all wavelengths). However, once a measurement is made in either quanta or energy it cannot be converted to the other quantity without complete knowledge of the spectral distribution of the measured flux. In what follows we will be concerned with photon scalar irradiance as defined in Eq. (1), measured in quantum units, which satisfies requirement (1) by being a well-defined and standardized physical quantity.

Measurement of Photon Scalar Irradiance

In the process of designing and building an instrument to measure the radiance distribution of radiant energy underwater,[3] it has become evident that photon scalar irradiance h can be measured by a method not previously utilized. This technique lends itself to precise and accurate measurements and can conceivably be employed by means of a simple reliable seagoing or laboratory instrument.

The Nikkon 8-mm $f/8$ fish-eye lens, manufactured by Nippon Kogaku K.K. (Japan Optical Industry Company), has collection properties that allow scalar irradiance to be measured. This lens is not an extension of a wide-angle lens[19] and does not exhibit the usual distortions associated with such a lens. However, the fish-eye lens has an inherent distortion which should not be considered an aberration but rather the result of projecting a hemisphere onto a plane.

Consider the fish-eye lens as a device that projects a hemisphere in object space onto a circle in image space (Fig. 1) and assume for the moment that losses due to reflection, absorption, and scattering while traversing the lens are negligible. Then, since energy is conserved, a narrow beam of radiant flux from the direction (θ,ϕ) in object space, passing through the entrance aperture of the lens to image space in the direction (θ',ϕ'), obeys the following relation[20]:

$$n^2 N(p,\theta,\phi)d\Omega(\theta,\phi)dS \cos\theta$$
$$= n'^2 N'(p,\theta',\phi')d\Omega(\theta',\phi')dS'(r',\phi') \cos\theta'. \quad (3)$$

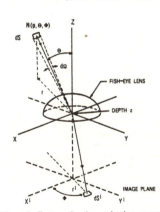

Fig. 1. Schematic diagram showing projection geometry of fish-eye lens for measuring scalar irradiance.

Here primed quantities refer to image space; unprimed, to object space. $N(\theta,\phi)$ is the field radiance incident on the lens, n the index of refraction, and $d\Omega$ an element of solid angle. If the object forming an image is far away compared to the square root of the entrance aperture (i.e., $r \gg \sqrt{a}$), which obtains for all practical situations, then

$$dS \cos\theta = a, \qquad (4)$$

where a is the area of the lens aperture which defines the narrow beam of flux under consideration. The irradiance in the image plane due to this narrow beam of flux is, by definition of irradiance,

$$H'(r',\phi') = N'(p,\theta',\phi') \cos\theta' \; d\Omega'. \qquad (5)$$

Substituting Eqs. (3) and (4) into Eq. (5), we have

$$H'(r',\phi')dS'(r',\phi') = \left(\frac{n}{n'}\right)^2 aN(p,\theta,\phi)d\Omega(\theta,\phi). \qquad (6)$$

Integration of the right-hand side of Eq. (6) over the complete upper hemisphere and the left-hand side over the area in image space to obtain the total energy per unit time $E_{\frac{1}{2}}$ projected from the object hemisphere onto the image plane gives

$$E_{\frac{1}{2}} = \int_{r'=0}^{r'_{max}} \int_{\phi=0}^{2\pi} H'(r',\theta')dS(r',\phi') = \left(\frac{n}{n'}\right)^2 a$$

$$\times \int_{\theta=0}^{\pi/2} \int_{\phi=0}^{2\pi} N(p,\theta,\phi)d\Omega(\theta,\phi). \qquad (7)$$

If we now consider a second fish-eye lens, oriented so as to project the opposite hemisphere (along the Z axis in Fig. 1) onto the image plane and sum the energy incident on the image plane from both directions, we have

$$E = E_{\frac{1}{2}} + E_{-\frac{1}{2}} = \left(\frac{n}{n'}\right)^2 a\left[\int_{\theta=0}^{\pi/2}\int_{\phi=0}^{2\pi}\right.$$

$$\times N(p,\theta,\phi)d\Omega(\theta,\phi) + \left.\int_{\theta=\pi/2}^{\pi}\int_{\phi=0}^{2\pi} N(p,\theta,\phi)d\Omega(\theta,\phi)\right],$$

$$E = \left(\frac{n}{n'}\right)^2 a\int_{\theta=0}^{\pi}\int_{\phi=0}^{2\pi} N(p,\theta,\phi)d\Omega(\theta,\phi). \qquad (8)$$

Finally, using the definition of scalar irradiance, Eq. (1),

$$E\left(\frac{photons}{time}\right) = \left(\frac{n}{n'}\right)^2 a \; (area) \; h(p) \left(\frac{photons}{time \cdot area}\right), \qquad (9)$$

where we have agreed to measure energy in quantum units.

Thus two fish-eye lenses, oriented back-to-back, having the property that they project a hemisphere in object space onto a circle in image space, provide a means of measuring scalar irradiance. It is necessary only to record the energy incident onto the image plane from both directions E to obtain a measure of $h(p)$.

The specific projection properties of the fish-eye lenses used in the radiance distribution camera system are known[19] and have been checked in detail.[3] How-

ever, the exact projection is not so important for measuring scalar irradiance as are losses in the lens system. To measure h accurately, it is necessary that transmission losses in the lens (not to be confused with flux density changes due to the lens projection properties) not be a function of incident angle. This has been measured for our lenses and found to be constant within 5%. More important than particular results, however, are the desired properties of an ideal photon scalar irradiance collector and techniques for testing the fidelity of a particular detector against the ideal. The two principal aspects of such a test, the spectral response and the geometrical response, are discussed below.

Absolute In-Water Calibration

Relative Spectral Response

In order for a detector to measure total quanta between 350 nm and 700 nm, it must have a uniform quantum sensitivity over this region and zero sensitivity outside this region. Or in the more familiar energy domain, it must have twice the sensitivity at 700 nm as it has at 350 nm, have linear sensitivity between these limits, and zero sensitivity outside (see Fig. 2). Jerlov and Nygard[15] discuss in some detail their spectral calibration procedures. Their careful work deserves consideration by anyone constructing an instrument to measure radiant energy underwater.

An alternative technique used by the Visibility Laboratory to determine relative spectral response is a Cary monochromator altered so that it can be used to compare the spectral output of any photosensitive surface with that of a blackbody. With this facility the relative spectral response of photodetectors (plus filters) can be rapidly measured. The objective of the relative spectral calibration is to measure the spectral response and to adjust this response, if necessary, until it corresponds as closely as practical to the ideal response shown in Fig. 2.

Relative Geometrical (Angular) Response

Since we are discussing instruments to make in-water measurements, the relative geometrical response must be measured in water. The fish-eye lens, protected with a polished Plexiglas hemisphere centered at the protection origin of the lens, has a geometrical response in water that is effectively the same as in air. The

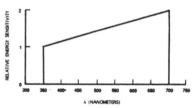

Fig. 2.   Ideal spectral response of quantum meter.

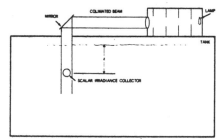

Fig. 3. Experimental arrangement to measure geometrical response and immersion effect.

procedure outlined here for calibrating a scalar irradiance collector is similar to the calibration of an underwater spectral irradiance collector which has been described in detail.[21]

In the laboratory a steady, uniform collimated beam of radiant energy is directed upon the collector through a flat, calm air–water interface and through a depth (z) of water as shown in Fig. 3. An ideal scalar irradiance collector will have a response that is independent of its angle of rotation in the beam. Most conveniently expressed on a polar plot (Fig. 4), an ideal collector will yield a perfect circle about the circumference of such a diagram. Imperfect collectors will, of course, depart from this ideal, so the direction of maximum response should be normalized to unity. The departure from unity at other angles can then be used to calculate or estimate the accuracy of a measurement for a given or estimated radiance distribution. This direction of maximum response is the orientation used to calculate the immersion effect and to make the final absolute calibration.

A procedure for measuring the immersion effect correction for a photon scalar irradiance collector is essentially the same as that described for irradiance collectors.[21] This correction, usually between 10% and 30%, allows in-water measurements to be related to an absolute calibration in air.

Absolute Calibration

Absolute calibration of an in-water photon scalar irradiance collector requires (1) a knowledge of the relative spectral responses (Fig. 2), (2) a knowledge of the geometrical response in water (Fig. 4), (3) an immersion effect correction factor, (4) a detector and related electronic circuitry that must be linear over the full working range of energy levels and spectral inputs. Once the above information is known, an absolute calibration can be carried out (see also Ref. 15). Without this spectral and geometrical response information of the collector, the accuracy of a measurement cannot be assessed satisfactorily.

Use of a standard of spectral irradiance[22] is the most accurate and straightforward technique for absolute calibration, provided that the relative spectral and geometrical responses are accurately known and are

reasonably close to the ideal response. It is only necessary to integrate the known spectral output (in quanta) of the standard lamp by the relative spectral response of the collector to obtain an absolute calibration in air. In practice great care must be exercised to avoid *spectral leakage* problems at the boundaries (350 nm and 700 nm) of the spectral region.

Another indirect calibration technique is direct comparison of the collector response with that of a pyrheliometer.[23] Here, as above, ability to assign an absolute value to the calibration will depend in no small measure on the fidelity of the spectral and geometrical responses to the ideal. Knowledge of the actual responses, however, allows one to determine the accuracy of the absolute calibration and to assess the magnitude, and perhaps the direction, of errors in experimental measurements.

Second-order calibration problems, such as the immersion effect correction's being a function of wavelength or geometry, have not been discussed. In general, efforts to calibrate or correct for such second-order effects are not commensurate with the increased accuracy obtained.

Summary

The long-standing need for an instrument to measure photosynthetically useful radiant energy in natural waters has been briefly reviewed. Photon scalar irradiance was presented as the optimum measure for such energy. A new technique for measuring photon scalar irradiance has been described.

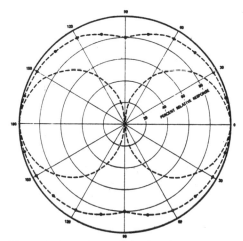

Fig. 4. Angular response of ideal scalar collector (*outer circle*). Inner (*dashed*) circles show ideal response of cosine collectors placed back-to-back. Most actual scalar irradiance collectors have responses whose curves lie somewhere between these ideal curves. An example of this is shown (*dotted curve*, x's data points) for an inexpensive fish-eye lens. Lenses used in the radiance distribution camera system have ideal responses within the accuracy of the measurement (5%).[2]

Finally, since there are several techniques in the literature for measuring scalar irradiance, principal considerations for the absolute calibration of in-water instruments have been outlined. Such absolute calibration would allow all radiant energy measurements in the laboratory or at sea to be directly intercompared in a meaningful manner.

This work was supported by National Science Foundation grant GA-19738A.

References

1. J. E. Tyler, Limnol. Oceanog. **13**, 1 (1968).
2. J. E. Tyler and R. C. Smith, *Measurements of Spectral Irradiance Underwater* (Gordon and Breach, New York, 1970).
3. R. C. Smith, R. W. Austin, and J. E. Tyler, Appl. Opt. **9**, 2015 (1970).
4. W. R. G. Atkins and H. H. Poole, J. Marine Biol. Assoc. U. K. **24**, 271 (1940).
5. W. R. G. Atkins and H. H. Poole, J. Int. Council Exploration Sea **23**, 327 (1958).
6. R. I. Currie, in *Symposium on Radiant Energy in the Sea*, Helsinki (International Union of Geodesy and Geophysics, 1960), pp. 107–112.
7. W. S. Maddux, Limnol. Oceanog. **11**, 136 (1966).
8. T. Sasaki, G. Oshiba, and M. Kishino, J. Oceanog. Soc. Japan **22**, 123 (1966).
9. P. H. Rich and R. G. Wetzel, Limnol. Oceanog. **14**, 611 (1969).
10. Working Group 15 (Moscow, 5–9 October 1964), UNESCO Tech. Pap. Marine Sci. No. 2 (1965).
11. E. Steemann Nielsen, Physiol. Plantarum, **5**, 334 (1952).
12. J. M. Pickett and J. Myers, Plant Physiol. **41**, 90 (1966).
13. R. C. Smith, Limnol. Oceanog. **13**, 423 (1968).
14. Working Group 15 (Karuizawa, 15–19 August 1966), UNESCO Tech. Pap. Marine Sci. No. 5 (1966).
15. N. G. Jerlov and K. Nygard, Univ. Copenhagen, Inst. Phys. Oceanog. Rept. 10 (1969).
16. J. E. Tyler and R. W. Preisendorfer, in *The Sea*, M. N. Hill, Ed. (Interscience, New York, 1962), Vol. 1, pp. 397–451.
17. J. J. Murray, F. E. Nicodemus, and I. Wunderman, Appl. Opt. **10**, 1465 (1971).
18. R. E. Craig, Photochem. Photobiol. **3**, 189 (1964).
19. K. Miyamoto, J. Opt. Soc. Am. **54**, 1060 (1964).
20. M. Born and E. Wolf, *Principles of Optics* (Macmillan, New York, 1964).
21. R. C. Smith, J. Marine Res. **27**, 341 (1969).
22. R. Stair, W. E. Schneider, and J. K. Jackson, Appl. Opt. **2**, 1151 (1963).
23. N. Robinson, *Solar Radiation* (Elsevier, New York, 1966).

Part VII

INTERDISCIPLINARY APPLICATIONS OF LIGHT IN THE SEA

Editor's Comments
on Papers 26 Through 32

OCEANIC PHOTOSYNTHESIS

One of the most important and influential papers on the relationship between light in the sea and oceanic photosynthesis was published by E. Steemann Nielsen and E. Aabey Jensen in the *Galathea Report* (1957, Volume 1). These authors used C^{14} as a tracer (Steemann Nielsen, 1952), to determine the rate of photosynthesis of oceanic phytoplankton and related their results to incident radiant energy. They documented light inhibition at high levels of radiant energy and reported the levels of incident energy at the "compensation" depth, that is, the light level for which the net rate of photosynthesis became zero.

They also established a general relationship between gross production and depth (as the latter relates to the available natural radiation) and worked out a method for estimating the maximum potential rate of photosynthesis at any location from radiant energy measurements, a result that can be used to estimate the maximum possible productivity of the world's oceans. In a more recent paper, E. Steemann Nielsen and E. G. Jørgensen (Paper 26) have reviewed the relationship between marine phytoplankton and incident radiant energy.

Much of the early research on the relationship between marine photosynthesis and radiant energy made use of photopic measurements of the available radiant energy. In 1964 the Scientific Committee on Oceanic Research (SCOR), together with IAPSO and UNESCO, recognizing that photopic measurements were of questionable application to photosynthesis, appointed a working group to determine the most appropriate measurement of radiant energy for correlation with oceanic photosynthesis. The unanimous recommendations of this working group were accepted and published by SCOR in 1974 (Paper 27).

Since the publication of these recommendations, A. Morel and R. C. Smith (1974) have determined appropriate conversion factors between a measurement of total quanta in the ocean and a simultaneous measurement of total energy (these being the primary and secondary recommendations, of the working group).

Production Equations

An important aspect of the interaction of radiant energy with oceanic phytoplankton is the development of mathematical models for predicting gross daily photosynthetic productivity. T. T. Bannister (Paper 20) has pointed out the need for production equations that do not contain parameters or coefficients that are dependent on the phytoplankton concentration, and has transformed the recent production equations of Steele (1962), Vollenweider (1965), and Fee (1969) to meet this requirement. Bannister (1974) has also developed a general theory of steady-state phytoplankton growth, and Crill (1974) has developed an analog model of the photosynthesis-light curve.

Mass Culture

Mass culture of phytoplankton has been studied in laboratory experiments but has not yet been undertaken as a profitable enterprise largely because of the cost of harvesting. The significance of light in mass cultures had been demonstrated by J. Myers and J.-R. Graham (Paper 29).

Image Transmission

The work of S. Q. Duntley and coworkers on contrast transmittance through ocean water is reviewed in Paper 2 and in Paper 32. More recently, Duntley (1974) has examined resolution limitations imposed by various types of water and has experimentally shown that the normal temperature and salinity gradiants found in the ocean do not appreciably attenuate high spacial image frequencies.

The Deep Scattering Layer

The vertical migration of the deep scattering layer in the ocean has attracted considerable interest. J. H. S. Blaxter and R. I. Currie (Paper 30) have directly demonstrated the influence of light on the deep scattering layer, and J. D. Isaacs et al. (Paper 31) have offered a logical tactic for this migration.

REFERENCES

Bannister, T. T. 1974. A general theory of steady state phytoplankton growth in a nutrient saturated mixed layer. *Limnol. Oceanogr.* **19**(1):13–30.

Crill, P. A. 1974. *An analog model of the photosynthesis-light curve and algal adaptation.* Thesis, San Diego State University.

Duntley, S. Q. 1974. Underwater visibility and photography. In N. G. Jerlov and E. Steeman Nielsen, eds., *Optical Aspects of Oceanography,* pp. 135–149. Academic Press, London, 494 pp.

Fee, F. J. 1969. A numerical model for the estimation of photosynthesis production, integrated over time and depth, in natural waters. *Limnol. Oceanogr.* **14**:906–911.

Morel, A., and R. C. Smith. 1974. Relation between total quanta and total energy for aquatic photosynthesis. *Limnol. Oceanogr.* **19**(4):591–600.

Steele, J. H. 1962. Environmental control of photosynthesis in the sea. *Limnol. Oceanogr.* **7**:137–150.

Steeman Nielsen, E. 1952. The use of radioactive carbon (C^{14}) for measuring organic production in the sea. *Perm. Inst. Cons. Explor. Mer J. du Conseil* **18**:117.

———— and E. A. Jensen. 1957. Primary oceanic production, the autotrophic production of organic matter in the oceans. In A. Fr. Bruun, Sv. Greve, and R. Spärck, eds., *Galthea Report, Volume 1: Scientific Results of the Danish Deep-Sea Expedition Round the World, 1950–1952.*

Vollenweider, R. A. 1965. Calculation models of photosynthesisi-depth curves and some implications regarding day rate estimates in primary production measurements. In C. R. Goldman, ed., *Primary Productivity in Aquatic Environments,* pp. 425–457. Ist. Idrobiol. Mem. 18 (suppl.).

26

Copyright © 1968 by the Scandinavian Society of Plant Physiology

Reprinted from *Physiol. Plant.* **21**:401–413 (1968)

The Adaptation of Plankton Algae
I. General part

By

E. Steemann Nielsen and Erik G. Jørgensen

Botanical Department, Royal Danish School of Pharmacy, København

(Received September 14, 1967)

Abstract

The various aspects of the adaptation of plankton algae to light and temperature are discussed. The shape of a light intensity–photosynthesis curve is shown to be an important means of describing the physiological adjustment of an algal population.

If the algae are not exposed to adverse influences such as poisons, pronounced nutrient deficiency or light shocks, the rate of real photosynthesis per mg chlorophyll *a* at 1 klux (incandescent light) should be about 0.4–0.6 mg C/hour. Hence this rate presents an excellent means of judging the quality of experiments.

Experiments are presented where *Chlorella pyrenoidosa* was adapted to light intensities between 0.32 klux and 21 klux. This alga adapts to different light intensities by varying the amount of pigments per cell. Algae grown at 1 klux have about 10 times more chlorophyll per cell than those grown at 21 klux. Other species of algae — but by no means all — are shown to behave in the same way.

The problem of algal resistance to photo-oxidation at high light intensities is discussed. Adaptation is shown to be one of the mechanisms which make the algae resistent. "Chlorophyll inactivation" is another.

Experiments with the diatom *Skeletonema costatum* concerning adaptation to different temperatures have been performed. The fact that the alga has essentially the same rate of photosynthesis per cell at all light intensities at 20°C and 7°C, may be attributed to an increase of all the enzymes at the low temperature. The amount of protein per cell was twice as high at 7°C as at 20°C.

1. Introduction

The adaptation of plankton algae to various ecological factors has now been in the focus of interest for several years. Two factors in particular have been studied, *i.e.,* rate of illumination and temperature; cf. *e.g.,* Steemann

321

Nielsen *et al.* (1959, 1961, 1962, 1964), Ryther *et al.* (1959), Ichimura (1960), Ichimura *et al.* (1962), Saijo and Ichimura (1962), Jørgensen (1964), Aruga (1965), Jørgensen and Steemann Nielsen (1965) and Talling (1966).

Investigations of adaptation have partly been made with cultures of unicellular algae in the laboratory, partly by the study of natural populations. In order to fully understand the importance for the algae of adapting to the different ecological conditions, it is necessary to examine all relevant facts of adaptation.

In the present article the word "adaptation" is of course used in the physiological sense, *i.e.* as a physiological adjustment to the surrounding conditions. Plant physiologists have used the word adaptation exclusively with such a meaning at least since the end of the last century. Biologists working with genetics on the other hand, understand by adaptation an hereditary alteration adjusting the organisms to the surroundings. For physiologists then adaptation is "phaenotypical" in contrast to the "genotypical" meaning as used in genetics.

It seems necessary to stress this difference in the meaning of the word. In a recent article Yentsch and Lee (1966) criticize workers dealing with adaptation without having recognized that these interpret the response of organisms as physiological adjustment and not as an hereditary alteration.

The photosynthetical mechanism is of course not the only mechanism which adapts according to the ecological condition. Respiration must also be mentioned. As will be shown in a subsequent paper, — Steemann Nielsen and Jørgensen (1968) — respiration and photosynthesis adapt concordantly. In plants growing harmoniously the two processes are adjusted in such a way that they match each other. However, in the present paper only photosynthesis is considered.

2. Light Intensity - Photosynthesis Curves

Two kinds of processes are taking part in photosynthesis, *i.e.*, photochemical processes and enzymatical processes. The rates of the photochemical processes depend on the light absorption by the photosynthetic pigments — and therefore on both the concentration of the pigments and of the light intensity. On the other hand, the rates of the enzymatic processes depend on the concentration of active enzymes and on the temperature. The rate of the over-all process may be limited either by the rate of the photochemical or by the rates of the enzymatic processes. Perhaps the most important consequence of adaptation — but not the only one — is the possibility of matching to some extent the photochemical and the enzymatical processes under the prevailing ecological conditions.

The shape of the curve presenting the rate of photosynthesis as a function of illumination — cf. *e.g.*, Figure 2 — gives important information about the adaptation. The slope of the initial part of the curve is a function of the photochemical part of photosynthesis. On the other hand, the horizontal part represents the maximum rate of the enzymatic processes. Therefore the light intensity at which the initial slope and the horizontal part of the light intensity–photosynthesis curve intersect describes, to a certain degree, the ratio between the two kinds of processes. This light intensity, introduced as I_k by Talling (1957), is thus an important means of describing the physiological adjustment of an algal population.

In laboratory work it is easy to present the rate of photosynthesis per definite units, such as number of cells or weight of chlorophyll. In Figure 2 the rate of photosynthesis is given as a function of light intensity in *Chlorella pyrenoidosa* at 20°C, grown at either 0.32, 1, 3 or 21 klux, the number of cells being the unit in 2 A and the weight of chlorophyll in 2 B. It is obvious that the curves vary *inter se* according to the unit applied.

Another way to present light intensity–photosynthesis curves is to normalize each curve with respect to the maximum rate of photosynthesis. When working with plankton collected in nature it is difficult in many cases to obtain reliable units to which the rate of photosynthesis can be related. Numbers of cells are unrealistic, if the plankton as is usual consists of many different species of varying sizes. Chlorophyll can often be used with success (cf. *e.g.*, Ichimura *et al.* 1962), but in many cases chlorophyll measurements in natural plankton collections are rather doubtful, either too high due to "dead" chlorophyll or too low due to difficulties in extracting the pigment.

As mentioned above the light intensity–photosynthesis curve is an important means of describing the physiological adjustment of a plant. It is, however, very important to keep in mind that it only gives an instantaneous measure of the ability of the plant to react to various experimental conditions of short duration. Most important for a plant is the rate of photosynthesis at the light intensity at which it grows.

All the following experiments have been made by the carbon-14 technique. The values are corrected to real photosynthesis; cf. Steemann Nielsen (1958).

3. Rate of Photosynthesis versus Chlorophyll Concentration as a Guide to the Quality of Photosynthesis Experiments

Harder (1923), Gabrielsen (1948) and Steemann Nielsen (1961) have all stressed the importance of studying the influence of chlorophyll concentration on the rate of photosynthesis at low light intensities where the rate of the overall photosynthesis process is limited by the rate of the photochemical part of the processes. As shown by Gabrielsen (1948) chlorophyll is ordinarily present in considerable excess in leaves of higher plants. Investigations concerning the influence of chlorophyll concentration can hence only be made using extremely chlorophyll-poor leaves. Dilute suspensions of plankton algae are still better suited for such investigations, as the pigments here are never found in real excess; cf. Steemann Nielsen (1961). In *Chlorella vulgaris* grown at either 3 or 30 klux (20°C), was found a rate of photosynthesis at 1 klux (incandescent light) of respectively 0.37 and 0.40 mg C per mg chlorophyll a and b. Computed instead per mg chlorophyll a, the rates were 0.50 and 0.57 mg C. The photosynthesis experiments were made with the carbon-14 technique. Steemann Nielsen (1965) has shown that the standardization of the C-14 ampoules was defective in the original carbon-14 'technique due to a wrong extrapolation of the self-absorption curve. The original numbers published have been corrected accordingly.

During the last few years a considerable number of planktonic green algae and diatoms have been investigated in this laboratory. When cultivated under ordinary conditions at 1 klux incandescent light they have given, per mg chlorophyll a, a rate of photosynthesis between 0.40 and 0.60 mg C/h (on average, about 0.50). Most of these experiments have not been published, but Jørgensen (1964 b) presented 9 single experiments with the diatom *Cyclotella Meneghiniana*. Further the experiments with Skeletonema presented in Figure 4 B show 0.52 mg C per mg chlorophyll a at 1 klux and even Chlorella pyrenoidosa cultured at the extremely low light intensity of 0.32 klux — *i.e.* just above the compensation point — had the quite normal rate of photosynthesis of 0.48 mg C per mg chlorophyll a at 1 klux.

Some exceptions to the rule have been found, however. *Chlorella vulgaris* grown at a low light intensity (3 klux) but provisionally transferred to a high light intensity for a few hours suffers inactivation of a part of the photochemical mechanism whereby the rate of photosynthesis at 1 klux decreases considerably; cf. Steemann Nielsen (1962).

Another exception to the rather constant rate of photosynthesis per mg chlorophyll a at 1 klux is found in extremely nutrient deficient cultures. In *Skeletonema costatum* where Jørgensen (in preparation) in normal experiments has obtained an average rate of 0.50 mg C per mg chlorophyll a at 1 klux, the rate was 0.19 in a culture being pronouncedly deficient in phosphorus, and 0.09 in a culture being pronouncedly deficient in nitrogen.

Finally, poisons may lower the rate per mg chlorophyll a at both high and low light intensities as shown, *e.g.*, by Weller and Franck (1941).

We have preferred to use incandescent light when relating the rate of photosynthesis to chlorophyll a because the main part of such light is in the spectral region where chlorophyll a absorbs light in green algae and diatoms. — The same is also true to some extent of chlorophyll b in green algae. Blue green algae absorb a considerable part of the incandescent light via the pigment phycocyanin. Hence such algae give quite different values for the photosynthetic rate per unit of chlorophyll a; cf. Steemann Nielsen (1961 p. 873).

It is impossible, even for green algae and diatoms, to give a precise value for the rate of photosynthesis per mg chlorophyll a at 1 klux in incandescent light. The variation of the value given in this article is due not only to the experimental scatter but also to the variation in the concentration of the other pigments (especially chlorophyll b). Finally, the content of chlorophyll a in the individual alga may have some effect; cf. Steemann Nielsen (1961). The mutual shading by the single pigment molecules inside a chloroplast must be expected to be more pronounced in chloroplasts rich in pigment than in those poor in pigments. In the latter, we therefore must expect the relatively highest value for the rate of photosynthesis per mg chlorophyll and hour at 1 klux, as normally observed in this laboratory.

Some variation in this value must therefore be expected. Large deviations from the value of 0.5 mg C per mg chlorophyll a and per hour, however, are indications of either faults in the experiments or influences of *e.g.* nutrient deficient cultures; cf. above.

Yentsch and Lee (1966) have presented experiments with the green alga *Nannochloris* normally grown at 10 klux on a 12 hour-light, 12 hour-dark regimen at 20°C by a turbistatic method. Prior to photosynthesis experiments, samples were drawn off from the turbistats and preconditioned for 24 hours at either different temperatures or different light intensities. The results of the photosynthesis experiments which were presented by Yentsch and Lee in their Figures 4 and 5 are very different from experiments made in this laboratory. This cannot be attributed to the use of another species. We have in fact obtained consistent results with all the species (more than 10) used.

It is easiest to explain the experiments on the temperature dependence of photosynthesis. At 20°C the I_k is 11.75 klux and the rate of photosynthesis per mg chlorophyll at 1 klux is 0.40 mg C/h. The latter value is in relatively good accord with the value obtained in this laboratory. The same is in fact also true for the experiments at 10°C. The slightly lower value for the rate of photosynthesis per mg chlorophyll at 1 klux could be due even to the drawing of the curve. However, at 5°C only 0.17 mg C was assimilated per mg chlorophyll at 1 klux, a reduction by a factor of more than two. The transfer directly from 20°C to 5°C has presumably given the algae a temperature shock, from which they had not recovered during the preconditioning period of only 24 hours.

More difficult to explain are the experiments presented in Yentsch and Lee's Figure 5 showing the response of photosynthesis to light in Nannochloris at 20°C for 24 hours at different light intensities. Whereas the curve for the algae from 25 klux is in agreement with similar curves obtained in this laboratory, the curves for the algae both grown at 6 and 2 klux are completely different. For algae grown at 6 and 2 klux the rates of photosynthesis per mg chlorophyll at 1 klux were 0.15 and 0.09 mg C/h respectively, the latter value being reduced by a factor of more than four compared with the value for the algae grown at 25 klux.

Little doubt exists that the experiments with the algae grown at 6 and 2 klux are defective. The cause of this is more difficult to ascertain. A theory of a toxic effect of some kind is proposed as the most possible cause. It is extremely unlikely that deficiency in nutrients could be significant. Such a deficiency should, of course, also have influenced the algae preconditioned at 25 klux.

In their Figure 7 Yentsch and Lee (1966) give some light intensity–photosynthesis curves for tropical phytoplankton at depths of 10 and 75 m's, respectively. The rates are given as apparent photosynthesis. We have corrected them to present real photosynthesis. The rates of photosynthesis (real) per mg chlorophyll at 1 klux for the plankton at 10 m was 0.60 and 0.43 mg C per hour, respectively, and in good agreement with our values, while the rates at 75 m were only 0.20 and 0.18 mg C per hour, respectively. These very low values are most likely due to a high concentration of "dead chlorophyll" at this depth. The experiments presented in Figure 2 shows that algae adapted even to extremely low light intensities have a normal rate of photosynthesis per mg chlorophyll.

Figure 6 in Yentsch and Lee (1966) shows a figure taken from Steemann Nielsen (1962). The text to the lower curve, however, is wrong. The algae are said to have been grown at 30 klux. They were grown at 3 klux, however, and had been transferred to 30 klux only for 3 hours. The experiment was given by Steemann Nielsen as an example of the "chlorophyll inactivation" which takes place in *Chlorella vulgaris* cells adapted to a low light intensity when the algae receive a light shock by being transferred to a much higher light intensity.

4. Adaptation to Different Light Intensities

The relevant literature concerning the adaptation of plankton algae were cited in the Introduction. When shifting cultures of algae from one condition of illumination to another, an adaptation of the photosynthetic apparatus takes place. Some species such as *Chlorella vulgaris* and *C. pyrenoidosa* are able to adapt very extensively, whereas other species — such as the diatom *Cyclotella Meneghiniana*; cf. Jørgensen (1964 a) — adapt much less extensively. For the stabilization of a new state of adaptation it is necessary that at least one new generation has been established; cf. Steemann Nielsen *et al.* (1962), Jørgensen (1964 a).

During the period when the algae adapt to a new light situation certain provisional situations may occur. They are most pronounced when shifting from a low to a high light intensity — including "inactivation of the chlorophyll" — but considerable variation is found from one species to another; cf. Steemann Nielsen (1962) and Jørgensen (1964 a), and section 4 b.

a. *Experiments with Chlorella pyrenoidosa*

As a general rule we may state that algae adapted to a high light intensity have increased the maximum rates of the enzymatic processes relative to the potential rates of the photochemical processes (which in principle is the same as the concentration of the photosynthetical pigments). In *Chlorella pyrenoidosa* and *C. vulgaris* the adaptation is mainly effected by varying the pigment content per cell. Figure 1 presents the concentration of chloro-

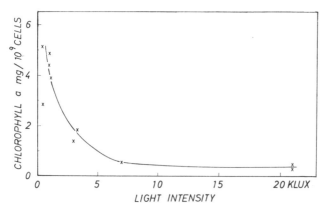

Figure 1. *The content of chlorophyll a per cell in Chlorella pyrenoidosa (211/ 8 b) as a function of the light intensity at which the algae were grown.* Continuous incandescent light, 20°C.

phyll *a* in mg per 10⁹ cells of C. pyrenoidosa in cultures grown in continuous incandescent light between 0.32 and 21.0 klux. A variation by a factor of about 10 is observed. At about 0.32 klux, which is just slightly above the compensation point, the variation was relatively considerable between two cultures both grown for 7 days at this light intensity. At all other light intensities the variation was only slight.

Some variation is always to be expected, however. A culture of Chlorella in continuous light is to a certain extent — but by no means completely — non-synchronized. The relative numbers of small and large cells are not constant.

The very low chlorophyll concentration at the relatively moderate light intensity of 21 klux is not complicated by a temperature of 20°C. This temperature is relatively low for the present species. At 29°C and the same light intensity, 0.30 mg chlorophyll *a* per 10⁹ cells was found. This is practically the same as at 20°C. Neither is the kind of illumination important. Experiments using other light sources — to be published later — have given the same results.

Altogether 6 series of experiments at 21 klux have shown the same low chlorophyll concentration per cell number. It is thus out of question that this can be due to the presence of only just divided cells. This was moreover not noticed during the counting of the cells.

Figure 2 in Steemann Nielsen *et al.* (1962) presents light intensity–photosynthesis curves from *Chlorella vulgaris* grown at either 30 or 3 klux. As the rate of photosynthesis was given per three different units — cell number, mg chlorophyll *a* or dry weight — it was possible to state that this species adapts to different light intensities by varying the concentration of pigment per cell. In Figure 2 A and B, similar light intensity–photosynthesis curves are given for *Chlorella pyrenoidosa* grown in continuous light of 21 klux, 3 klux, 1 klux and 0.32 klux, showing the adaptation processes to continue down to light intensities near the compensation point.

At the two lowest light intensities used for growing *Chlorella pyrenoidosa*, the cells are growing abnormally large before dividing. They produce 16–32 autospores in contrast to the normal 4–8. The very high chlorophyll concentration per cell is at least partly due to this circumstances. The very high rate of light-saturated photosynthesis per cell number in the algae grown at 1 klux — cf. Figure 2 A — is probably also a consequence of the large size of the cells. In the algae grown at the still lower light intensity of 0.32 klux the same was not observed. Here, the alga is near to the lowest illumination rate, where it can grow at all.

However, *Chlorella pyrenoidosa* adapts to different light intensities by varying the concentration of pigment per cell. It has been shown in this laboratory by Jørgensen (unpublished) that many other species of algae adapt in the same way. We may mention the green algae *Ankistrodesmus falcatus*, *Chlamydomonas moevusii*, *Monodus subterraneus*, *Scenedesmus obliquus*, and the blue green alga *Synechococcus elongatus*.

In contrast to the species mentioned above, the diatom *Cyclotella Meneghiniana* has the same chlorophyll content per cell when grown by a high (30 klux) and a relatively low (3 klux) light intensity. The only way for this

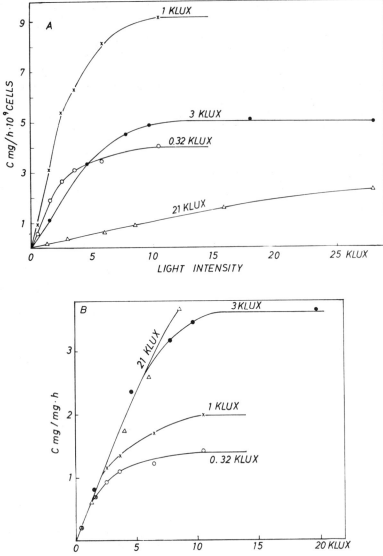

Figure 2. *The rate of photosynthesis as a function of the rate of illumination in Chlorella pyrenoidosa* (211/8 b) *grown in continuous incandescent light of 0.32, 1, 3 and 21 klux and 20°C. A: the rate per cell number and hour. B: the rate per mg chlorophyll a and hour. The light saturated rate of photosynthesis for algae grown at 21 klux was 10 mg C per mg chlorophyll a and hour.*

alga to adapt to a low and a high light intensity is by varying the maximum rates of the enzymatical processes; cf. Jørgensen (1964 a and b).

The same type of adaptation as in Cyclotella has been found by Jørgensen in other diatoms when grown in incandescent light: *Skeletonema costatum,*

Nitzschia closterium and *N. palea* and in the green alga *Scenedesmus quadricauda.*

b. *Means to protect the cells against too high light intensities*

One advantage of the adaptation of the algae to a high light intensity — and probably the most important one — is the ability of algae under such conditions to better resist very high light intensities. If the dark reactions in photosynthesis are unable to keep pace with the photochemical reactions the latter may induce photooxidations which again may destroy compounds in the chloroplasts such as the enzymes and thus decrease the overall rate of photosynthesis or even kill the algae. The lower the I_k is, the lower is the light intensity at which photooxidation can be expected.

In several bacteria it is evident that carotenoids function as photochemical buffering agents, cf. Burnett (1965). A similar system is proposed by Krinsky (1964, 1966) for higher green plants and algae. The carotenoid pair, antheraxanthin–zeaxanthin could fulfil the requirements of being "chemical buffer" to protect the cells against lethal photosensitized oxidations.

During the period when the algae adapt to a new and higher light intensity certain provisional situations may occur. In *Chlorella vulgaris* "chlorophyll inactivation" was found when cells adapted to 3 klux were transferred to 30 klux: cf. Steemann Nielsen (1962). As the mechanisms are not functioning momentarily, immediately after the transfer to high light intensity photooxidation destroys *e.g.* some of the enzymes found in the chloroplasts, whereby the rate of light-saturated photosynthesis decreases.

However, not all plankton algae seem to "inactivate the chlorophyll". The diatom *Cyclotella Meneghiniana* grown at 3 klux is thus able to tolerate a transfer even to the extremely high light intensity of 100 klux without effecting "chlorophyll inactivation" and without showing any decrease in the rate of light-saturated photosynthesis; cf. Jørgensen (1964 a). On the contrary this rate increased. It must be assumed that this alga has another mechanism protecting the cells against photooxidation.

5. Adaptation to different temperatures

When discussing the influence of temperature on the photosynthesis — and respiration — of plankton algae the time course is extremely important. Immediately after a shift of the temperature we observe a simple effect on all chemical processes, including the enzymatic processes in respiration and photosynthesis. In the latter process it is only the rate at light-saturation which will be influenced, enzymatic processes here limiting the over-all processes. At low light intensities, where temperature-insensitive photochemical part processes limit the rate of photosynthesis, temperature is of no importance. Several workers have convincingly shown this, *e.g.*, Aruga (1965 b). Figure 3 presents some of our own experiments with the diatom *Skeletonema costatum.*

However, for longer periods this simple situation will not last. The algae gradually adapt to the new situation. The adaptation will ordinarily be

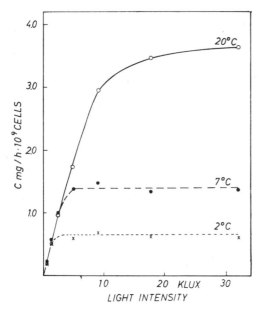

Figure 3. *The rate of photosynthesis as a function of light intensity in Skeletonema costatum grown at 3 klux, 20°C, but transferred for 30 minutes to 20°C, 7°C and 2°C.*

accomplished in some few days. Working with the influence of temperature on the photosynthesis or growth of plankton algae, for some species at least, it is necessary to make the alteration of the temperature in small steps. In this laboratory we use an alteration of about 5°C during each step; cf. Jørgensen and Steemann Nielsen (1965). Otherwise the algae may suffer harm.

In Figure 4 A and B, the rate of photosynthesis of the diatom *Skeletonema costatum* grown in continuous illumination of 3 klux at 20°C, 8°C and 2°C is presented as a function of light intensity. In 4 A the number of cells is used as the unit, in 4 B the weight of chlorophyll *a*. In the last case the initial slopes of all three curves are the same. However, if the number of cells is used as the unit, only the initial slopes of the curves for 20°C and 8°C are identical; a lower concentration of chlorophyll per cell at 2°C causes the less steep slope of this curve.

Whereas the light-saturated rate of algae adapted to 20°C immediately after transferring from 20°C to 8°C decreases to about one third (cf. Figure 3) after the termination of the adaptation to 8°C, the light-saturated rate is practically the same as at 20°C. The decrease by 12°C has, as shown above, the effect that the rate of enzymatic processes at constant enzyme concentration decreases to about one third. The rate of enzymatic processes is also dependent on the concentration of the enzymes, however.

In earlier publications we have advanced the hypothesis, that the concentration of enzymes per cell increases when the temperature is low; cf. Steemann Nielsen and Hansen (1959) and Jørgensen and Steemann Nielsen (1965). This is the most easy way of explaining the ability of the algae to maintain rates of enzymatic processes when these should have decreased due to the decrease in temperature. Further support for the hypothesis is

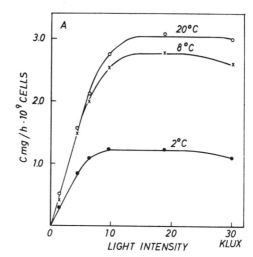

Figure 4. *The rate of photosynthesis as a function of light intensity in Skeletonema costatum grown in continuous illumination of 3 klux at 20°C, 8°C and 2°C. The temperature during the experiment the same as the growth temperature. A: the rate per cell number and hour. B: the rate per mg chlorophyll a and hour.*

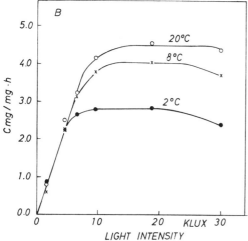

the fact, presented by Jørgensen and Steemann Nielsen (1965), that the growth rate pronouncedly decreased when the temperature decreased, although the rates of photosynthesis and respiration did not decline. The enzymes represent a major part of the protein in the cell and the protein constitutes a considerable part of the total organic matter in a plankton alga. Therefore, if the concentration of the enzymes increases — *e.g.*, by a factor of about three on lowering the temperature from 20°C to 8°C — much more organic matter has to be produced in order to increase the number of cells by a factor of two, the growth rate thus being lower.

The simplest way of proving the correctness of the hypothesis is to investigate the content of protein per cell number in algae adapted either to a high or to a low temperature. This has now been done; cf. Jørgensen (1968). The amount of protein per Skeletonema cell was twice as high at 7°C as at 20°C.

As shown by Jørgensen and Steemann Nielsen (1965) the shift between light period and dark period is of decisive importance when Skeletonema is grown at a very low temperature. The light intensity–photosynthesis curves presented in Figure 4 (this paper) were all made with algae grown in continuous light. However, if the algae at 2°C had been grown in periods of 9 hours' light (3 klux) and 15 hours' dark the initial slope of the curve had been exactly the same as the slopes of the curves at the higher temperatures; cf. Figure 2 and 3 in the paper mentioned. The rate of photosynthesis per cell up to 4.5 klux was the same as the rate at the higher temperatures.

It was further observed by Jørgensen (1968) that in Skeletonema cultures grown at low temperature (7°C) and continuous light a few per cent of the cells were abnormal, being 4–5 times higher than the average cell height. This was not found in cultures grown in a light-dark regimen. Some of the cells in the culture in continuous light thus seem to be unable to start the cell division. This is a further confirmation of the fact that the combination of a low temperature and continuous light is unfavourable for the growth of this particular diatom species.

In Danish waters Skeletonema is found at low temperature in winter, when the days are short and the light intensities low. It has never been found in Arctic waters during summer, probably due to the fact that the days are too long and perhaps the light intensities too high.

References

Aruga, Y.: Ecological studies of photosynthesis and matter production of phytoplankton I. Seasonal changes in photosynthesis of natural phytoplankton. — Bot. Mag. Tokyo 78: 280–288. 1965 a.
— Ecological studies of photosynthesis and matter production of phytoplankton II. Photosynthesis of algae in relation to light intensity and temperature. — Ibid. 78: 360–365. 1965 b.
Burnett, J. H.: Functions of carotenoids other than in photosynthesis. — In Chemistry and Biochemistry of Plant Pigments (T. W. Goodwin, ed.), pp. 381–403. Academic Press. 1965.
Gabrielsen, E. K.: Effects of different chlorophyll concentrations on photosynthesis in foliage leaves. — Physiol. Plant. 1: 5–37. 1948.
Harder, R.: Über die Bedeutung von Lichtintensität und Wellenlänge für die Assimilation farbiger Algen. — Z. Bot. 15: 305–355. 1923.
Ichimura, S.: Photosynthesis pattern of natural phytoplankton relating to light intensity. — Bot. Mag. Tokyo 73: 458–467. 1960.
— Saijo, Y. & Aruga, Y.: Photosynthetic characteristics of marine phytoplankton and their ecological meaning in the chlorophyll method. — Ibid. 75: 212–220. 1962.
Jørgensen, E. G.: Adaptation to different light intensities in the diatom Cyclotella Meneghiniana Kütz. — Physiol. Plant. 17: 136–145. 1964 a.
— Chlorophyll content and rate of photosynthesis in relation to cell size of the diatom Cyclotella Meneghiniana. — Ibid. 17: 407–413. 1964 b.
— The adaptation of plankton algae II. Aspects of the temperature adaptation of Skeletonema costatum. — Ibid. 21: 423–427. 1968.
— & Steemann Nielsen, E.: Adaptation in plankton algae. — Mem. Ist. Ital. Idrobiol., 18 Suppl.: 37–46. 1965.
Krinsky, N. I.: Carotenoid de-epoxidations in algae I. Photochemical transformation of antheraxanthin to zeaxanthin. — Biochim. Biophys. Acta 88: 487–491. 1964.
— The role of carotenoid pigment as protective agents against photosensitized oxidation in chloroplasts. — In Biochemistry of Chloroplasts (T. W. Goodwin, ed.) 1: 423–430. Academic Press. 1966.

Ryther, J. H. & Menzel, D. W.: Light adaptation by marine phytoplankton. — Limnol. Oceanogr. 4: 492–497. 1959.

Saijo, Y. & Ichimura, S.: Some considerations on photosynthesis of phytoplankton from the point of view of productivity measurement. — J. Oceanog. Soc. Japan 20: 687–693. 1962.

Steemann Nielsen, E.: Experimental methods for measuring organic production in the sea. — Rapp. et Proc.-Verb. 144: 38–46. 1958.

— Chlorophyll concentration and rate of photosynthesis in Chlorella vulgaris. — Physiol. Plant. 14: 868–876. 1961.

— Inactivation of the photochemical mechanism in photosynthesis as a means to protect the cells against too high light intensities. — *Ibid*. 15: 161–171. 1962.

— On the determination of the activity in ^{14}C-ampoules for measuring primary production. — Limnol. Oceanogr. 10, Suppl.: R.247–R.252. 1965.

— & Hansen, V. K.: Light adaptation in marine phytoplankton populations and its interrelation with temperature. — Physiol. Plant. 12: 353–370. 1959.

— — Influence of surface illumination on plankton photosynthesis in Danish waters (56°N) throughout the year. — *Ibid*. 14: 595–613. 1961.

— — & Jørgensen, E. G.: The adaptation to different light intensities in Chlorella vulgaris and the time dependence on transfer to a new light intensity. — *Ibid*. 15: 505–517. 1962.

— & Jørgensen, E. G.: The adaptation of plankton algae III. With special consideration of the importance in nature. — *Ibid*. 21: 647–654. 1968.

— & Park, T. S.: On the time course in adapting to low light intensities in marine phytoplankton. — J. Cons. Int. Explor. Mer. 29: 19–24. 1964.

Talling, J. F.: Photosynthetic characteristics of some freshwater plankton diatoms in relation to underwater radiation. — New Phytol. 56: 29–50. 1957.

— Photosynthetic behaviour in stratified and unstratified lake populations of a planktonic diatom. — J. Ecol. 54: 99–127. 1966.

Weller, S. & Franck, J.: Photosynthesis in flashing light. — J. Physic. Chem. 45: 1359–1373. 1941.

Yentsch, C. S. & Lee, R. W.: A study of photosynthetic light reactions, and a new interpretation of sun and shade phytoplankton. — J. Mar. Res. 24: 319–337. 1966.

27

Reprinted from pp. 37–42 of *Sci. Com. Oceanic Res. Proc.* **10**(1):1–83 (1974)

PHOTOSYNTHETIC RADIANT ENERGY:
RECOMMENDATIONS

SCOR Working Group 15

The terms of reference given to Working Group 15 by SCOR were:

(i)   To identify exactly what measurement of irradiance is required by biological oceanographers;

(ii) To recommend apparatus and procedure for measuring the variable defined above.

The Working Group has confined its activity to the measurements of radiant energy as it relates to the determination of oceanic primary productivity by the C^{14} incubation technique.

With respect to term of reference (i):

The basic recommendation of Working Group 15 is that biologists should measure the total-quanta available for photosynthesis within the wavelength limits 350 to 700 nm. The units of this measurement would be quanta per unit area per sec.

Because of the technical problems associated with the measurement of spherical irradiance (i.e., a measurement giving equal weight to quanta arriving at a point from all possible directions) Working Group 15 recommends a measurement of total down-welling irradiance (in quanta per unit area per sec) on a horizontal flat surface, and within the wavelength limits 350 to 700 nm.

In the event that no quanta meter is available, the subordinate recommendation of Working Group 15 is that biologists should measure total downwelling irradiance (in watts per unit area) on a horizontal flat surface and within the wavelength limits 350 to 700 nm.

With respect to term of reference (ii), Working Group 15 recommends that:

Apparatus recommendation:

I.   The radiant-flux sensor for measuring downwelling quantum irradiance should incorporate:

a)   A photosensitive device having suitable spectral-sensitivity characteristics and adequate response for the planned application.

b)   An optical filter having spectral transmittance such that the product of its spectral transmittance multiplied by the spectral sensitivity of the photosensitive device (multiplied (when necessary) by the spectral transmittance of the collector in C below) imparts equal quantum sensitivity to the combination, within the wavelength limits 350 nm and 700 nm and substantially zero sensitivity at other wavelengths.

c)   A white irradiance collector, that is, a radiant flux collector that performs the integration specified in the equation below and that exhibits minimum selective absorption, especially in the spectral region 350 nm to 700 nm.

$$E = 2\pi \int_{0°}^{90°} B \sin \theta \cos \theta \, d\theta \qquad (1)$$

In the equation, E represents the total collected quanta per unit area per sec and B represents the quanta per unit solid angle per unit area per sec arriving from various directions.

II.   In the event that no quanta-meter can be made available, the subordinate recommendation of Working Group 15 is that the photosensitive device be fitted with an optical filter (in place of the filter discussed in (b) above) having spectral transmittance such that the product of its spectral transmittance multiplied by the spectral sensitivity of the photosensitive device (multiplied (when necessary) by the spectral transmittance of the white collector), imparts equal energy sensitivity to the combination, within the wavelength limits 350 nm and 700 nm and substantially zero sensitivity at other wavelengths.   Recommendations (a) and (c) in this section remain the same.

Procedural recommendations:

I.   For measuring downwelling quantum-irradiance in the ocean, Working Group . 15 recommends the following procedural details:

a)   "Downwelling irradiance" as used herein implies that the underwater irradiance collector is horizontally oriented.   When necessary, precautions should be taken to insure horizontal orientation.

b)   Measurements of irradiance should be obtained in the vicinity of the phytoplankton samples that are being studied.   Both the quanta meter and the phytoplankton samples must be far away from any perturbation of the normal light distribution, due to the presence of the ship or other interfering obstacle.

c)   At each location and depth, the measurement of quanta should be sufficiently prolonged to obtain a useful average value.

d) Measurements with a quanta meter should be corrected for the optical immersion effect.

e) The underwater quanta meter should be equipped with a suitable depth transducer.

II. For measuring downwelling quantum-irradiance at deck level, and/or in an incubator, Working Group 15 recommends:

a) A horizontally oriented quantum-irradiance meter mounted in gymbals and located in an open area as free as possible from interference from shadows cast by the ship's rigging or equipment.

b) The quantum-meter should be constructed so that it can be immersed in each section of the deck incubator so that measurements of total quanta available for simulated-in-situ carbon fixation can be made.

c) Measurements of quantum irradiance in each section of a deck incubator, and the incubation of all photosynthesis samples, should be conducted under lighting conditions that closely simulate the natural lighting conditions encountered by the phytoplankton in the ocean at the depth from which they were collected.

Comments on Recommendations

In circulating these formal recommendations to members of the Working Group for signature, the following comments have been offered by individual members.

H. Jitts has commented that Procedural Recommendation I(c) is "weak and unclear". Professor Ivanoff has suggested the inclusion of a numerical specification of accuracy in the measurement of quanta above the surface. N. G. Jerlov and E. Steemann Nielsen have jointly advocated greater technical detail.

These criticisms are in large part corrected by the information given in the section headed, Discussion of State of the Art, which follows.

Y. Ochakovsky has pointed out that under term of reference (ii) (section on Apparatus Recommendation), if no quanta meter can be made available, the energy units in Equation 1, section I(c), will be E, in watts/area and B, in watts/unit solid angle area. Also, in Section I and II of the Procedural Recommendations, if measurements in watts are made, all references to quanta meters and quantum units must be changed to energy units (i.e., watts). Dr Ochakovsky is correct.

Discussion of State of the Art
by
John E. Tyler - Chairman W.G. 15

The concepts and procedures set forth in the above Working Group 15 recommendations were employed successfully during the SCOR DISCOVERER Expedition and an adequate technology exists.

The purpose of this addendum to the Working Group 15 recommendation is to relate and reference this technology.

The conceptual objective of recommendation (i) is to measure the quanta (or energy) available to the phytoplankton.   Radiant flux will be "available" to oceanic phytoplankton only if the wavelength distribution of the flux lies within the spectral sensitivity range of the phytoplankton and then only if that flux can be successfully collected, in a geometrical sense, by the phytoplankton.

In order to fulfill this latter conceptual requirement, it is necessary to have detailed information regarding the directional collecting properties of phytoplankton populations.   This information is not available and it was therefore necessary for the Working Group to consider measurements that would closely approximate the directional collecting properties of the phytoplankton.   The Working Group would have preferred to recommend a measurement of spherical quantum irradiance (as implied in recommendation (i)) but felt that spherical irradiance collectors were not sufficiently well developed.

The radiometric concepts of irradiance, spherical irradiance (and others) are defined and discussed in detail in references (1), (2), (3), and (7).   Simple instruments for measuring irradiance and quantum irradiance in the ocean have been described in references (4), (5), and (6).

Detailed data on the spectral quanta (and energy) that penetrates into various types of ocean water are available in the Data Report of the SCOR DISCOVERER Expedition (reference 9, Sections F and G) and in reference (8)).   These data are useful in determining the sensitivity requirements of submersible irradiance meters.   It has been estimated that irradiance meters for this purpose should be calibrated with an accuracy better than $\pm 5\%$, should exhibit a precision of 1% and that experimental measurements of quanta/$m^2 \cdot$ sec (or watts/m^2) in the ocean should have an accuracy better than $\pm 10\%$.

The measurement of irradiance (in any inuts) from shipboard is made difficult by the presence of the ship which casts a large shadow, by the motion of the air-ocean interface which superimposes noise on the measurement and by variations in surface flux due to clouds.   No quantitative evaluation of the effects of these problems has been made.   Section I(c) of the Prcoedural Recommendations alludes to these problems.

It has been suggested that a continuous record of surface quanta should be made and that underwater measurements of quanta should be averaged over a period at least as long as one ocean-swell period.

Methods for correcting for the optical immersion effect referred to in Procedural Recommendations I(d) are detailed and illustrated in reference (4).   The depth transducer referred to in Procedural Recommedation I(e) should have an accuracy consistent with the objectives of the research.

For the experiment on the SCOR DISCOVERER Expedition, a continuous recording of total quanta/m^2 (350-700 nm) was obtained in the simulated-in-situ incubator (under a blue-green filter--see comments below on Section II(c) of Procedural Recommendations), a continuous recording of total quanta/m^2 (350-700 nm) was also obtained at a depth of one of the in-situ incubation bottles, and total quanta/m^2 (350-700 nm) was measured as a function of depth in the ocean in order to locate the in-situ incubation bottles at depths where the available quanta/m^2 was the same as measured under blue-green filters in the various sections of the deck incubator.   Additional details of this procedure are given in the Data Report of the SCOR DISCOVERER Expedition, reference (9), Sections A and Introduction.

Section II(c) of the Procedural Recommendations requires that the spectral and spatial distribution of radiant flux in each incubator section should simulate the natural spectral and spatial distribution of the radiant flux available to the phytoplankton in the natural environment. The spatial distribution of underwater radiant flux has been measured and the data published in references (10), (11), and (12). The spectral distribution of underwater radiant flux is recorded in references (8) and (9) - Sections F and G.

The incubator used in the SCOR DISCOVERER Experiment was divided into sections, each one of which was painted to simulate the measured spatial distribution of underwater radiant flux. (Data for this purpose is available in references (10), (11), and (12).) Each section of the incubator was covered with one or more blue-green glass filters chosen to simulate the spectral distribution of the underwater radiant flux at different depths. Additional details of this procedure are recorded in reference (9), Section B.

Additional references relevant to the DISCOVERER Experiment will be found at the end of the text that precedes each section of the Data Report of the SCOR DISCOVERER Expedition (reference (9)).

Brief reports of the SCOR Working Group 15 planning meetings in 1964 and 1966 and of the sea trials in 1968 are published in references (13), (14), and (15).

REFERENCES

1. Judd, D. B. (ed.). J. O. S. A., vol. 52, p. 490 (1962).

2. Preisendorfer, R. W. In IUGG Monographie #10, Symposium on Radiant Energy in the Sea, N. G. Jerlov (ed.), pp. 11-30 (June 1961).

3. Jerlov, N. G. Optical Oceanography (Chapt. I). Elsevier Publishing Co. Amsterdam (1968).

4. Smith, R. C. An underwater Spectral Irradiance Collector. Journ. of Marine Research, vol. 27, pp. 341-351 (1969).

5. Jerlov, N. G. and Nygård, K. A Quanta and Energy Meter for Photosynthetic Studies. Kø/benhavns Univ., Inst. for Fysisk Oceanografi., Report #10 (1969)

6. Prieur, Louis. Photometre Marin Mesurant Un Flux de Photons (Quanta-Metre). Cahier Oceanographie, vol. 22, pp. 493-501 (1970).

7. Tyler, J. E. and Preisendorfer, R. W. In The Sea: Ideas and Observations on Progress in the Study of the Sea, Vol. I, Section IV, Chapter 8, "Light". Interscience Publishers, New York (1962).

8. Tyler, J. E. and Smith, R. C. Measurements of Spectral Irradiance Underwater. Gordon and Breach Science Publishers, New York (1970).

9.   Tyler, J. E., ed.   Data Report of the SCOR DISCOVERER Expedition
           SIO   Ref. Rep. 73-t16, 1000 pp. (June 1973).

10.   Tyler, J. E.   Radiance Distribution as a Function of Depth in an Underwater
           Environment.   University of California, Bulletin of the Scripps
           Institution of Oceanography, Vo. 7, #5, pp. 363-412 (1960).

11.   Jerlov, N. G. and Fukuda, M.   Radiance Distribution in the Upper Layers of
           the Sea.   Tellus, vol. 12, pp. 348-355 (1960).

12.   Sasaki, T.   On the Instruments for Measuring Angular Distributions of Under-
           water Daylight Intensity.   In Physical Aspects of Light in the Sea,
           J. E. Tyler, ed. Univ. of Hawaii Press, pp. 19-24 (1964).

13.   Tyler, J. E. (Chairman).   Report of the first meeting of the joint group of experts
           on photosynthetic radiant energy.   UNESCO technical papers in marine
           science, # (1965).

14.   Tyler, J. E. (Chairman).   Report of the second meeting of the joint group of
           experts on photosynthetic radiant energy.   UNESCO technical papers
           in marine science, #5 (1966).

15.   Tyler, J. E. (Chief Scientist).   Technical report of sea trials conducted by the
           Working Group on Photosynthetic Radiant Energy.   UNESCO technical
           papers in marine science, #13 (Aug. 1969).

28

Reprinted from *Limnol. Oceanogr.* **19**(1):1–12 (1974)

Production equations in terms of chlorophyll concentration, quantum yield, and upper limit to production

T. T. Bannister
Department of Biology, University of Rochester, Rochester, New York   14627

Abstract

For eventual incorporation into a general theory of phytoplankton dynamics, an equation for gross daily production in a water column is needed which is an explicit function of phytoplankton concentration and which does not contain parameters or coefficients dependent on phytoplankton concentration: the production equations of Steele, of Vollenweider, and of Fee are transformed to meet this requirement. It is also desirable, in order to simplify theoretical analysis, that the parameters in production equations be chosen so as to minimize the range of values needed to describe the variation of photosynthesis in nature. This can be achieved by choosing the quantum yield of photosynthesis in weak light as parameter instead of the maximum rate in saturating light. In lakes, the value of the quantum yield in weak light is likely to be about 0.06 moles carbon per einstein absorbed. An additional advantage of the transformed equations is to make evident a simple relation between actual daily production and the upper limit to production.

Equations for gross daily production in a water column are important both in analysis of field data and for eventual development of a general theory of phytoplankton dynamics. A number of treatments of daily production have appeared. Vollenweider (1965) has pointed out shortcomings of the earlier work, including the nonequational form of the expression of Ryther (1956*a*), the insufficiently accurate light curve equations adopted by Riley (1946), Tamiya et al. (1953), and Oorschot (1955), and the inaccuracy of the approximate depth integral proposed by Talling (1957*b*). These deficiencies have been overcome in the more recent production equations of Steele (1962), Steele and Menzel (1962), Vollenweider (1965), and Fee (1969). Nevertheless, these equations are still not ideal in two respects.

First, the equations have not been given in the most fundamental and general form. That is, they have not been written as functions of those factors which most directly determine production in a column. These factors are:

i) the illumination I_0 incident on the column; I_0 is an environmental variable dependent on latitude, season, time of day, and local weather;

ii) the intensity of light absorption by the water and its nonphytoplankton components; this intensity, which can be expressed by an extinction coefficient k_w, is at least in part an environmental variable dependent on the content of silt, dissolved pigments, and other nonalgal matter in the water;

iii) the concentration of phytoplankton pigments (most conveniently measured as chlorophyll concentration C); the product of chlorophyll concentration and a phytoplankton extinction coefficient k_c is a measure of the intensity of light absorption by phytoplankton; together k_w, k_c, and C determine the fraction of the incident light absorbed by phytoplankton;

iv) the efficiency with which light absorbed by phytoplankton generates photosynthetic products (e.g. reduced carbon); this efficiency is a function related to the photosynthetic light curve (the characteristic dependence of the instantaneous rate of photosynthesis on illumination); the efficiency function, therefore, contains light curve parameters.

Also, previous production equations have not included the most fundamental light curve parameters. The often used parameter p_{max} (the instantaneous rate of photosynthesis per unit volume of water at saturating illumination) varies widely and is dependent on chlorophyll concentration, temperature, and the state of physiological adaptation of the algae. A more nearly constant parameter—and therefore a more fundamental one—is the quantum yield of photosynthesis in weak light; it is independent of chlorophyll concentration and temperature and is less sensitive to adaptive state. The most general and fundamental equations for daily production will be, therefore, functions of the two environmental variables I_0 and k_w, the chlorophyll concentration C, the phytoplankton extinction coefficient k_c, and light curve parameters which are independent of chlorophyll concentration and have been chosen for greatest possible constancy.

Secondly, the manner in which a production equation will be eventually incorporated into a general dynamic theory also generates a demand for an equation that is an explicit function of chlorophyll concentration and that does not contain parameters dependent on chlorophyll concentration. This requirement arises from the following considerations. Even for the simple case of a well mixed, vertically homogeneous column, several interdependent simultaneous equations will be needed to describe gross production, losses of several kinds, nutrient supply rate, the behavior of primary consumers, and the characteristic dependence

of phytoplankton composition and parameters on growth rate and adaptive state. Since several of the equations will involve the phytoplankton concentration (most often expressed as the chlorophyll concentration) as a dependent variable, the simultaneous solution of the equations will be simplified if the production equation is an explicit function of chlorophyll concentration. Similarly, the mathematical treatment will be simplified by choosing the most nearly constant light curve parameters.

No previous treatment of daily production has given attention to the choice of most nearly constant parameters. Some attention has been devoted to the relation between daily production and chlorophyll concentration, but, up to now, no general equation, which is an explicit function of chlorophyll concentration has been published. Ryther and Yentsch (1957) showed how chlorophyll concentration could be taken into account in Ryther's (1956a) graphical estimation of daily production. Murphy (1962) was first to "partition" the total extinction coefficient into chlorophyll-dependent and chlorophyll-independent components, but he dealt only with relative production and his mathematical treatment had weaknesses (*see* Patten 1968). Steele (1962) and Steele and Menzel (1962) clearly knew how a production equation might be expressed as an explicit function of chlorophyll concentration. They recognized the means of partitioning the extinction coefficient and they showed how the photosynthetic parameter p_{max} could be expressed in terms of chlorophyll concentration. They did not, however, give a general equation but rather one in which specific numerical values of parameters had been substituted. Furthermore, their treatment was complicated by an attempt to take account of adaptation.

The equations most recently advanced for daily production are two proposed by Vollenweider (1965). The first, identified as "Vollenweider's equation," neglected photoinhibition at high light. The second, referred to as "Fee's production equation," since Fee (1969) elaborated on it, did take account of photoinhibition. Neither expressed their equations in terms of chlorophyll concentration. Subsequently, M. Lorenzen (1972) derived an equation for the time rate of change of phytoplankton concentration; into this "growth equation" was incorporated Vollenweider's production equation recast in terms of phytoplankton concentration. The recast production equation implied by the growth equation, but not actually written, has two deficiencies. First, it does not specify how phytoplankton concentration is to be measured. Secondly, it includes a parameter K_{max} defined as the component of the growth rate constant contributed by gross production. This parameter is not appropriate for a production equation since it is not a basic photosynthetic parameter, it is not easily evaluated experimentally, and it is the ratio of daily production (the quantity a production equation is intended to give!) and the phytoplankton carbon concentration in the water column. Thus, like Steele and Steele and Menzel, Lorenzen knew how to recast a production equation in terms of phytoplankton concentration, but did not actually write the production equation in its most general and fundamental form.

Under appropriate conditions, the equations of Steele, Vollenweider, and Fee are all potentially useful representations of daily production in a water column. My purpose here is to show how these equations can be rewritten in forms that are the most general and fundamental and that will be most useful in developing a general dynamic theory. After a short review of the properties of light curves and their parameters, I show how their equations can be recast as functions of chlorophyll concentration and the often used but widely variable parameter—the assimilation number. I then show that the production equations can be written in terms of the more nearly constant parameter —the quantum yield of photosynthesis in weak light. Finally, I point out that the transformed equations make evident simple relations between actual daily production and the upper limit to daily production, discussed by Ryther (1959), and also between

this upper limit and the maximum daily production per unit chlorophyll, a quantity used by Megard (1972).

Light curves

Light curves (also called "P vs. I curves") are graphs or equations describing the characteristic dependence of the instantaneous rate p of photosynthesis on the incident illumination I. Light curves always refer to photosynthesis in an optically thin layer— thin enough so that the incident illumination is essentially unattenuated by the layer and all algae are uniformly illuminated. Photosynthetic light curves have well known properties. Over a low illumination range, the rate of photosynthesis rises almost linearly with illumination. At higher illuminations, the rate rises more slowly and eventually reaches a maximum rate p_{max} at light saturation. At yet higher illuminations, the rate generally declines due to photoinhibition.

Directly related to the characteristic dependence of rate on illumination is the equally characteristic dependence of the quantum yield ϕ of photosynthesis on illumination. As defined by physiologists (*see* Kok 1960), the quantum yield ϕ is the ratio of the rate of photosynthesis (in moles of oxygen evolved or of carbon incorporated per unit time) and the rate (in einsteins absorbed per unit time) at which light is absorbed by the phytoplankton. The quantum yield has the dimensions moles oxygen or carbon per einstein absorbed. Over the low illumination range of a light curve, the quantum yield has an essentially constant, maximum value ϕ_{max}, and this value is directly proportional to the slope of the light curve. [If a light curve were graphed as rate (in moles carbon per unit time) versus absorbed illumination (in einsteins absorbed per unit time), then ϕ_{max} would be equal to the slope.] Above the linear region of the light curve, the quantum yield ϕ steadily declines as illumination increases and approaches a limit of zero at infinite illumination.

Up to light saturation, light curves of

photosynthesis can be crudely represented by the equation of a rectangular hyperbola (Tamiya et al. 1953; Oorschot 1955) and more accurately by the equation of a nonrectangular hyperbola (Rabinowitch 1951) or by the equation of Smith (1936). The latter may be written as

$$p(\text{g C m}^{-3} \text{ day}^{-1}) = p_{\max}I[(I_{0.7})^2 + I^2]^{-\frac{1}{2}}. \quad (1)$$

Here p is the instantaneous rate of photosynthesis and I is the illumination incident on an optically thin layer. Of the two parameters, $p_{\max}$ is the light-saturated rate, and $I_{0.7}$ is the illumination at which the rate p reaches 70% of the light-saturated rate.

To represent light curves both below and above saturation (i.e. to take account of photoinhibition at high light), Steele (1962) used an exponential function that can be written as

$$p(\text{g C m}^{-3} \text{ day}^{-1}) = p_{\max}(I/I_{\max}) \, e^{1-(I/I_{\max})}. \quad (2)$$

Here, the parameter $p_{\max}$ has exactly the same meaning as in Smith's equation; it is the maximum rate at light saturation. The parameter $I_{\max}$ is the illumination at which the light-saturated rate $p_{\max}$ occurs. An analysis of equation 2 shows that the rate p declines as the illumination I rises above $I_{\max}$. Steele's equation can provide an accurate fit of light curves up to light saturation, but, as Vollenweider (1965) pointed out, may provide only a rough representation in the range of photoinhibition at high light.

To obtain a better fit to experimental data at high light, Vollenweider (1965) proposed a modification of Smith's equation:

$$p(\text{g C m}^{-3} \text{ day}^{-1}) = p'_{\max}I[(I_K)^2 + I^2]^{-\frac{1}{2}}[1 + (aI/I_K)^2]^{-n/2}. \quad (3)$$

In this equation, the two parameters a and n characterize photoinhibition. The parameters $p'_{\max}$ and I_K are analogous to, but different from, $p_{\max}$ and $I_{0.7}$ in Smith's equation. If Smith's and Vollenweider's equations were fitted to the same set of light curve data (Smith's equation being fitted only to points at and below saturation), it

would be generally found that $p'_{\max} \neq p_{\max}$ and $I_K \neq I_{0.7}$. $p'_{\max}$ is not a rate ever actually attained, and I_K is not the illumination at which p reaches 70% of the maximum attained rate.

From a mathematical point of view, the choice of particular parameters in a light curve equation is an arbitrary one. For example, in place of $p_{\max}$ and $I_{0.7}$ in Smith's equation, one might choose as parameters the half-maximum rate $p_{0.5}$ and the illumination $I_{0.5}$ at which the half-maximum rate occurs. Smith's equation would then be written as

$$p(\text{g C m}^{-3} \text{ day}^{-1}) = 2p_{0.5}I[(3I_{0.5})^2 + I^2]^{-\frac{1}{2}}. \quad (4)$$

Relation between ϕ_{max} and p_{max} and p'_{max}

An alternative choice of parameter is that of the maximum quantum yield ϕ_{max} in place of $p_{\max}$ or $p'_{\max}$. The following derivation shows the relation between ϕ_{max} and $p_{\max}$ and $p'_{\max}$.

Consider an optically thin layer of algal suspension with a surface area of 1 m² and a shallow depth dz (m). The chlorophyll concentration in the layer is C (mg Chl m⁻³). The instantaneous incident illumination on the 1 m² surface is I (einst m⁻² day⁻¹). [1 einst m⁻² day⁻¹ of visible (400–700 nm) solar light corresponds to about 0.0084 g cal (total solar energy) cm⁻² min⁻¹. This equivalency is based on the close approximation that the average energy of an einstein of visible solar light is 52 kcal and that visible energy is 43% of the total (Withrow and Withrow 1956).] From the differential form of Beer's Law, the instantaneous rate dI_A at which light is absorbed by phytoplankton in the layer is

$$dI_A \text{ (einst absorbed m}^{-2} \text{ day}^{-1}) = -dI = k_cCIdz. \quad (5)$$

Here k_c is the average spectral extinction coefficient per unit of phytoplankton concentration, where phytoplankton concentration is expressed as the concentration of chlorophyll a in the water. The dimensions of k_c are m² mg⁻¹ Chl a. The instantaneous rate of photosynthesis, assuming Smith's equation, will be

$$pdz(\text{g C m}^{-2}\,\text{day}^{-1}) = p_{max}Idz[(I_{0.7})^2 + I^2]^{-\frac{1}{2}}, \quad (6)$$

or, in moles carbon,

$$pdz/12(\text{moles C m}^{-2}\,\text{day}^{-1}) = (p_{max}Idz/12)[(I_{0.7})^2 + I^2]^{-\frac{1}{2}}. \quad (7)$$

The quantum yield ϕ will be

$$\phi(\text{moles C einst}^{-1}\,\text{absorbed}) = pdz/12dI_A = (p_{max}/12k_cC)[(I_{0.7})^2 + I^2]^{-\frac{1}{2}}. \quad (8)$$

The quantum yield will reach a maximum in weak light, i.e. when I becomes small compared with $I_{0.7}$; therefore,

$$\phi_{max} = \phi_{I \to 0} = p_{max}/12k_cCI_{0.7}, \quad (9)$$

or

$$p_{max} = 12\phi_{max}I_{0.7}k_cC. \quad (10)$$

An analogous derivation, for Steele's light curve (equation 2), leads to

$$p_{max} = 12\phi_{max}I_{max}k_cC/e. \quad (11)$$

Similarly, for the Vollenweider light curve (equation 3),

$$p'_{max} = 12\phi_{max}I_Kk_cC. \quad (12)$$

Substitutions of equations 10, 11, and 12 into the respective light curve equations transforms the latter into functions of the parameter ϕ_{max}.

Production equations of Steele, Vollenweider, and Fee

Vollenweider (1965) derived an equation for daily production in a totally absorbing, homogeneous water column. On the assumptions that Beer's Law

$$I = I_0 e^{-\epsilon z} \quad (13)$$

applies in the column, and that the light curve is represented by Smith's equation, Vollenweider showed that integration over depth and time leads to the production

$$\Pi(\text{g C m}^{-2}\,\text{daily}) = (p_{max}\lambda/\epsilon)\int_{-\frac{1}{2}}^{+\frac{1}{2}} \sinh^{-1}(I_0/I_{0.7})\,dt'. \quad (14)$$

This equation is hereafter identified as "Vollenweider's production equation." In this equation λ is the time in days between sunrise and sunset, and $t'\,(=t/\lambda)$ is a dimensionless time of day such that $t' = -\frac{1}{2}$ at sunrise, zero at noon, and $+\frac{1}{2}$ at sunset. To calculate the approximate production on a sunny day, Vollenweider assumed that the incident illumination I_0 varied during the day according to

$$I_0 = 0.5I_{0\,max}\,(1 + \cos 2\Pi t'). \quad (15)$$

$I_{0\,max}$, the maximum illumination at noon, varies with season and latitude.

Smith's equation applies accurately only below light saturation; it might be expected, therefore, that Vollenweider's production equation would be limited to situations where I_0 does not exceed $I_{0.7}$. However, an analysis by Vollenweider showed that, for a totally absorbing layer, equation 14 overestimates daily production by only about 15% even though photoinhibition is neglected and even if the noonday illumination $I_{0\,max}$ is 10 to 20 times larger than $I_{0.7}$. For many applications, a 15% error may be acceptable.

For a column that is not totally absorbing, photoinhibition at high light becomes more important, and it is necessary to adopt a light curve such as Steele's (1962) or Vollenweider's (1965). Steele's light curve (equation 2) is easily integrated over depth. For a homogenous, partially absorbing layer, daily production is given by

$$\Pi(\text{g C m}^{-2}\,\text{daily}) = (p_{max}\lambda e/\epsilon)\int_{-\frac{1}{2}}^{+\frac{1}{2}} (e^{-TI_0/I_{max}} - e^{-I_0/I_{max}})\,dt'. \quad (16)$$

Here $T\,(=e^{-\epsilon d})$ is the fraction of the incident illumination I_0 transmitted out of the bottom of the layer, located at $z = d$. Equation 16 is hereafter referred to as "Steele's Production equation," although Steele did not in fact write it in this form.

The Vollenweider light curve (equation 3) cannot be integrated analytically over either depth or time; an equation of daily production must therefore remain in the form of the double integral. For a homogeneous, partially absorbing layer in which Beer's Law holds, daily production is

$\text{II}(\text{g C m}^{-2} \text{ daily}) =$

$$(p'_{max}\lambda/\epsilon) \int_{-\frac{1}{2}}^{+\frac{1}{2}} \int_{TI_0}^{I_0} [(I_K)^2 + I^2]^{-\frac{1}{2}} \times$$
$$[1 + (aI/I_K)^2]^{-n/2} \, dI \, dt'. \quad (17)$$

This equation, first stated by Vollenweider and later discussed by Fee (1969), is hereafter referred to as "Fee's production equation."

Production equations in terms of ν, k_c, and k_w

Production equations 14, 16, and 17 are not explicit functions of the chlorophyll concentration, and they contain the parameters p_{max} or p'_{max} and the absorption coefficient ϵ, which are dependent on chlorophyll concentration. One way to convert these production equations to the required form is to replace the parameters p_{max} and p'_{max} with the appropriate functions of the chlorophyll concentration and the maximum assimilation number ν (g C mg^{-1} Chl day^{-1}), which is a chlorophyll-independent parameter. Also, since Beer's Law can be written as

$$I = I_0 e^{-(k_c C + k_w)z}, \quad (18)$$

it is apparent that the chlorophyll concentration-dependent coefficient ϵ can be replaced by $(k_c C + k_w)$. As stated earlier, C is the chlorophyll a concentration (mg Chl m^{-3}) due to phytoplankton in the water, and k_c is the average spectral extinction coefficient (m^2 mg^{-1} Chl) for the absorption of light by phytoplankton pigments, per unit of chlorophyll concentration. k_w is the average spectral extinction coefficient (m^{-1}) for absorption by water and its nonphytoplankton components. k_c can be regarded as a constant to a good approximation; its value appears to be about 0.016 m^2 mg^{-1} Chl (see p. 10). In contrast, k_w will vary widely in different waters depending on the content of silt, dissolved pigments, and other nonphytoplankton components.

In the production equations of Vollenweider (equation 14) and of Steele (equation 16), the parameter p_{max} is the instantaneous rate actually attained at light saturation; p_{max} may therefore be replaced

by the product of the maximum assimilation number ν and the chlorophyll concentration. This substitution, together with that for ϵ, transforms Vollenweider's equation into

$$\text{II}(\text{g C m}^{-2} \text{ daily}) = [\nu C\lambda/(k_c C + k_w)] \times$$
$$\int_{-\frac{1}{2}}^{+\frac{1}{2}} \sinh^{-1}(I_0/I_{0.7}) \, dt'. \quad (19)$$

The same substitutions transform Steele's production equation into

$$\text{II}(\text{g C m}^{-2} \text{ daily}) = [\nu C\lambda e/(k_c C + k_w d)] \times$$
$$\int_{-\frac{1}{2}}^{+\frac{1}{2}} (e^{-TI_0/I_{max}} - e^{-I_0/I_{max}}) \, dt'. \quad (20)$$

In the production equation of Fee (equation 17), the same substitution for ϵ may be made, but the replacement of p'_{max} by νC would be incorrect. The difficulty, as mentioned before, is that p'_{max} is not the maximum rate actually attained. The latter quantity Fee termed p_{opt}, and he derived a production equation in terms of p_{opt}. Since $p_{opt} = \nu C$, his transformed equation can be rewritten as

$\text{II}(\text{g C m}^{-2} \text{ daily}) =$

$$[\nu C\lambda\delta/(k_c C + k_w)] \int_{-\frac{1}{2}}^{+\frac{1}{2}} \int_{TI_0}^{I_0} [(I_K)^2 + I^2]^{-\frac{1}{2}} \times$$
$$[1 + (aI/I_K)^2]^{-n/2} \, dI \, dt'. \quad (21)$$

Here, δ is a complicated function of the parameters a and n:

$$\delta = [1 + (I_{opt}/I_K)^2]^{\frac{1}{2}} \frac{[1 + (aI_{opt}/I_K)^2]^{n/2}}{I_{opt}/I_K}, \quad (22)$$

where

$$I_{opt}/I_K = \langle (1-n)/2n$$
$$+ \{[(1-n)/2n]^2 + 1/na^2\}^{\frac{1}{2}}\rangle^{\frac{1}{2}}. \quad (23)$$

Equations 22 and 23, in which I_{opt} is the illumination at which the maximum rate p_{opt} occurs, are correct; in Fee (1969) the analogous equations erred in dropping I_K. Although the transformed production equation (equation 21) is in the form required for general theory, the need for the two auxiliary equations (22 and 23) is a nuisance. It will be seen shortly that this complication is avoided when the maximum quantum yield ϕ_{max} is chosen as parameter instead of ν.

Advantage of the parameter ϕ_{max}

The parameters p_{max}, p'_{max}, p_{opt}, and ν, and also the light saturation parameter $I_{0.7}$, I_{max}, and I_K, are functions of temperature and of the state of algal adaptation. The temperature dependence of these parameters has been demonstrated many times by both physiologists (*see* Rabinowitch 1951) and limnologists (e.g. Talling 1957a). Concerning adaptive changes, Steele (1962) pointed out that p_{max} and $I_{0.7}$ vary over an at least fourfold range in adaptation to strong or weak illumination. Myers and Graham (1971) reported an equally large variation in p_{max} and $I_{0.7}$ when the growth rate of *Chlorella* in well nourished turbidostatic cultures varied in response to illumination. A wide variation in the assimilation number also occurs; experimental values compiled by Rabinowitch (1951) indicate as much as a 50-fold range. It is evident that a single value of ν will describe with reasonable accuracy only a limited segment of the full range of variation in nature.

In contrast, available evidence indicates that the maximum quantum yield is much more nearly constant. It has long been recognized by physiologists (*see* Rabinowitch 1951; Kok 1960) that ϕ_{max} is not a function of temperature. This has also been demonstrated by limnologists; for example, Talling (1957a) showed the temperature independence of the slope of light curves in the low light range where the slope is proportional to ϕ_{max}. As yet there are insufficient data to determine the extent of change in ϕ_{max} in algal adaptation. On the one hand, the data of Myers and Graham (1971) show that ϕ_{max} did not change at all while p_{max} and $I_{0.7}$ varied fourfold. On the other hand, in cultures synchronized by light-dark cycles, ϕ_{max}, although relatively constant during the light period, does fall off about 50% during a part of the dark period (Govindjee et al. 1968; Wang 1968; Senger and Bishop 1967). Oorschot (1955) claimed an about 70% decline of ϕ_{max} in batch cultures in which growth had ceased as a result

of nitrogen exhaustion. However, the severe loss of chlorophyll under these conditions raises questions about the accuracy of the yield determination.

Because ϕ_{max} is independent of temperature and does not change in at least some cases of adaptation (when p_{max} and $I_{0.7}$ do change), it is certain that ϕ_{max} is more constant than p_{max}, ν, and $I_{0.7}$. Therefore, to the same degree of accuracy, a single value of ϕ_{max} will be able to represent a larger segment of the full range of variation in nature than can a single value of p_{max} or ν.

To write the production equations of Vollenweider (equation 14), Steele (equation 16), and Fee (equation 17) in terms of ϕ_{max}, I make use of the previously derived equations (10, 11, and 12) relating ϕ_{max} to p_{max} or p'_{max}. After substituting for p_{max} and ϵ, Vollenweider's equation becomes

$$\Pi(\text{g C m}^{-2}\text{ daily}) = \frac{12\phi_{max}I_{0.7}k_cC\lambda}{k_cC + k_w} \times$$
$$\int_{-\frac{1}{2}}^{+\frac{1}{2}} \sinh^{-1}(I_0/I_{0.7})\, dt', \qquad (24)$$

Steele's equation becomes

$$\Pi(\text{g C m}^{-2}\text{ daily}) = \frac{12\phi_{max}I_{max}k_cC\lambda}{k_cC + k_w} \times$$
$$\int_{-\frac{1}{2}}^{+\frac{1}{2}} (e^{-TI_0/I_{max}} - e^{-I_0/I_{max}})\, dt', \qquad (25)$$

and Fee's equation becomes

$$\Pi(\text{g C m}^{-2}\text{ daily}) = \frac{12\phi_{max}I_K k_cC\lambda}{k_cC + k_w} \times$$
$$\int_{-\frac{1}{2}}^{+\frac{1}{2}}\int_{TI_0}^{I_0} [(I_K)^2 + I^2]^{-\frac{1}{2}} \times$$
$$[1 + (aI/I_K)^2]^{-n/2}\, dI\, dt'. \qquad (26)$$

Note that Fee's equation can be written in terms of ϕ_{max} without the added complexity of the auxiliary equations (22 and 23).

For calculation of Π by equations 24, 25, and 26, the light saturation parameter ($I_{0.7}$, I_{max}, and I_K) before the integral sign must be expressed in einsteins m^{-2} day^{-1}; this is necessitated by ϕ_{max} being expressed as moles carbon per einstein absorbed. Since the integrals are dimensionless, the light saturation parameters and the incident illumination, appearing *after* the integral sign, can be expressed in any convenient units.

Estimated value of ϕ_{max}

Kok (1960) has reviewed the many laboratory determinations which, taken together, indicate a maximum quantum yield of about 0.10 moles oxygen evolved per einstein absorbed. This value is the approximate upper limit for the yield in low light when the wavelength of illumination and the physiological state of the organisms are both optimized. Several factors can be expected to lead to a lower time-average value for the maximum quantum yield of carbon incorporation in lakes. First, quantum yield and action spectra (Emerson and Lewis 1942, 1943; Haxo and Blinks 1950; Tanada 1951; Haxo 1960) indicate that some phytoplankton accessory pigments (e.g. carotenoids) sensitize photosynthesis less efficiently than does chlorophyll; these studies indicate that the average spectral yield is commonly 10 to 20% less than that at wavelengths absorbed solely by chlorophyll. Secondly, as already mentioned, studies of synchronous algal cultures show that the maximum quantum yield is depressed during part of the dark period (Wang 1968; Govindjee et al. 1968; Senger and Bishop 1967); asynchrony in lake phytoplankton could lead to a time-average yield in lakes 10–20% less than that characteristic of the light period. Thirdly, as Ryther (1956b) pointed out, the yield for carbon is likely to be about 85% of the yield for oxygen, as a result of nitrate reduction and the formation of some carbon compounds more reduced than carbohydrates. Taking these reductions together, it is likely that ϕ_{max} will have a time-average value in lakes of about 0.06 moles C einst^{-1} absorbed.

This estimate is admittedly a rough one, and there may be some real variation in lakes. On the other hand, with the exception of the claim of Oorschot (1955), available evidence suggests a fairly narrow range for ϕ_{max}. The value cannot surpass 0.10 moles C einst^{-1} absorbed, since the carbon yield must generally be less than the highest yield of oxygen evolution. Also, ϕ_{max} cannot fall much below 0.04 moles C einst^{-1} absorbed, at least in highly productive lakes in which daily production can exceed 5 g C m^{-2} daily. With a lower yield, Vollenweider's production equation (24) predicts a lower value of production than is actually measured, despite the fact that the equation neglects photoinhibition and therefore overestimates production. In sum, I believe that a value of ϕ_{max} of 0.06 will be correct, to within a factor smaller than 1.5, for most if not all lakes.

Upper limit to production Ψ

The form of the production equations (24, 25, and 26), expressed in terms of ϕ_{max}, has a physical significance not apparent in the original production equations of Vollenweider, Steele, and Fee. Each of the transformed equations contains the factor $k_cC/(k_cC + k_w)$; this factor is easily shown to be the fraction of the total light absorbed due to phytoplankton. [From the differential form of Beer's Law, the total light absorbed in a thin layer is $I(k_cC + k_w)\,dz$, while the light absorbed by phytoplankton only is $Ik_cC\,dz$. Then the fractional absorption by phytoplankton is $k_cC/(k_cC + k_w)$.]

The relation between daily production and an upper limit to production can be deduced by considering a mixed layer of a lake which is just barely totally absorbing [i.e. $T = \exp - (k_cC + k_w)d \simeq 0.01$], and in which absorption by water and nonphytoplankton components is a significant contribution to the total absorption [e.g. $k_cC/(k_cC + k_w) = 0.5$]. Daily production in the layer will be given by equation 24. Now, imagine that the phytoplankton concentration were to increase until the value of $k_cC/(k_cC + k_w)$ became unity, corresponding to all incident light being absorbed by phytoplankton. At this point, daily production would reach an unsurpassable upper limit. Inspection of equation 24 shows that this upper limit, defined as Ψ, is given by

$$\Psi(\text{g C m}^{-2}\text{ daily}) =$$
$$12\phi_{max}I_{0.7}\lambda\int_{-1\frac{1}{2}}^{-\frac{1}{2}}\sinh^{-1}(I_0/I_{0.7})\,dt'. \quad (27)$$

By defining Ψ in this way, equation 24 for daily production can be rewritten in a much simplified form:

$$\Pi(\text{g C m}^{-2}\ \text{daily}) = \Psi k_c C/(k_c C + k_w). \quad (28)$$

Ψ is not a function of the chlorophyll concentration, but rather only of incident illumination I_0, daylength λ, and the photosynthetic parameters ϕ_{max} and $I_{0.7}$. Actual daily production Π in the layer depends on phytoplankton concentration solely through the factor $k_c C/(k_c C + k_w)$. Analogous expressions for Ψ can be obtained from the production equations of Steele and Fee (equations 25 and 26), provided these equations are modified (by setting $T = 0$) so that they apply to a totally absorbing layer.

Ψ defined in this way is identical to the upper limit to production discussed by Ryther (1959). To calculate values of the upper limit to production, he used a graphical procedure which, though correct, was complicated and unsuited to taking account of adaptive changes in the photosynthetic parameters. In contrast, the upper limit is easily calculated by equation 27, for any parameter values.

For a partially absorbing layer $(T > 0)$, one can formally define Ψ_T as the limit to production when $k_c C/(k_c C + k_w)$ is unity. However, Ψ_T has a different meaning from that of Ψ for a totally absorbing layer. Ψ_T is the hypothetical upper limit to production in a layer the transmission of which is constant and larger than 0.01, when phytoplankton account for all, and water and nonphytoplankton none, of the light absorption. This situation is probably an unreal one and, therefore, the concept of Ψ_T as an upper limit is likely to be of little use. Furthermore, unlike Ψ, Ψ_T is dependent on the transmission T and, hence, on the chlorophyll concentration in the layer.

Ψ and the maximum daily production per unit chlorophyll

For lake layers which are totally absorbing and essentially homogeneous, experimental graphs of daily production (g C m^{-3} daily) versus depth commonly exhibit a maximum at a depth z' (m) at which the

illumination is about 10% of that at the surface (Rodhe 1965). Therefore,

$$z' = 2.3/(k_c C + k_w). \quad (29)$$

The ratio of the maximum daily production at the depth z' and the chlorophyll concentration C Megard (1972) termed the maximum specific daily production P_{max} (g C mg^{-1} Chl daily). Talling (1957b) and Rodhe (1965) showed that daily production in a homogeneous, totally absorbing column can often be quite accurately estimated as

$$\Pi(\text{g C m}^{-2}\ \text{daily}) = P_{max} C\, z'. \quad (30)$$

From equations 28, 29, and 30, it is easily shown that

$$\Psi = 2.3\, P_{max}/k_c. \quad (31)$$

This simple relation between P_{max} and the upper limit to production Ψ, in a totally absorbing layer, does not appear to have been recognized previously. The relation should be useful for estimating the upper limit attainable in a given lake. Like Ψ, P_{max} is a function of incident illumination, daylength, and the parameters ϕ_{max} and $I_{0.7}$.

Final remarks

In simplest terms, what is accomplished in the foregoing is to show how the best of the published equations for daily production can be written as explicit functions of chlorophyll concentration and how the widely variable parameter p_{max} (or p'_{max}) can be replaced by the more constant parameter ϕ_{max}. The transformation is simple, and it is surprising that the transformed equations have not been presented previously. It is also surprising that the simple relation between daily production and the upper limit to production, and also between the upper limit and the maximum daily production per unit of chlorophyll, have not been pointed out heretofore. Apparently, neither the relation between ϕ_{max} and p_{max}, nor the relative constancy of ϕ_{max}, nor the concept of fractional absorption has been generally appreciated. Or, perhaps, the possibility of mathematical simplification of a later dynamic theory has been a less pressing concern, up to now, than that the forms of production equations be directly applicable to commonly available field data.

The transformed production equations are not as easily applied to field data as the original production equations. The reason is that p_{max} and ϵ are more commonly and easily measured than ϕ_{max}, k_c, and k_w. Also the transformed equations require measurement of the chlorophyll concentration. However, if p_{max} (or p'_{max}) and $I_{0.7}$ (or I_{max} or I_K) are available (as required by the original production equations), and since C is easily measured, ϕ_{max} could be calculated if only k_c were available. It is really the difficulty of determining k_c and k_w that makes the transformed equations more difficult to use.

The problem of measuring k_c and k_w in a lake is not yet satisfactorily resolved. The problem is exacerbated by the fact that, in production equations, k_c must refer specifically to light absorption by the pigments of live, photosynthetically competent phytoplankton, while k_w must refer to light absorption not only by water and its nonphytoplankton components, but also by any pigments of dead and decaying algae. Recent efforts to distinguish between chlorophyll and pheophytin have an obvious bearing in this regard (Lorenzen 1967; Yentsch 1965; Moss 1967). One approach to the problem is to accept arguments (which I will state elsewhere) that k_c is approximately constant and that its value is close to 0.016 m^2 mg^{-1} Chl. [This is the approximate mean of the estimates, ranging from 0.013 to 0.020, of Megard (1972), Talling (1970), M. Lorenzen (1972), and C. J. Lorenzen (1972).] If the mean value is accepted, and if values of ϵ and C are measured, then k_w is calculable. Another approach was that of Megard (1972) who estimated k_c and k_w from production measurements in areas of varying chlorophyll concentration in Lake Minnetonka. His method required a large number of measurements and assumed that the values of k_c and k_w were the same in all areas. The lack of good methods for k_c and k_w will understandably discourage use of the transformed production equations in some kinds of fieldwork. On the other hand,

a dynamic theory relating the properties of a phytoplankton crop to environmental characteristics must take into account that daily production is fundamentally dependent on the fractional absorption by phytoplankton—hence on k_c and k_w. As seen by a phytoplankton population, the light absorption coefficient k_w is an environmental property no less important than incident illumination and nutrient supply. Thus, for theory, the transformed equations are required.

Finally, it may be noted that the transformed production equations retain some of the weaknesses of the equations from which they were derived. First, the equations apply to a homogeneous water column. In principle, this is not a serious deficiency since vertical inhomogeneity can be treated by subdividing the column into shallow, essentially homogeneous layers, and by applying the production equations (with appropriate new limits for the depth integral) to each layer separately. This possibility was realized by Ryther (1956a) and was recently restated by Fee (1969). Secondly, constancy of chlorophyll concentration over a day has been implicitly assumed. In fact, during periods of rapid net growth, and even in a steady state determined by balanced production and losses over a 24-hr period, chlorophyll concentration may vary appreciably during daylight hours (Yentsch and Ryther 1957; Shimada 1958). One way to overcome this difficulty might be to redefine C in the production equations as the time-average chlorophyll concentration during daylight hours, possibly weighted in favor of the hours when production is greatest. Thirdly, the often encountered asymmetry of production in morning and afternoon (Doty and Oguri 1957; Lorenzen 1963; Newhouse et al. 1967; Vollenweider 1965; Tilzer 1973), possibly related to synchrony or to nutrient depletion, is not taken into account.

References

DOTY, M. S., AND M. OGURI. 1957. Evidence for a photosynthetic daily periodicity. Limnol. Oceanogr. **2**: 37–40.
EMERSON, R. E., AND C. M. LEWIS. 1942. The

photosynthetic efficiency of phycocyanin in *Chroococcus*, and the problem of carotenoid participation in photosynthesis. J. Gen. Physiol. **25**: 579–595.

———, AND ———. 1943. The dependence of the quantum yield of *Chlorella* photosynthesis on wave length of light. Amer. J. Bot. **30**: 165–178.

FEE, E. J. 1969. A numerical model for the estimation of photosynthetic production, integrated over time and depth, in natural waters. Limnol. Oceanogr. **14**: 906–911.

GOVINDJEE, R., E. RABINOWITCH, AND GOVINDJEE. 1968. Maximum quantum yield and action spectrum of photosynthesis and fluorescence in *Chlorella*. Biochim. Biophys. Acta **162**: 539–544.

HAXO, F. T. 1960. The wavelength dependence of photosynthesis and the role of accessory pigments, p. 339–360. *In* M. B. Allen [ed.], Comparative biochemistry of photoreactive systems. Academic.

———, AND L. R. BLINKS. 1950. Photosynthetic action spectra of marine algae. J. Gen. Physiol. **33**: 389–422.

KOK, B. 1960. Efficiency of photosynthesis, p. 563–633. *In* W. Ruhland [ed.], Handbuch der Pflanzenphysiologie, v. 5, Part 1. Springer.

LORENZEN, C. J. 1963. Diurnal variation in photosynthetic activity of natural phytoplankton populations. Limnol. Oceanogr. **8**: 56–62.

———. 1967. Determination of chlorophyll and pheo-pigments: spectrophotometric equations. Limnol. Oceanogr. **12**: 343–346.

———. 1972. Extinction of light in the ocean by phytoplankton. J. Cons., Cons. Int. Explor. Mer **34**: 262–267.

LORENZEN, M. 1972. The role of artificial mixing in eutrophication control. Ph.D. thesis, Harvard Univ.

MEGARD, R. O. 1972. Phytoplankton, photosynthesis, and phosphorus in Lake Minnetonka, Minnesota. Limnol. Oceanogr. **17**: 68–87.

MOSS, B. · 1967. A spectrophotometric method for the estimation of percentage degradation of chlorophylls to pheo-pigments in extracts of algae. Limnol. Oceanogr. **12**: 335–340.

MURPHY, G. I. 1962. Effect of mixing depth and turbidity on the productivity of fresh-water impoundments. Trans. Amer. Fish. Soc. **91**: 69–76.

MYERS, J., AND J. GRAHAM. 1971. The photosynthetic unit of *Chlorella* measured by repetitive short flashes. Plant Physiol. **48**: 282–286.

NEWHOUSE, J., M. S. DOTY, AND R. T. TSUDA. 1967. Some diurnal features of a neritic surface plankton population. Limnol. Oceanogr. **12**: 207–212.

OORSCHOT, J. L. P. VAN. 1955. Conversion of light energy in algal culture. Meded. Landbouwhogesch. Wageningen **55**: 225–276.

PATTEN, B. C. 1968. Mathematical models of plankton production. Int. Rev. Gesamten Hydrobiol. **53**: 357–408.

RABINOWITCH, E. 1951. Photosynthesis and related processes, v. 2, Part 1, p. 964–1006. Interscience.

RILEY, G. A. 1946. Factors controlling phytoplankton populations on George's Bank. J. Mar. Res. **6**: 54–73.

RODHE, W. 1965. Standard correlations between pelagic photosynthesis and light, p. 365–381. *In* C. R. Goldman [ed.], Primary productivity in aquatic environments. Mem. Ist. Ital. Idrobiol. **18** (suppl.), *also* Univ. Calif. 1966.

RYTHER, J. H. 1956a. Photosynthesis in the ocean as a function of light intensity. Limnol. Oceanogr. **1**: 61–70.

———. 1956b. The measurement of primary production. Limnol. Oceanogr. **1**: 72–84.

———. 1959. Potential productivity of the sea. Science **130**: 602–608.

———, AND C. S. YENTSCH. 1957. The estimation of phytoplankton production in the ocean from chlorophyll and light data. Limnol. Oceanogr. **2**: 281–286.

SENGER, H., AND N. I. BISHOP. 1967. Quantum yield of photosynthesis in synchronous *Scenedesmus* cultures. Nature **214**: 140–142.

SHIMADA, B. M. 1958. Diurnal fluctuations in photosynthetic rate and chlorophyll content of phytoplankton from eastern Pacific waters. Limnol. Oceanogr. **3**: 336–339.

SMITH, E. L. 1936. Photosynthesis in relation to light and carbon dioxide. Proc. Nat. Acad. Sci. U.S. **22**: 504–511.

STEELE, J. H. 1962. Environmental control of photosynthesis in the sea. Limnol. Oceanogr. **7**: 137–150.

———, AND D. W. MENZEL. 1962. Conditions for maximum primary production in the mixed layer. Deep-Sea Res. **9**: 39–49.

TALLING, J. F. 1957a. Photosynthetic characteristics of some freshwater plankton diatoms in relation to underwater radiation. New Phytol. **56**: 29–50.

———. 1957b. The phytoplankton population as a compound photosynthetic system. New Phytol. **56**: 133–149.

———. 1970. Generalized and specialized features of phytoplankton as a form of photosynthetic cover, p. 431–435. *In* Prediction and measurement of photosynthetic productivity. Proc. IBP/PP Tech. Meeting, Třeboň (Czechoslovakia). Centre Agr. Publ. Doc., Wageningen.

TAMIYA, H., AND OTHERS. 1953. Kinetics of growth of *Chlorella*, with special reference to its dependence on quantity of available light and on temperature, p. 204–232. *In* J. S. Burlew [ed.], Algal culture from laboratory to pilot plant. Carnegie Inst. Wash. Publ. 600.

TANADA, T. 1951. The photosynthetic efficiency of carotenoid pigments in *Navicula minima*. Amer. J. Bot. **38**: 276–283.

TILZER, M. M. 1973. Diurnal periodicity in the phytoplankton assemblage of a high mountain lake. Limnol. Oceanogr. **18**: 15–30.

VOLLENWEIDER, R. A. 1965. Calculation models of photosynthesis-depth curves and some implications regarding day rate estimates in primary production measurements, p. 425–457. *In* C. R. Goldman [ed.], Primary productivity in aquatic environments. Mem. Ist. Ital. Idrobiol. **18** (suppl.), *also* Univ. Calif. 1966.

WANG, R. 1968. Dependence of photosynthesis on life cycle stage in *Chlorella pyrenoidosa*. Ph.D. thesis, Univ. Rochester.

WITHROW, R. B., AND A. P. WITHROW. 1956. Generation, control, and measurement of visible and near-visible radiant energy, p. 125–258. *In* A. Hollaender [ed.], Radiation biology, v. 3. McGraw-Hill.

YENTSCH, C. S. 1965. Distribution of chlorophyll and phaeophytin in the open ocean. Deep-Sea Res. **12**: 653–666.

———, AND J. H. RYTHER. 1957. Short-term variations in phytoplankton chlorophyll and their significance. Limnol. Oceanogr. **2**: 140–142.

Submitted: 12 January 1973
Accepted: 26 October 1973

AUTHOR'S NOTE

J. F. Talling has kindly pointed out errors in the prose on page 9 under the heading "Ψ and the maximum daily production per unit chlorophyll." Correction is achieved with the following.

Talling (1975) showed that the depth integral of the instantaneous rate p (g C m^{-3} day^{-1}) of production is approximately given by

$$\int_0^d p \ dz \simeq p_{max} \ z'$$

where p_{max} is the maximum rate, usually occurring where illumination is 30 percent to 70 percent of that at the surface, and z' is the depth where illumination is about 10 percent that at the surface. Rodhe (1965) and Megard (1972) showed that daily production π is approximately given by an analogous equation

$$\pi(\text{g C m}^{-2} \text{ daily}) = \int \int_0^d p \ dz \ dt \simeq \mathscr{P}_{max} \ z'$$

where $\mathscr{P}_{max}$ (g C m^{-3} daily) is the maximum daily production. Megard (1972) introduced the quantity P_{max}, the ratio of $\mathscr{P}_{max}$ and the concentration C (mg Chl a m^{-3}) of chlorophyll, whence

$$\pi \simeq P_{max} \ C \ z'. \tag{30}$$

Since z' is the depth for 10 percent of surface illumination (Rodhe 1965),

$$z' = \frac{2.3}{(k_c C + k_w).} \tag{29}$$

From equations 28, 29, and 30, it is easily shown that, for a well-mixed, totally absorbing layer,

$$\Psi = \frac{2.3 \ P_{max}}{k_c.} \tag{31}$$

My first sentence included errors, but the equations and the conclusions of the final sentences I believe to be correct. It may be added that the 10 percent transmission at depth z′ refers to the most penetrating spectral component (Rodhe 1965). The average spectral transmission would be somewhat less and the factor of 2.3 in equation 29 should then be somewhat larger. According to Talling, J. F. 1971 The underwater light climate as a controlling factor in the production ecology of freshwater photoplankton. *Mitt. Int. Verein. Limnol. (Int. Assoc. Theor. Appl. Limnol. Commun.)* **19**:214–43.

References

Megard, R. O. 1972. Phytoplankton photosynthesis, and phosphorus in Lake Minnetonka, Minnesota. *Limnol. Oceanogr.* **17**:68–87.

Rodhe, W. 1965. Standard correlations between pelagic photosynthesis and light. *In* C. R. Goldman, ed., *Primary Productivity in Aquatic Environments*, pp. 365–381. Mem. Ist. Ital. Idrobiol. **18** (suppl), *also* Univ. Calif., 1966.

Talling, J. F. 1957. The phytoplankton population as a compound photosynthetic system. *New Phytol.* **56**:133–149.

Talling, J. F. 1971. The underwater light climate as a controlling factor in the production ecology of freshwater photoplankton. *Mitt. Int. Verein. Limnol. (Int. Assoc. Theor. Appl. Limnol. Commun.)* **19**:214–43.

Reprinted from *Plant. Physiol.* **36**(3):342–346 (1961)

ON THE MASS CULTURE OF ALGAE. III. LIGHT DIFFUSERS; HIGH VS LOW TEMPERATURE CHLORELLAS [1,2]

JACK MYERS & JO-RUTH GRAHAM

DEPARTMENTS OF BOTANY & ZOOLOGY, UNIVERSITY OF TEXAS, AUSTIN

The present work is introduced by a previous report (4). The yield of algae under sunlight irradiance is limited by the characteristic of light-saturation. The Chlorellas and all other algae studied to date become light-saturated at values of irradiance far less than that of full sunlight. The resulting efficiency of light utilization is considerably lower than that achievable under low incident irradiance. For mass culture under sunlight illumination attempts to increase yield become attempts to minimize or circumvent the limitations due to light-saturation. The present work is concerned with two possible approaches: A, the use of a cone diffuser as a physical means of light attenuation and B, the use of an alga with a higher temperature optimum and a presumed higher irradiance for light-saturation.

METHODS

We have chosen to study yield (mg/day) in cultures of Chlorella maintained under controllable conditions in the laboratory. Most of the methods used have been described in detail (4). Continuous light from a tungsten lamp, filtered through water and copper sulfate, was presented to the open horizontal surface of a culture as an approximately collimated beam of irradiance equivalent to full sunlight. The culture vessel was a flat-bottomed glass cylinder of 67 mm ID containing 1,010 ml of algal suspension at a normal working depth of 267 mm. It was aerated with 5 % CO_2 in air and stirred just sufficiently to prevent cell sedimentation. Further details of the culture vessel and subsequent modifications are presented in figure 1.

The information sought required comparison of cultures, or series of cultures, each illuminated at approximately the same incident irradiance but differing in terms of one defined variable. Each culture was operated under a chemostat (6) system of a constant rate of dilution with fresh medium and an equal rate of withdrawal of algal suspension. Such management provided a steady-state system similar to that which would be a method of choice in any practical large-scale culture. Unfortunately, the low specific growth rates of algae make the chemostat system sluggish in reaching a steady-state and in practice each culture had to be maintained for 7 to 12 days.

DIFFUSING CONE

One attack upon the problem of light-saturation is to find some physical method of attenuating the high irradiance of sunlight without energy loss, i.e., by spreading it out over a greater area as viewed by the cells of the culture. Of the several methods which have been suggested (3, 9) the most attractive is the use of diffusing cones held base-up and projecting into a deep culture. The culture vessel was designed to compare yields in a culture with and without a diffusing cone. In initial control experiments without use of the cone it became apparent that yield was dependent upon cell concentration. The preceding paper (4) describes yield as a function of cell concentration and therefore provides the control or base line for study of effects of adding the diffusing cone.

Our first cone was turned from a solid lucite rod, leaving a very finely turned thread to give a diffusing surface. The diameter of the base was 50 mm and the height of the conical portion was 244 mm. The cone was held in an opaque diaphragm of 50 mm diameter. The diffusing conical surface for light output had an area of 192 cm² as compared to a base input area of 19.6 cm². The intent was to achieve a 10 : 1 attenuation of input irradiance. We attempted to obtain a profile of the output irradiance over the surface of the cone but were able to do this only in a relative fashion. When immersed in water and illuminated through the base most of the light emerged from the lower third of the cone.

Since an algal suspension is itself a diffusing layer it appeared that an inverted cone might give similar results even with a clear surface. Accordingly, our second model was a hollow glass cone. It was held under an opaque diaphragm of 47 mm diameter in order to restrict the input beam to the ID of the top opening. The OD was 50 mm and the height of the conical portion was 227 mm. The light input area was 17.35 cm² and the output area 178 cm².

The arrangement of the cones in the culture chamber is shown in figure 1. By use of the cones the culture volume was decreased by about 150 ml from the original volume of 1,010 ml.

[1] Received December 19, 1960

[2] This work was supported by a grant from the Rockefeller Foundation.

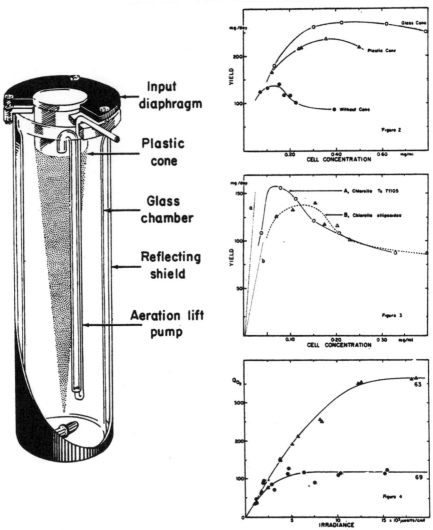

FIG. 1 (*left*). The growth chamber with plastic cone in place. The cutaway drawing shows details of the cone, the top input diaphragm, the glass chamber, and the surrounding reflecting shield. The inserted glass tubes made a lift pump delivering aerated suspension to the top of the culture with minimum splashing.

FIG. 2 (*top right*). Effects of the plastic and glass cones. The lower curve, showing performance without the cone, is taken from the preceding paper (4) with data plotted only for the last seven experiments. *Chlorella ellipsoidea* at 25° C.

FIG. 3 (*center right*). Comparison of yields without a cone for Chlorella Tx71105 at 35° C (A) and *Chlorella ellipsoidea* at 25° C (B). The dotted curves (a) and (b) represent the respective limits imposed by maximum specific growth rates taken as 6.4 days⁻¹ for Chlorella Tx71105 (a) and 3.0 days⁻¹ for *Chlorella ellipsoidea* (b).

FIG. 4 (*bottom right*). Rate of photosynthesis, Qo_2 in μl/mg-hr, vs. irradiance as measured in Warburg buffer 0.190 M $KHCO_3$ plus 0.010 M Na_2CO_3 for cells from cultures 63 and 69 of Chlorella Tx71105 at 35° C.

TABLE I
SUMMARY DATA

EXPT.	IRRADIANCE kcal/day	CELL CONC mg/l	YIELD mg/day	EFFICIENCY* %	SPECIFIC GROWTH RATE** day^{-1}	RATE RESPIRATION μl O_2/mg-hr	CHLOROPHYLL CONTENT %
Chlorella ellipsoidea at 25° C; plastic cone; input area 19.64 cm²; volume 850 ml							
51	14.5	0.122	166	6.2	1.6	11.6	2.7
52	14.2	0.239	217	8.3	1.1	7.7	4.4
55	13.8	0.251	218	8.6	1.0	5.8	3.7
54	14.0	0.357	236	9.1	0.78	3.9	4.2
53	13.8	0.494	219	8.6	0.52	...	4.8
Chlorella ellipsoidea at 25° C; glass cone; input area 17.35 cm²; volume 856 ml							
57	12.9	0.137	180	7.5	1.5	9.4	2.7
56	13.3	0.305	262	10.6	1.0	4.5	4.1
59	13.2	0.424	270	11.1	0.74	3.6	4.3
58	13.3	0.620	266	10.8	0.50	2.7	4.9
60	12.8	0.784	251	10.6	0.37	1.8	4.6
Chlorella Tx71105 at 35° C; without cone; input area 19.64 cm²; volume 1010 ml							
65	13.7	0.038	108	4.3	2.9	...	...
70	13.6	0.080	156	6.2	2.0	18.2	...
66	14.0	0.112	144	5.6	1.3	16.7	...
63	13.8	0.152	121	4.7	0.79	12.1	...
67	14.4	0.207	108	4.1	0.53	8.8	...
69	14.2	0.330	87	3.3	0.27	7.4	...

* Yield $\times$ 0.0054/irradiance; heat of combustion of cells produced taken as 5.4 cal/mg.
** Yield/vol. $\times$ cell concentration.

Summarized data obtained with the cones are presented in table I [cf. (4), table II] and the yields compared to those previously obtained without the cone are shown in figure 2. Two advantageous effects of the cones are clearly evident. First, maximum yield or efficiency in utilization of input light is increased about twofold. Second, the optimum cell concentration is displaced toward higher values at which harvesting becomes more economical in terms of volume of suspension to be processed.

USE OF HIGHER TEMPERATURE ALGA. All of the experiments reported above were done with *Chlorella ellipsoidea* Gerneck at 25° C, which is close to the optimum temperature. A second series of experiments was designed to make a comparative evaluation of an alga of higher temperature tolerance. For this purpose we chose Chlorella Tx71105 (8) maintained at a temperature of 35° C. Culture management in all other respects was similar to that used in the preceding report (4) and without introduction of a cone.

The results summarized in table I and figure 3 provide comparison of performance of the two strains at temperatures close to their respective optima. Chlorella Tx71105 at 35° C gave a maximum yield about 15 % higher and achieved at lower cell concentration. Auxiliary measurements of characteristics of the cells and the cultures of Chlorella Tx71105 were attempted but were less satisfactory than those previously reported for *Chlorella ellipsoidea*. Our lead

sulfide cell was more difficult to use at the higher temperature and the characteristic of light penetration into the culture could not be determined satisfactorily. Chlorophyll analyses were attempted but proved unsatisfactory for this strain because of incompleteness of extraction by boiling methanol.

Irradiance curves for the harvested cells were measured in a light beam of the same spectral character used for the cultures. The single light beam available required use of the Warburg carbonate-bicarbonate buffers and repeated measurements on a single vessel at randomly ordered values of irradiance. In spite of lack of precision, which reflects known limitations in the procedure, the irradiance curves followed the same trend and demonstrated the same phenomenon previously observed with *Chlorella ellipsoidea*. Two curves presented in figure 4 were chosen from the set because of their greater reliability; each curve contains data from duplicate experiments on different days. The curve for culture 69, describing cells grown at high cell concentration and low average irradiance per cell, shows a low irradiance for light-saturation and a low maximum rate of photosynthesis.

DISCUSSION

The experimental conditions were purposely chosen to study the special or limiting case of an algal culture under continuous sunlight irradiance. Evaluation in terms of practical application requires additional consideration of limitations in the experimental conditions used.

DIFFUSING CONE: The results demonstrate a twofold increase in yield and efficiency of input light utilization by a diffusing cone of arbitrarily chosen geometry. As applied to the practical case of a large culture under diurnal solar illumination there are opposing factors which would enhance or reduce the observed gain in yield. Obviously the gain will be reduced under diurnal sunlight since the most serious effects of light-saturation occur only during a portion of the day. On the other hand a large culture, having its surface covered with close-packed cones, would have little loss of light from the sides of the culture and could take more complete advantage of the cone principle. We have noted previously (4) the consequences of the unmeasured light losses which occurred from the sides of our culture vessel (edge effects). These are considered significant even though minimized by a surrounding reflecting shield. We failed in attempts to obtain a quantitative measure of these losses and can only record that they appeared considerably larger with the cone than without it.

The term "diffusing cone" is intended in a generic sense to describe a static optical device for distributing the incident surface irradiance over a larger area within a deep culture. The choice of optimum geometry has not been examined. For a large culture the use of pyramids rather than cones would allow closer packing and complete use of the surface. And it is not clear that either pyramids or cones should have straight sides for optimum effects. For attainment of maximum yield the intent is only that the incident light be so distributed that no cell of a culture is ever light-saturated. Visual inspection of our experimental cones showed that the cone surface did not radiate uniformly. Hence the desired 10:1 attenuation was not achieved and some areas of light-saturation did occur.

In short, the observed twofold gain in yield by use of the cone was obtained in spite of increased edge losses and in spite of lack of attainment of uniform light distribution. It would appear that practical test of the diffusion cone principle is merited for mass cultures under diurnal sunlight illumination. For the proposed use of algal cultures as gas exchangers in space vessels the diffusing cone principle should be an even greater advantage.

HIGH-TEMPERATURE STRAIN: The small gain in optimum yield observed with Chlorella Tx71105 at 35° C as compared with *Chlorella ellipsoidea* at 25° C is considerably less than might have been expected. Furthermore, even this small advantage was obtained only at low values of cell concentration. These findings suggest re-examination of the premises made for use of high temperature algal strains in mass culture.

In early attempts at mass culture the algae chosen were *Chlorella pyrenoidosa, Chlorella ellipsoidea,* and species of Scenedesmus, all with temperature optima at about 25° C (low-temperature strains). The first isolation by Sorokin (8) of a Chlorella with a temperature optimum as high as 39° C provided a new and important alga for mass culture. Subsequently a variety of strains of Chlorella, Scenedesmus, and other genera with temperature optima in the range of 35 to 40° (high-temperature strains) have been isolated and used in various laboratories (9). The higher temperature optima presented an important practical advantage in minimizing the requirement of cooling of cultures under sunlight illumination. Of this advantage no question can be raised. It was envisioned also that the higher rates of metabolism and growth of the high-temperature strains would allow greater yields in mass culture. The two possible bases for such an expectation should be re-examined.

As compared to the low-temperature strains, the high-temperature strains have maximum specific growth rates about threefold higher. Specific growth rate, k, is defined as dN/N dt where N is a measure of cell quantity and t is time. For purposes of cell production in mass culture the criterion of performance is production rate or yield which for a given culture is dN/dt or kN. In general yield will be referred to some measure of culture size such as illuminated area giving dimensions such as g/m^2-day. However, for the present work the culture geometry and illumination were held constant and comparisons of yield can be made in terms of cell mass produced per day.

From experience with other microorganisms it might be reasoned that the higher maximum specific growth rates of the high-temperature strains should result in higher yield. However, the specific growth rate of an alga is governed also by the effective irradiance per cell and its maximum value is attained only at light-saturation. All the cells of a culture cannot be maintained at light-saturation without considerable transmission and loss of light. Furthermore those cells which are maintained at light-saturation cannot be working at maximum efficiency. All experimental work has confirmed the early observation of Ketchum et al (1) that "the theory of the optimum catch" applies to algal cultures, that maximum yield is attained under conditions at which the cells are growing at less than their maximum specific growth rate. Maximum yield is to be expected at a cell concentration such that most of the incident light is absorbed. It follows that yield is governed by efficiency of light utilization rather than by specific growth rate.

A second reason to anticipate advantage of the high-temperature strains under sunlight illumination lies in an expected higher point of light-saturation. For example, suppose that the point of light-saturation for the high-temperature strains lies at about 1,200 ft-c as compared to about 400 ft-c for the low-temperature strains. Then under the high illuminance of sunlight the losses attributable to light-saturation will be lower for the high-temperature strains. The basis of this argument has been presented in the preceding paper (4).

Unfortunately the expectation of a consistently high point of light-saturation for Chlorella Tx71105 is denied by the experimental findings. The illuminance curves obtained by Sorokin (7) make it clear that cells grown at 39° C and 60 ft-c reach saturation at an illuminance about one-fourth as great as for cells grown at 400 ft-c. The data of figure 4, though incomplete and far less precise, confirm Sorokin's observations. The same effects are observable, even under an incident irradiance of full sunlight, when cell concentrations are so high as to give a low average irradiance per cell. It is now abundantly clear that a fixed irradiance curve of photosynthesis is not an intrinsic character of an algal species (2, 4, 5, 8).

The data of figure 3 show for Chlorella Tx71105 at 35° C a maximum yield only about 15 % higher than for *Chlorella ellipsoidea* at 25° C. This is a significant advantage but not one which would by itself dictate the choice of alga for mass culture. No greater advantage in yield under steady-state conditions has yet been shown for any high-temperature strain under any chosen conditions of illumination.

SUMMARY

Yields (mg/day) of two strains of Chlorella were studied in steady-state cultures under a continuous visible irradiance equivalent to full sunlight. For *Chlorella ellipsoidea* at 25° C the maximum production rate was increased about twofold by use of plastic or glass diffusing cones designed to supply the incident energy over a greater surface at reduced irradiance. The diffusing cone principle appears to be a practicable means of circumventing the limitations of light-saturation under sunlight illumination. In cultures without the cones Chlorella Tx71105 at 35° C showed a maximum production rate only about 15 % higher than that obtained with *Chlorella ellipsoidea* at 25° C. The bases for earlier expectation of much higher production rates obtainable with high-temperature strains of algae have been re-examined.

LITERATURE CITED

1. KETCHUM, B. H., LOIS LILLICK, & A. C. REDFIELD. 1949. The growth & optimum yields of unicellular algae in mass culture. J. Cell. Comp. Physiol. 33: 267–280.
2. MYERS, J. 1946. Influence of light intensity on the photosynthetic characteristics of Chlorella. J. Gen. Physiol. 29: 429–440.
3. MYERS, J. 1958. Algal growth: Processes & products. Transactions of the Conference on the Use of Solar Energy, E. F. Carpenter, ed. Vol. iv. Univ. of Arizona Press, Tucson. Pp. 1–11.
4. MYERS, J. & JO-RUTH GRAHAM. 1959. On the mass culture of algae. II. Yield as a function of cell concentration under continuous sunlight irradiance. Plant Physiol. 34: 345–352.
5. MYERS, J. & W. A. KRATZ. 1955. Relations between pigment content & photosynthetic characteristics in a blue-green alga. J. Gen. Physiol. 39: 11–22.
6. NOVICK, A. & L. SZILARD. 1950. Description of the chemostat. Science 112: 715–716.
7. SOROKIN, C. 1958. The effect of past history of cells of Chlorella on their photosynthetic capacity. Physiol. Plantarum 11: 275–283.
8. SOROKIN, C. & J. MYERS. 1953. A high-temperature strain of Chlorella. Science 117: 330–331.
9. TAMIYA, H. 1957. Mass culture of algae. Ann. Rev. Plant Physiol. 8: 309–334.

30

Reprinted from pp. 5–11 of *Zool. Soc. London Symp. No.* **19**:1–14 (1967)

THE EFFECT OF ARTIFICIAL LIGHTS ON ACOUSTIC SCATTERING LAYERS IN THE OCEAN

J. H. S. BLAXTER

Natural History Department, Marischal College, University of Aberdeen, Scotland

and

R. I. CURRIE

*National Institute of Oceanography, Wormley, Godalming, Surrey, England**

* Present address: The Marine Station, Millport, Isle of Cumbrae, Scotland.

[*Editor's Note:* In the original, material precedes this excerpt.]

Night

Simple initial experiments (nos. 1–4; see Appendix)

On two nights the 1 500 W diver's lamp was lowered into the top of a scattering layer and switched on. Within 2 min the top of the trace had apparently moved down to a depth 20 m below the light. On switching off, the status quo was restored within about 5 min (see Fig. 4). This sequence was repeated a number of times and gave the first clear indication that scattering layers would respond to underwater lights.

Next the 1 000 W free-flooding bulb was used with, and then without, its reflector at depths ranging from 50 to 150 m. With the reflector the reaction of the layer was negligible, presumably due to the much reduced sphere of influence. Without the reflector the results were more striking. Switched on at 50 m, for example, within a rather homogeneous trace, the light caused a clear area to appear on the recorder between 40 and 60 m. An attempt was made to force the trace much deeper by lowering the light to 150 m in 10 m stages over a period of 10 min. This proved unsuccessful though a slight lightening of the trace occurred from 100 to 180 m when the light was left on at 150 m. On switching off, the trace thickened again.

The effect of the intensity of the source

To assess this the light source was kept at a fixed depth and its intensity varied by changing the voltage with a variable transformer. The voltage was lowered only in so far as it did not appear (to the human eye) to reduce the colour temperature of the filament noticeably.

In the first experiment (no. 5) the 1 000 W free-flooding bulb was used at 55 m and then at 100 m and its intensity reduced in stages from a maximum to about 1/10 maximum. The intensity was monitored by an underwater light meter (Craig and Lawrie, 1962) (on loan from the Marine Laboratory, Aberdeen) fixed 20 m below the source.

Some idea of the irradiance produced by a 1 000 W underwater light can be given partly on the basis of actual measurements and partly on computed values (Fig. 2), the latter being derived in the following manner. If we simplify the problem and neglect the change in colour composition with distance from the source—resulting from selective absorption by the water—and further consider only the mono-path or non-scattered component E^0 of the irradiance, then the relation

$$E^0{}_x = \frac{I\,e^{-cx}}{x^2}$$

where I is the radiant intensity of the source in the direction considered

359

and c is the attenuation coefficient, can be used as an approximation to the irradiance produced at a distance x from the source. In this equation c is strictly speaking the attenuation coefficient for collimated light (Duntley, 1960), but a good enough approximation for present purposes can be made by using the value obtained for daylight attenuation, which averaged 0·05 for blue light in the upper layers of the area concerned.

The distance of the scattering layer from the light source when different voltages were used is plotted in Fig. 3A and the calculated intensity at the level of the layer in each instance given in Fig. 3B. These figures show how the layer moved away from the source as its power

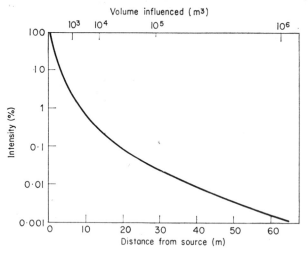

FIG. 2. Calculation of attenuation of light intensity from an underwater source given as a percentage with distance (lower abscissae), showing also the volume of water influenced (upper abscissae). Natural light is ignored; $c = 0·05$.

increased, apparently at the same time moving into a lower ambient light intensity.

In the second experiment (no. 9 (a)) the 1 000 W diver's lamp was used at 50 m and the effect of varying the voltage assessed in the same way. The light conditions in the region around the source were estimated by calculation as before but checked this time by monitor-readings from another underwater irradiance meter (provided by Dr. Boden and Dr. Kampa) about 50 m below the source.

An examination of the record showed a number of components, including a very insensitive one and one that made considerable movements in response to changes in voltage. The plots for this latter layer are also given in Fig. 3A and B as distance from the source against power of the source and as light intensity at the level of the trace in each part of the experiment. A similar result to the earlier experiment was obtained, though the range of movement and variation in light intensity at the depth of the trace were somewhat different.

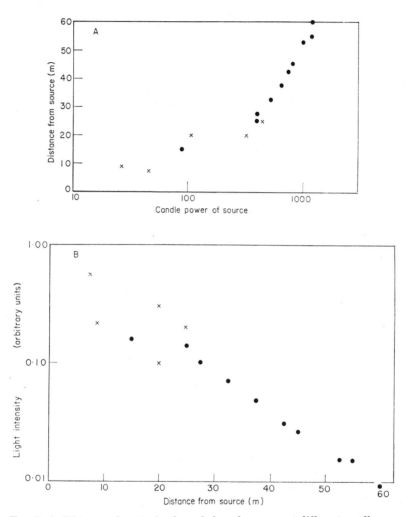

FIG. 3. A. Distance of scattering layer below the source at different candle powers of the source; ×, expt. 5; ●, expt. 9a. B. Light intensity at the depth of the layer during different phases of expts. 5 (×) and 9a(●). The arbitrary units of light intensity are equivalent to metre candles using the Aberdeen light meter (Craig and Lawrie, 1962). One arbitrary unit is also roughly equivalent to 8 μW/cm² using the Boden and Kampa photometer in the "open" position, i.e. with the full response of a S.4 RCA 931–A multiplier phototube.

Forcing down the layer by night

Attempts to do this were included in a number of experiments, the most successful being no. 6. Here the 1 000 W free-flooding bulb was switched on at the top of the layer by night at 55 m and then lowered in stages of about 50 m to a final depth of 250 m during a period of 40 min. The record from the 67 kc/s echo-sounder (Fig. 5) shows that the layer was apparently forced down from 60 to 300 m during this time.

Dusk

On four occasions lights were successfully used to delay scattering layers from making their dusk ascent. In the first instance (no. 7) the 1 000 W free-flooding bulb was switched on near dusk at 350 m just as one of the deeper layers had started to rise. There was an immediate disappearance of the trace which seemed to re-form at 410 m, remaining there until the filament of the bulb burned out. Later in the cruise (no. 10) a similar experiment was done. The light, again at a depth of 350 m, was switched on as the middle of the layer came up to it, diffusing the layer so that it disappeared within about 2 min. There is a suggestion in the record (see Fig. 6) that some components of the layer may have continued their upward movement but for the most part it faded out. On switching off the light some 8 min later, the layer appeared to re-form and continue its upward movement. The light was then rapidly raised to a depth of 108 m where it was switched on again. Once more an effect was produced and the upward movement of the layer arrested. After $11\frac{1}{2}$ min the light was extinguished and immediately the layers resumed their rise to the surface.

The diver's lamp was also used twice on traces at 100 m at dusk (nos. 8 and 9b). A comparison of one experiment with its control (Fig. 7) shows that the ascent was retarded by 20 min, until the normal ambient conditions were restored by switching off the light.

Day

The daytime experiments were attended by a series of frustrating failures. To reach the necessary depths a cable had to be improvised from various assorted lengths available on board ship. Great difficulty was experienced in overcoming leakages in this cable and its connexions, but in the end two experiments proved possible (nos. 11 and 12). Experiment 11 was interrupted by a sudden invasion of the 100–200 fm layer by larger individual scatterers which confused the results. In expt. 12 the 1 000 W free-flooding light was lowered to 450 and then

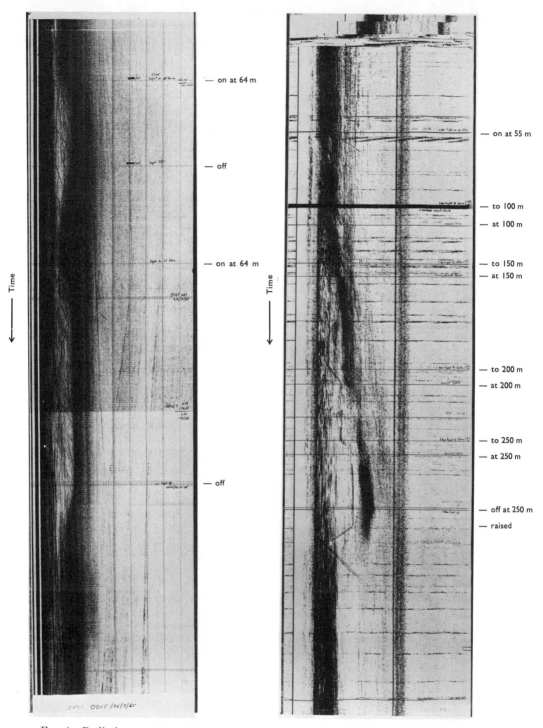

— on at 64 m

— off

— on at 64 m

— off

— on at 55 m

— to 100 m
— at 100 m

— to 150 m
— at 150 m

— to 200 m
— at 200 m

— to 250 m
— at 250 m

— off at 250 m
— raised

Time

Time

FIG. 4. Preliminary attempt to influence a scattering layer by switching the light source on and off (1 500 W diver's lamp). Record from the 10 kc/s echo-sounder; expt. 1. Scale width of record is 200 fm (366 m).

FIG. 5. Forcing down the scattering layer from 60 to 300 m by lowering the 1 000 W free-flooding bulb from 55 to 250 m; 67 kc/s echo-sounder; expt. 6. Scale width of record is 400 fm (732 m).

363

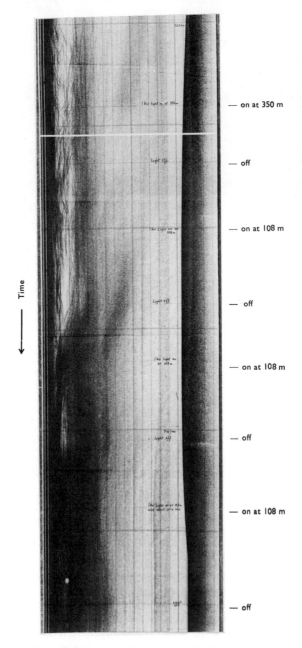

FIG. 6. Delaying the dusk rise of a deep scattering layer using 1 000 W free-flooding bulb; 10 kc/s echo-sounder; expt. 10. Scale width of record is 400 fm (732 m).

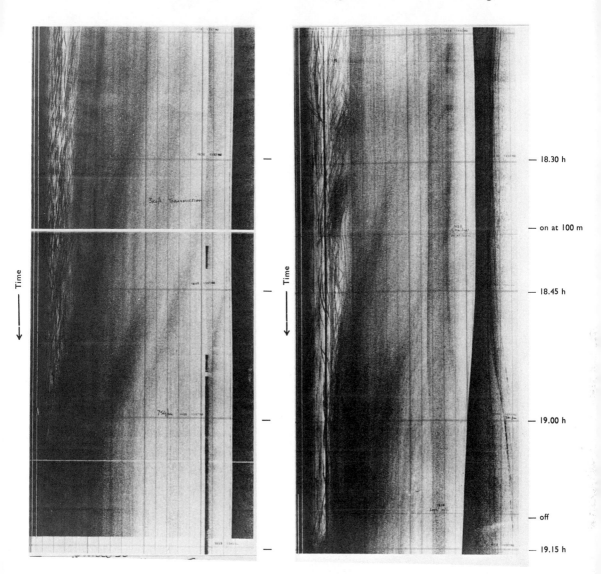

FIG. 7. Delaying the dusk rise of shallow
scattering layer using 1 500 W diver's lamp;
10 kc/s echo-sounder; expt. 8 below; control
(13/10) above. Scale width of record is
400 fm (732 m).

500 m. The record, when the light was switched on, showed definite indications of dispersion of the layer at that depth, but unfortunately this was confused by a wet patch on the record paper which made it impossible to discern any subsequent detail. Furthermore, bad wire angles were experienced during the experiment. It seemed likely that the light may have influenced too small a volume of the echo-sounder beam to create any very pronounced effect.

DISCUSSION

Composition of the scattering layers

In the Introduction it was noted that there is still some doubt how far the sonic scattering layers represent accumulations of particular organisms, or merely exist because in the layers the populations of organisms have a characteristic which brings about the scattering of sound of the frequency used. Different transmitted acoustic frequencies may be scattered at similar depths or at different depths, as the case may be. In Fig. 8 are given the superimposed simultaneous records from three echo-sounders operating at 10, 36 and 67 kc/s respectively, showing clearly that while some of the scattering layers are common to all three frequencies, others were present at only one of the frequencies used. Furthermore, the nightly coalescing of discrete daytime layers, and the confusion of them with the surface reverberation, makes the identification of any particular layer almost impossible at the time when most of the experiments were conducted.

The daytime layers thus offer better prospects of identification. Indeed, these experiments were conducted in conjunction with an intensive programme of biological and acoustical sampling from which it is hoped to gain a clearer insight into the relationships between the scattering and the animal populations at different depths. This material, however, must inevitably undergo detailed analysis and it will be some time before the results are available.

Suffice it to say at the moment, perhaps, that on theoretical grounds the most probable sources of the scattering appear to be organisms containing gas bladders of some kind. This does not mean that organisms not containing gas are unable to scatter sound, for if they are of sufficient size or reach sufficient abundance they will do so. The scattering coefficient is, however, essentially a function of the difference in density and compressibility between the organism and the medium, and the gas-containing organism is potentially a much better acoustic scatterer, in fact by several orders of magnitude.

The diffuse character of the scattering layers leads one to suppose that the scattering is being brought about by the smaller organisms and it seems that in most localities these are unlikely to reach sufficient abundance to bring about scattering, unless they do contain gas bladders (Hersey and Backus, 1962). Thus it appears that potentially the

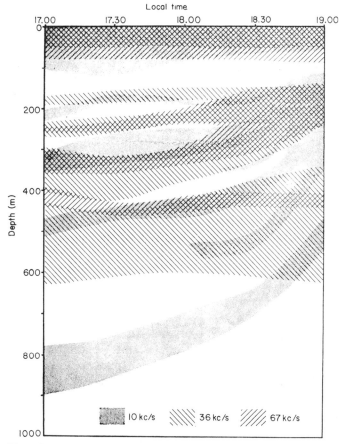

Fig. 8. Plot of upward movements of scattering layers at dusk with superimposed simultaneous records from different echo-sounders (9 October 1965).

bathypelagic fish, numerically the most abundant organisms with gas bladders, are a likely cause of the sound scattering, and it is interesting to note that in the area where the studies were made comparatively large populations of these, particularly of the lantern fishes (*Diaphus* sp. and *Lampanyctus* sp.), were taken.

Another point which should be made is that the concentration of the scatterers in these layers may not necessarily be very large. Indeed rough calculations of the volume scattering coefficient would be consistent with a maximum concentration, of the size of fish common in the Isaacs-Kidd trawls, of the order of 1 per 20 m^3.

Effect of intensity of artificial light

While distinct reactions have already been noted in relation to changes in the intensity of the light sources, it is interesting to compare the ambient light conditions before the experiment with the light intensity in which the organisms remain after reacting to the artificial light stimuli. In general the light seems to be of the order of ten times greater after the lights have been switched on (see Table I).

TABLE I

Ambient light before and during experiments

Expt.	Ambient intensity in layers before expt. (natural light, arbitrary units*)	Intensity at new depth of layer after light switched on (natural plus artificial light, arbitrary units*)
5	$6 \cdot 5 \times 10^{-3}$ (moonlight at 55 m)	3×10^{-1} (organisms at 75 m, 20 m below light)
5	6×10^{-4} (moonlight at 100 m)	$1 \cdot 2 \times 10^{-2}$ (organisms at 125 m, 25 m below light)
9	$1 \cdot 3 \times 10^{-4}$ (starlight at 40 m)	$1 \cdot 6 \times 10^{-1} - 9 \times 10^{-3}$ (organisms at 65–110 m, 15–60 m below light, varying with its intensity)

* The same arbitrary units are used as in Fig. 3B. They are roughly equivalent to metre candles or $8\mu W/cm^2$.

[Editor's note: Material has been omitted at this point.]

REFERENCES

Craig, R. E. and Lawrie, R. G. (1962). An underwater light intensity meter. *Limnol. Oceanogr.* **7**, 259–261.

Duntley, S. Q. (1960). Measurement of the transmission of light from an underwater point source. U.S. Navy Bureau of Ships, Contract No. bs-72039, Task 5, Report No. 5–11, 1–16 (unpublished manuscript).

Hersey, J. B. and Backus, R. H. (1962). Sound scattering by marine organisms. *In* "The Sea", Vol. 1, pp. 498–539. Interscience, New York and London.

31

Reprinted from *Deep-Sea Res.* **21**:651–656 (1974)

Deep Scattering Layers: vertical migration as a tactic for finding food

JOHN D. ISAACS,* SARGUN A. TONT* and GERALD L. WICK*

(*Received* 15 *October* 1973; *in revised form* 26 *March* 1974; *accepted* 28 *March* 1974)

Abstract—Their daily migrations lead the animals in the Deep Scattering Layer to food. The animals' response to light and their interaction with ocean currents maintain them within regions of high phytoplankton standing crop and transport them away from unproductive regions.

NUMEROUS investigators have searched for the advantages to the Deep Scattering Layer (DSL) organisms that counterbalance the energy expended in their daily excursions. Several hypotheses have been advanced and none of these seems to pre-empt the rest. Most likely the advantages are combinations of several factors. In this paper we elucidate and expand upon an idea originating with HARDY (1953), who proposed that the migrating organisms are transported greater horizontal distances as a consequence of their vertical migration than would be possible if the organisms employed the same energy to move only in a horizontal direction. This greater translocation is possible because the shear currents of the ocean *relative to the surface* generally increase with depth, and as an organism descends deeper it will experience more horizontal displacement relative to the surface. Thus, as HARDY (1953, p. 121) wrote, "It is as if each tiny creature is given ten-league boots to set it striding through the sea". Although Hardy mentioned that by controlling its depth a creature could conceivably seek congenial environments for surface feeding, he did not postulate a mechanism that would enable the creature to exercise this power. We propose that by responding to light intensity, most vertically migrating marine creatures are directed to food.

Throughout the world's oceans, several types of organisms have been suggested as the main constituents of Deep Scattering Layers. Dependent upon location of the station, fish with gas-filled swimbladders, euphausiids, sergestid prawns, and physonectid siphonophores are among the most frequently mentioned (HERSEY and BACKUS, 1962; BARHAM, 1963, 1966). At times a migrating layer will divide and some of the parts will show static or only partially migratory characteristics. In this paper we deal only with fully migrating layers, that is, the layers that move downward during dawn and approach the surface at sunset and which display horizontal continuity over distances of the order of 100 km or more. Records obtained during the following cruises have been used in this study: STEP I, September–December 1960; SCOR (Scientific Committee for Oceanic Research) DISCOVERER Expedition, May 1970 (Fig. 1). Our results do not depend on the identity of the species participating in the migration.

The theories suggesting the 'reasons' for the vertical migration of Deep Scattering Layers were critically evaluated by MCLAREN (1963). In our study we examined the

*Scripps Institution of Oceanography, University of California, San Diego, La Jolla, California 92037, U.S.A.

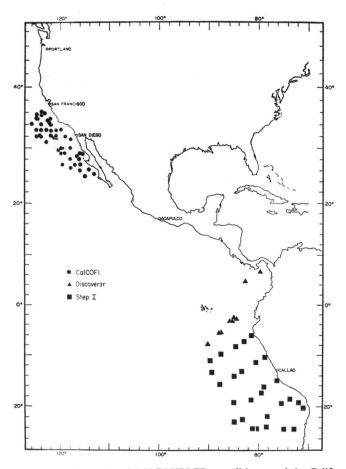

Fig. 1. Cruise stations of STEP I and DISCOVERER expeditions, and the California Current region stations monitored for the data used in Fig. 4.

evidence relating to a new thesis originally suggested by one of us in a semi-popular article that by migrating deeper in clear water than in turbid water, the marine animals statistically are transported *out* of the areas of low standing crop of phytoplankton *into* areas of high standing crop (ISAACS, 1969). In areas of low standing crops of phytoplankton, daylight penetrates further into the ocean causing the migrating animals to descend deeper. In the turbid water associated with high standing crop, the migrating forms remain closer to the surface. Their interaction with the current shear will generally result in their transport out of areas of low standing crop and transport into (or maintenance within) areas of high standing crop and, ordinarily, of high productivity. Insolation, and hence cloud cover, also affect the depth of migration, but cloud cover is not so important as water transparency.

There can be little doubt that the migrant Deep Scattering Layers respond to the *in situ* light levels in some way, probably principally to the absolute light levels and secondarily to light gradients. Using all available complete data, our analysis of both STEP I and DISCOVERER data clearly shows that the depths of the Deep

Scattering Layer in daytime are correlated with the surface layer isolumes (Figs. 2 and 3). In both figures the statistical significance $P < 0.05$. Acoustic records were made with precision depth recorders operating at a dominant frequency of 12 kHz. The distance between the ocean surface and the top of the layer, during local noon hour, has been taken as the daytime depth.

KAMPA (1970), summarizing her investigations of Deep Scattering Layers in the northeastern Atlantic and in the Gulf of California, stated that the intensity of irradiance encountered by the mesopelagic animals in the scattering layers varied at most by a factor of six despite the remoteness of the regions studied and despite variation of hundreds of meters in the midday depths sought by the animals. Her irradiance values ranged approximately between 10^{-4} and 10^{-5} μW cm^{-2} nm^{-1} for wavelength of 480 nm. CLARKE and BACKUS (1964) reported that in the slope water southeast of New York during the daylight hours the layers are associated with specific isolumes ranging from 10^{-1} to 10^{-4} μW cm^{-2} over a range of 320 to 650 nm centred at 480 nm.

The irradiances at Deep Scattering Layer depths used in this study were calculated from the upwelling spectral irradiance data obtained at 480 nm by SMITH (1973) using the Scripps spectroradiometer during the SCOR DISCOVERER expedition. The data were all taken within 100 m of the surface. To determine the light irradiance values at the Deep Scattering Layer depths of several hundred meters, we need to use the diffuse attenuation coefficients for irradiance at these greater depths. The extrapolation was performed in two steps. First, the deepest irradiance value measured at each station was extrapolated to 100 m using the attenuation coefficient calculated for the interval immediately above the deepest measured value. These values were further extrapolated to the Deep Scattering Layer depth using 0·04 m^{-1} for the attenuation coefficient in that interval. Fortunately, the attenuation coefficients below 100 m measured for most of the deep ocean are quite uniform and fall in the range of 0·03 to 0·04 m^{-1}. (JERLOV, 1968; JOHN TYLER, personal communication). Our cal-

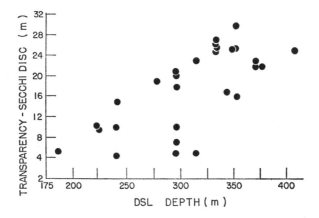

Fig. 2.   Relationship between the average surface water transparency and the Deep Scattering Layer depth for STEP I expedition. $P < 0.05$. Eighty-five per cent of the data were obtained under overcast· skies.

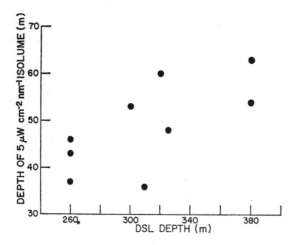

Fig. 3. Relationship between depth of 5 μW cm⁻² nm⁻¹ isolume and the Deep Scattering Layer depth for DISCOVERER expedition. $P < 0.05$.

culated light irradiance values are shown in Table 1 together with the surface irradiances. The essential point of this section is not that the Deep Scattering Layer organisms necessarily maintain themselves at a specific isolume, but that their response to light maintains them approximately in the same light environment. Even with this less stringent requirement the main point of our paper is supported.

If the daytime depth of the Deep Scattering Layer were associated with a specific isolume, then this depth would be a function of transparency of the water, insolation, cloud cover, and the sea surface conditions. At all but high latitudes and near sunset and sunrise (i.e. low sun angle) the subsurface light intensity is only slightly affected by the sea surface condition. At low latitudes, where our data were recorded, the noon light intensity fluctuations attributable to the sea surface are 3% at most (JERLOV, 1968). Among all of the factors, transparency is the most important in determining noon-time depth of the Deep Scattering Layer. Solar irradiance, I, in the ocean decreases from the surface according to the equation

$$I = I_0 \, e^{-Kz}, \tag{1}$$

where I_0 is the solar irradiance incident on the sea surface (dependent upon the cloud cover and the relative orientation of the surface to the sun), K is the diffuse attenuation coefficient for irradiance of seawater (dependent upon the transparency), and z is the depth below the surface. If the effect of the sea surface condition is negligible, the only important factors are cloud cover and transparency in low latitudes at noon during a limited season of the year, as was the case for the cruise data used here.

We evaluated how well the relationship between Deep Scattering Layer daytime depth, transparency, and incident solar radiation data from the DISCOVERER expedition fitted equation (1). The coefficient K for each station was averaged from the weighed values of K_i determined from irradiance measurements at regular depth intervals beneath the sea surface to a maximum of 100 m. Below 100 m, K_i was taken

Table 1.   DISCOVERER *Expedition noon-time data. Surface irradiance and irradiance calculated at the Deep Scattering Layer depth for 480 nm.*

Sta. no.	Surface irradiance (μW cm^{-2} nm^{-1})	Irradiance at Deep Scattering Layer depth (μW cm^{-2} nm^{-1})
9	$5\cdot5 \times 10^3$	$1\cdot7 \times 10^{-5}$
10	$4\cdot3 \times 10^3$	$2\cdot1 \times 10^{-4}$
11	$8\cdot3 \times 10^3$	$2\cdot7 \times 10^{-4}$
12	$7\cdot7 \times 10^3$	$1\cdot4 \times 10^{-5}$
13	$6\cdot5 \times 10^3$	$8\cdot6 \times 10^{-5}$
14	$9\cdot1 \times 10^3$	$1\cdot4 \times 10^{-4}$
15	$8\cdot4 \times 10^3$	$5\cdot8 \times 10^{-6}$
16	$7\cdot6 \times 10^3$	$2\cdot6 \times 10^{-4}$

as $0\cdot04$ m^{-1}. Using a value of I of 4×10^{-4} μW cm^{-2} nm^{-1}, equation (1) is satisfied for Deep Scattering Layer depth with a statistical significance of $P < 0\cdot04$.

HARDY (1953) proposed the idea that migrating animals use the current gradient to their advantage. We are presenting a specific consequence of this aspect of vertical migration. Using geostrophic flow in the California Current region (CALIFORNIA COOPERATIVE FISHERIES INVESTIGATIONS, 1966), we have calculated the daily displacement of organisms at the stations shown in Fig. 1 and found a striking correlation with transparency. The horizontal displacement is relative to the surface for organisms at their midday depths. In this current system Deep Scattering Layer organisms spend less time in regions of high transparency and more time in regions of low transparency (Fig. 4). This relationship is statistically significant at $P < 0\cdot05$. Since low transparency is associated with regions of high phytoplankton content and vice versa, this consequence of vertical migration is clearly beneficial to the organisms, as it leads them to primary food and the associated consumer organisms.

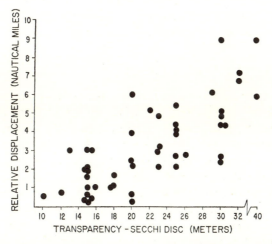

Fig. 4.   The daily displacement relative to the surface of Deep Scattering Layer organisms (calculated from geostrophic flow) plotted against Secchi disc depth.

This conclusion is based on the average flow as related to average phytoplankton conditions over a period of time. It will be interesting to investigate some of the possibilities of day-to-day migrations in regions with various degree of phytoplankton patchiness. Under a highly populated patch, Deep Scattering Layer organisms would be closer to the surface than they would under the general conditions. Assuming that the patch moves with the surface water and that current speed relative to the solid earth decreases with depth, then the organisms' chances of remaining with the patch will be greater than if they migrated to a greater depth. The spectrum of dimensions of phytoplankton patchiness is not well known, and we are unable to estimate the consequences of such inhomogeneities. However, the small-scale effects seem to add to the effectiveness of the general mechanism, and, to the degree to which herbivores are involved in the migration, even to give rise to increased grazing that could reduce patchiness of some intermediate dimensions.

Heavy cloud cover, which is often associated with upwelling and hence with subsequent blooms, may even allow the Deep Scattering Layer to anticipate regions of high productivity.

Acknowledgements—We thank T. E. Chase and J. E. Tyler, Scripps Institution of Oceanography, La Jolla; H. B. Stewart, Jr. and L. W. Butler, U.S. Department of Commerce, Atlantic Oceanographic and Meteorological Laboratories, Miami, Florida; and E. G. Barham and R. W. Owen, of U.S. Department of Commerce, National Marine Fisheries Service, La Jolla, for their help in completion of this study, which was supported in part by the Marine Life Research Group, Scripps, and by the Sea Grant Institutional Program, University of California.

REFERENCES

Barham E. G. (1963) Siphonophores and the Deep Scattering Layer. *Science*, **140**, 826–828.
Barham E. G. (1966) Deep Scattering Layer migration and composition: observations from a diving saucer. *Science*, **151**, 1399–1403.
California Cooperative Oceanic Fisheries Investigations (1966) *CalCOFI ATLAS, no.* 4 Geostrophic flow of the California Current at the surface and 200 meters. State of California Marine Research Committee.
Clarke G. L. and R. H. Backus (1964) Interrelations between the vertical migration of Deep Scattering Layers, bioluminescence, and changes in daylight in the sea. *Bulletin de l'Institut océanographique, Monaco*, **64**(1318), 2–23.
Hardy A. C. (1953) Some problems of pelagic life. In: *Essays in marine biology*, Oliver & Boyd, Edinburgh, pp. 101–121.
Hersey J. B. and R. H. Backus (1962) Sound scattering by marine organisms. In: *The sea*, M. N. Hill, editor, Wiley, New York, 1, pp. 498–539.
Isaacs J. D. (1969) The nature of oceanic life. *Scientific American*, **221**(3), 146–162.
Jerlov N. G. (1968) *Optical oceanography*, Elsevier, New York, 194 pp.
Kampa E. M. (1970) Photoenvironment and sonic scattering. In: *Proceeding of an International Symposium on Biological Sound Scattering in the Ocean*, G. B. Farquhar, editor, MC Report 005, U.S. Government Printing Office, Washington, D.C., No. 0851-0053, pp. 51–59.
McLaren I. A. (1963) Effects of temperature on growth of zooplankton, and the adaptive value of vertical migration. *Journal of the Fisheries Research Board of Canada*, **20**, 685–727.
Smith R. C. (1973) Scripps spectroradiometer data. In: *Data Report, SCOR—DISCOVERER Expedition May* 1970, Vol. II, John E. Tyler, editor, University of California, Scripps Institute of Oceanography, SIO Reference 73-16, pp. G1–G160.

32

Reprinted by permission from pp. 452–455 of *The Sea,* Vol. 1, M. N. Hill, ed., New York: Wiley, 1962, 864 pp.

UNDERWATER VISIBILITY

S. Q. Duntley

Nowhere in nature are the principles of protective coloration and camouflage better displayed than in the feeding-grounds of the sea, where predators and prey alike depend for survival upon their ability to see. When man invades the underwater world and peers through his face-plate at the new surroundings his success and his safety depends in large measure upon his visual capability.

1. Image Transmission

Visibility underwater is restricted in a manner somewhat analogous to the obscuration produced by dense haze or fog in the atmosphere, but the nature of image transmission by water differs importantly from that by the atmosphere because of the vastly greater space-rate of thermodynamically non-reversible energy transformation, i.e. the transformation of light into heat, chemical potential energy (as in photosynthesis), etc. This major effect, called *absorption*, causes all aspects of daylight in the sea to decrease so rapidly with depth that visual ranges along paths of sight inclined either upward or downward are profoundly affected in a manner quite different from the atmospheric case, where absorption is negligible except in clouds of dark smoke or dust.

Natural waters are usually composed of horizontal strata each of which is nearly uniform in its optical properties. When the path of sight is entirely within a uniform stratum, the spectral radiance, $N(z_1, \theta, \phi)$, measured at depth z_1 by a radiance photometer pointed in a direction having zenith angle θ and azimuth angle ϕ is found to be related to the corresponding spectral radiance $N(z_2, \theta, \phi)$ at depth z_2 by the approximation

$$N(z_2, \theta, \phi) = N(z_1, \theta, \phi) \exp\{-[K(z, \theta, \phi)](z_1 - z_2)\}, \tag{1}$$

where the z-axis is vertical and positive from the mean sea-surface *upward* and $K(z, \theta, \phi)$ is the attenuation coefficient for spectral radiance in the direction θ, ϕ at all depths between z_1 and z_2. This mathematical model introduces the justifiable approximation that the radiance K-function, $K(z, \theta, \phi)$, is the same at all points throughout the path of sight.

If equation (1) is represented by the differential equation

$$dN(z, \theta, \phi)/dr = -K(z, \theta, \phi) \cos \theta\, N(z, \theta, \phi), \tag{2}$$

where $r \cos \theta = z_1 - z_2$, and if the equation of transfer for spectral field radiance is written

$$dN(z, \theta, \phi)/dr = N_*(z, \theta, \phi) - \alpha(z)N(z, \theta, \phi), \tag{3}$$

and if the equation of transfer for the apparent spectral radiance, $_tN(z_t, \theta, \phi)$, of the visual target is written

$$d_tN(z, \theta, \phi)/dr = N_*(z, \theta, \phi) - \alpha(z)_tN(z, \theta, \phi), \tag{4}$$

375

wherein the depth of the target $z_t = z_1$ and the depth of the observer $z = z_2$, then (2), (3) and (4) can be combined and integrated throughout the path of sight to show that the apparent spectral radiance of the target, $_tN_r(z, \theta, \phi)$, is related to the inherent spectral radiance of the target, $_tN_0(z_t, \theta, \phi)$, by the equation

$$_tN_r(z, \theta, \phi) = {}_tN_0(z_t, \theta, \phi) \exp\{-\alpha(z)r\}$$
$$+ N(z_t, \theta, \phi) \exp\{+K(z, \theta, \phi)r\cos\theta\}[1 - \exp\{-\alpha(z)r + K(z, \theta,\phi)r\cos\theta\}], \quad (5)$$

wherein the first term on the right represents the residual image-forming light from the target and the second term represents radiance contributed by the scattering of ambient light in the sea throughout the path of sight.

If the target is seen against a background of inherent spectral radiance, $_bN_0(z_t, \theta, \phi)$, the apparent spectral radiance, $_bN_r(z, \theta, \phi)$, of the background will be given by an equation identical with (5) after replacing the presubscripts t by b. This equation and equation (5) can be combined with the defining relations for inherent spectral contrast, $C_0(z_t, \theta, \phi)$, and apparent spectral contrast $C_r(z, \theta, \phi)$, which are, respectively,

$$C_0(z_t, \theta, \phi) = [{}_tN_0(z_t, \theta, \phi) - {}_bN_0(z_t, \theta, \phi)]/{}_bN_0(z_t, \theta, \phi),$$

and

$$C_r(z, \theta, \phi) = [{}_tN_r(z, \theta, \phi) - {}_bN_r(z, \theta, \phi)]/{}_bN_r(z, \theta, \phi).$$

When this is done, the ratio of inherent spectral contrast to the apparent spectral contrast is found to be

$$C_0(z_t, \theta, \phi)/C_r(z, \theta, \phi)$$
$$= 1 - [N(z_t, \theta, \phi)/{}_bN_0(z_t, \theta, \phi)][1 - \exp\{\alpha(z)r - K(z, \theta, \phi)r\cos\theta\}]. \quad (6)$$

In the special case of an object suspended in deep water, $_bN_0(z_t, \theta, \phi) = N(z_t, \theta, \phi)$ so that

$$C_r(z, \theta, \phi) = C_0(z_t, \theta, \phi) \exp\{-\alpha(z)r + K(z, \theta, \phi)r\cos\theta\}. \quad (7)$$

When the observer's path of sight to an object seen against a background of water is horizontal the apparent spectral contrast is

$$C_r(z) = C_0(z) \exp\{-\alpha(z)r\}, \quad (8)$$

which indicates that the apparent contrast is independent of azimuth and depends only on the total attenuation coefficient, $\alpha(z)$, for image-forming light.

For many practical purposes it is a sufficient approximation to assume $K(z, \theta, \phi)$ to be independent of direction and to be of the same magnitude as the irradiance K-functions described in the previous chapter, i.e. to assume $K(z, \theta, \phi) = k(z, -) = K(z, -) = K(z, +) = K(z)$ in all of the foregoing equations, indicating thereby that the reduction of contrast in image transmission through water is virtually independent of azimuth. The principles described by the foregoing equations were first discovered in the course of experiments with an underwater telephotometer by Duntley (1949, 1950). The data provide verification of the contrast reduction equations and demonstrate the practical validity of the approximation (Duntley, 1951; Duntley and Preisendorfer, 1952; Preisendorfer, 1957).

2. Inherent Contrast

The inherent spectral contrast, $C_0(z_t, \theta, \phi)$, of objects under water presents a far more intricate analytical problem than does the contrast reduction effect discussed above. To be rigorous, all of the reflectance and gloss characteristics of both the target and its background must be known, and the three-dimensional configuration of target and background must be taken into account with respect to the underwater radiance distribution which irradiates their surfaces. No practical general procedure for meeting these requirements is available but research directed toward this goal is in progress. Two common and important special cases are, however, easily treated: (1) an object which appears as a dark silhouette, wherein the inherent contrast is -1, and (2) a horizontal matte surface of known submerged reflectance, wherein the inherent spectral contrast is controlled by the downwelling and upwelling spectral irradiances $H(z, -)$ and $H(z, +)$ (Duntley, 1960).

3. Sighting Range

Most underwater sighting ranges are so short that the visual angle subtended by ordinary objects is sufficient to make the exact angular size of the object unimportant. Underwater sighting ranges are, therefore, usually controlled by the contrast transmittance of the path of sight, i.e. by water clarity. This is not true of very small objects (e.g. small pebbles, grains of sand, etc.) nor is it true when semi-darkness prevails because of depth or low solar elevation. Nomographic charts for predicting underwater sighting ranges for objects of any size from data on $\alpha(z)$, $K(z)$, depth, solar altitude, target reflectance, bottom reflectance, etc. have been prepared by Duntley (1960) on the basis of equation (6) and visual threshold data by Taylor (1960).

Application of the nomographic visibility charts to a wide variety of underwater visibility problems in many kinds of natural water has resulted in the following useful rules-of-thumb:

1. Most objects can be sighted at 4 to 5 times the distance

$$1/[\alpha(z) - K(z) \cos \theta].$$

2. Large dark objects, seen as silhouettes against a water background, can be sighted at the distance $4/\alpha(z)$ when the path of sight is horizontal. (This rule can be used by swimmers for estimating $\alpha(z)$.)

3. In some natural waters $\alpha(z) = 2.7K(z)$; in such waters the downward sighting range of most objects $= \frac{7}{8}$ the horizontal sighting range of large, dark objects.

Exceptions to these rules-of-thumb are common. They seldom, for example, apply to a white Secchi disk in clear water, even if the disk is observed by a swimmer; this is because of the high inherent contrast of the white disk. No simple conversion exists between Secchi-disk data and the sighting ranges of other objects. The nomographic charts, however, provide correct sighting ranges for Secchi disks and other objects under virtually all circumstances.

References

Duntley, S. Q., 1949. Exploratory studies of the physical factors which influence the visibility of submerged objects. *Proc. Armed Forces-Nat. Res. Council Vision Comm.*, **23**, 123.

Duntley, S. Q., 1950. The visibility of submerged objects I. *Proc. Armed Forces-Nat. Res. Council Vision Comm.*, **27**, 57.

Duntley, S. Q., 1951. The visibility of submerged objects II. *Proc. Armed Forces-Nat. Res. Council Vision Comm.*, **28**, 60.

Duntley, S. Q., 1960. Improved nomographs for calculating visibility by swimmers, (natural light). Bureau of Ships Contract NObs-72039, Rep. 5–3, Feb.

Duntley, S. Q. and R. W. Preisendorfer, 1952. The visibility of submerged objects. Final Rep. Visibility Laboratory, N5ori 07864, Mass. Inst. Tech., Aug.

Preisendorfer, R. W., 1957. A model for radiance distributions in natural hydrosols. Scripps Institution of Oceanography, SIO Reference No. 58-42.

Taylor, J. H., 1960. Visual contrast thresholds for large targets. Part I. The case of low adapting luminances. SIO Reference No. 60-25, June 1960.

AUTHOR CITATION INDEX

SUBJECT INDEX

Absorption, 21, 38, 43, 56
Absorption (attenuation) data
   infrared, 143ff.
   spectral, 136ff.
   ultraviolet, 146ff.
Anton Dohrn, 7
Apparent optical properties. *See* Optical
   properties
Asymptotic radiance distribution, 5ff., 62
   equation for, 259ff.
   theory, 253ff.
Attenuation length, 21

Backward scattering, 49 (eq. 14), 56, 65ff.
Beam attenuation (absorption) coefficient
   artificial sea water, 138ff.
   distilled water, 135, 138ff.
   heavy water, 138ff.
   infrared, 144
   selected spectral data, 102, 138
   spectral, 135, 138ff.
   ultraviolet, 146
Beam transmissometer, 21, 104ff., 110,
   112, 113
   calibration of, 114ff., 117
Beam transmittance, 100ff.
   precision measurement, 101
   sources of error, 114, 115, 166

Calypso, 7
Collimated light, 28
Contrast transmittance, 34ff.

Danish Deep-Sea Expedition, 6
Deep scattering layer, 320
   recordings of effect of light, 363ff.
   vertical migration tactic, 369
Distribution functions, 43, 55

Equation of transfer, 48, 49 (eq. 8)
   for radiance, 104, 220ff.
      Monte Carlo solution, 222

Forward scattering, 49 (eq. 13), 56, 65ff.

Horizontal irradiance. *See* Irradiance,
   vector

Image transmission, 320, 375
   apparent contrast, 376
   inherent contrast, 376
   sighting range, 377
Inherent optical properties. *See* Optical
   properties
International Council for the Exploration
   of the Sea, 1936 conference on light
   in the sea, 2ff.
Irradiance, 27, 29, 55
   instruments, 289, 292ff., 299ff., 312ff.
   data, 306ff.
   quantum (photon), 312ff., 334ff.
   meter
      plane collectors, 52
      spherical collectors, 52
   photon scalar, 312ff.
      optical measurement, 313
   quantum, 288, 313, 334ff.
   scalar, 51
      calculation from radiance distribution,
         43
      collector, 288
      defined, 289, 313
      vs. spherical irradiance, 289
   spectral, instruments
      band width, 294–297
      calibration, 295
      immersion effect, 315
      spectroradiometer, 292ff.
      wavelength response, 293
   spherical
      calculation from radiance distribution,
         43
      defined, 289
      instruments, 52, 291
   vector, 52

About the Editor

JOHN E. TYLER was born in Boston, Massachusetts, November 11, 1911. He graduated from the Massachusetts Institute of Technology where he specialized in optics and completed his thesis in the field of spectrophotometry and colorimetry under Professor Arthur C. Hardy. He became a Research Associate at M.I.T. and, under the direction of Professor Martin J. Buerger, designed and constructed the first crystal pattern synthesizer.

In 1941, he joined the research staff of the National Research Corporation in Boston, where he engaged in research on optical instruments and the vacuum coating of optical components to produce low-reflecting films. In 1944, he transferred to the research laboratories of the Interchemical Corporation in New York City and became director of the Spectroscopy Laboratory. During the war years, he was active in optical research for the U.S. National Defense Research Committee. He was appointed to the research staff at the Scripps Institution of Oceanography in 1952, where he is currently a research physicist.

From 1964 to 1972 he was chairman of a Working Group for the Scientific Committee on Oceanic Research (SCOR) and worked directly with a distinguished group of international scientists on problems associated with oceanic photosynthesis, its measurement and the determination of in situ quantum efficiency. In 1970, he was chief scientist of the SCOR Discoverer Expedition during which the SCOR Working Group obtained essential data relating photosynthesis in the ocean to the available natural radiant flux.

Mr. Tyler is a fellow of the Optical Society of America and of the American Association for the Advancement of Science and is a member of Sigma Xi. Since 1939 he has a continuous record of published contributions to the scientific literature and is internationally recognized for his contributions in the field of hydrologic optics.